Chris

Slynr

Techno B

Telecommunications

Telecommunications

Second Edition

Warren Hioki

Prentice Hall
Upper Saddle River, NJ 07458

Library of Congress Cataloging-in-Publication Data

HIOKI, WARREN
    Telecommunications / Warren Hioki. — 2nd ed.
      p. cm.
    Includes bibliographical references and index.
    ISBN 0-13-123878-7
    1. Telecommunication. I. Title
TK5015.H56 1994
004.6.—dc20

94026757
CIP
AC

Acquisitions editor: *Charles Stewart*
Editorial/production supervision: *Kim Gueterman*
Cover design: *Bryant Design*
Cover photo courtesy of: *Index Stock Photography Incorporated*
Manufacturing buyer: *Ilene Sanford*

© 1990 1995 by Prentice-Hall, Inc.
A Simon & Schuster Company
Upper Saddle River, New Jersey 07458

Printed in the United States of America
10 9 8 7 6 5 4 3

ISBN 0-13-123878-7

Prentice-Hall International (UK) Limited, *London*
Prentice-Hall of Australia Pty. Limited, *Sydney*
Prentice-Hall Canada Inc., *Toronto*
Prentice-Hall Hispanoamericana, S.A., *Mexico*
Prentice-Hall of India Private Limited, *New Delhi*
Prentice-Hall of Japan, Inc., *Tokyo*
Simon & Schuster Asia Pte. Ltd., *Singapore*
Editora Prentice-Hall do Brasil, Ltda., *Rio de Janeiro*

To my loving wife, Andra

Contents

3   TERMINALS   37

4   SERIAL INTERFACES   61

5   THE UART   104

6　INTERFACING THE 8085A MICROPROCESSOR TO THE 8251A USART　124

7 THE TELEPHONE SET AND SUBSCRIBER LOOP INTERFACE   145

8 THE TELEPHONE NETWORK   183

Contents

10   PROTOCOLS   280

11 LOCAL AREA NETWORKS 330

Preface

Telecommunications, Second Edition, has been upgraded to include the latest standards and emerging technologies that have developed in recent years. The book is intended to provide the reader with the technical aspects and background material on the subject of telecommunications, one of the fastest growing industries in the world. A broad range of topics is covered without the intricate details of mathematical derivations and proofs. Instead, fundamental principles are emphasized in a simplified, yet comprehensive and practical manner. To achieve this, numerous sketches with detailed explanations are included throughout, all of which are aimed at providing the reader with the practical knowledge deemed necessary for today's telecommunications engineer and technician.

Intended Audience

This book is intended primarily for use as a college text for either a data communications or telecommunications program. The material has proved to be of great value to undergraduates, as well as graduates seeking to extend or review their discipline. A fundamental background in mathematics, electronics, and digital circuits is helpful. Instructors will find that the material is clear, concise, and written with sufficient depth. Experiments can be easily implemented to reinforce example problems presented in the text. This book can also serve as an excellent reference guide for those who work in industry and are concerned with keeping abreast of the subject matter. For those interested in achieving some degree of computer literacy, a wealth of technical terms and acronyms are defined and discussed throughout the text. They are also included in an extensive glossary at the end of the text.

Organization of the Text

The chapters in this text have been arranged in an order that has been successfully presented in a series of telecommunications courses over the years. It is by no means the suggested order in which the material should be taught, as some chapters may be covered independently of each other. For example, some instructors may choose to cover Chapter 9, Modems, prior to Chapter 8, The Telephone Network. The same may be true of Chapter 14, Fiber Optics, which is a subject in itself. Chapter organization is as follows:

Chapter 1 *Introduction:* An overview of the subject of telecommunications is presented in this chapter. A distinction between the analog and digital signal is made as well as the data communications system versus the telecommunications system.

Chapter 2 *Transmission Codes:* Several codes commonly used in communication systems are examined in this chapter.

Chapter 3 *Terminals:* In this chapter, various types of terminals and their features are discussed, with emphasis on the ASCII terminal. Examples of escape sequences and the use of control characters are presented.

Chapter 4 *Serial Interfaces:* This chapter distinguishes between synchronous and asynchronous serial transmission. RS-232-C is examined in depth, and the more recent RS-449 and its supporting standards RS-422-A and RS-423-A, are introduced.

Chapter 5 *The UART:* An extensive study of the UART is given in this chapter. Transmitted and received serial characters are examined. The conversion process of these characters from serial to parallel form and vice versa are examined.

Chapter 6 *Interfacing the 8085A Microprocessor to the 8251A USART:* This chapter is devoted to interfacing the 8085A microprocessor to the 8251A USART. An asynchronous 4800-baud program is discussed along with an interface schematic diagram.

Chapter 7 *The Telephone Set and Subscriber Loop Interface* Various types of telephones are discussed in this chapter, including the rotary dial, DTMF, and cordless telephones. Subscriber loop interface specifications are also covered and cellular telephony is also featured in this chapter.

Chapter 8 *The Telephone Network:* This chapter examines the telephone network's history and how it has evolved into the world's most sophisticated network of computers.

Chapter 9 *Modems:* The Bell family of modems and their characteristics are covered in this chapter. The latest CCITT (ITU) modem recommendations are also covered. Features such as data compression, scrambling and descrambling, loopback tests, and various modulation techniques are presented.

Chapter 10 *Protocols:* Synchronous serial link protocols are discussed in this chapter. These include BISYNC, SDLC, and HDLC. In addition, various standards organizations are described. These organizations have given rise to the development of the ISO/OSI seven-layer model. ISDN, ATM, SONET, and other emerging technologies are introduced.

Chapter 11 *Local Area Networks:* Common LAN topologies and access control methods are introduced in this chapter, with emphasis placed on the Ethernet protocol.

Chapter 12 *Error Detection, Correction, and Control:* This chapter presents an in-depth look at some of the common error detection and correction mechanisms used to today's telecommunications systems. Examples of computing the CRC block check character are given. Hamming codes are developed and used to correct single-bit errors.

Chapter 13 *Noise:* Noise is an inherent problem in any communication system. This chapter introduces the various types of noise, their characteristics, and their effects on the telecommunications system.

Chapter 14 *Fiber Optics:* With fiber optics rapidly becoming an integral segment of the telecommunications industry, this chapter is intended to provide a detailed look into this relatively new technology. The theory of light, including Snell's law, is presented prior to studying the characteristics of fiber.

Supplementary Aids for the Instructor

This text is supplemented with the new *Laboratory Manual for Telecommunications,* a lab manual that combines the unique feature of telephony related experiments with the option of ordering supporting hardware. Instructors often refrain from implementing lab experiments for their students due to lack of parts. Many experiments out of laboratory manuals are never implemented due to the large quantity of parts required by the experiment. The time necessary for students to build experiments, before even making any measurements, often exceeds the designated laboratory time. As a result, what could have been an effective lab job never gets performed. Now, for the first time, custom-designed circuit boards, tailored toward individual instructor needs, are available with quick turn-around time. These include:

— DACs and A/D Converters	— Fiber Optic Links
— Analog Multiplexers and switches	— Op Amp Circuits
— Sample and Hold Amplifiers	— AM and FM Circuits
— Phase Locked Loops	— 555 Timer Circuits
— UART Interface	— 4-bit DTMF Generators and Receivers
— Ring Generator Circuits	— Oscillator Circuits

Circuit boards and parts can be ordered to accommodate single or multiple laboratory experiments. Test points, power supply jacks, and silk screened reference designators for components are included. Boards can be ordered completely assembled and tested. They can also be ordered in kit form for students to build. For information contact:

Ångstrom Electronics
Las Vegas, NV
1-800-558-7731

Also available for the instructor is the *Telecommunications Instructor's Manual*. This manual provides the following:
— A total of 137 transparency masters of figures and tables taken from the text.
— Solutions to all end of chapter problems presented in the *Telecommunications* (Second Edition) text.
— Solutions to questions presented in *Laboratory Manual for Telecommunications*.

Acknowledgments

This text could not be written without the numerous contributions and hard work of many individuals. I wish to express my sincere gratitude by acknowledging those of you that have made this possible.

Beverly Bellows, *Digital Equipment Corporation*
Gregg Castro, *Pacific Bell*
Fred Dane, *San Jose City College*
Andra Hioki, *Ångstrom Electronics*
Sheldon Hochheiser, *AT&T Bell Laboratories*
Dick Jamison, *Hewlett-Packard Optical Communication Division*
Carl Jensen, Jr., *DeVry Institute of Technology*
Jim Lane, *Telco Systems Inc.*
Clay Laster, *San Antonio College*
Ruth Lombardi, *AT&T Bell Laboratories*
Ken Miller, *Concord Data Systems*
Benjamin Nelson, *Markem Corporation*
Tom Novicki, *Racal-Vadic*
R. Papannareddy, *Purdue University North Central*
Joy Parillo, *AT&T Archives*
Greg Pearson, *Microcom Inc.*
Preston Peek, *Northern Telecom Inc.*
John Quick, *DeVry Institute of Technology*
Ronald Rowe. *Adaptive Computer Technology*
Linda Simon, *Digital Equipment Corporation*
J.E. Stanley, *Hughes Aircraft Co.*
Ralph Tarrant, *IBM*
Bill Townsend, *Digital Equipment Corporation*
John Uffenbeck, *Wisconsin Indianhead Technical Institute*

Janet Wells, *AT&T Bell Laboratories*
Chuck Wojslaw, *San Jose State University*

I would like to especially thank Charles Stewart of Prentice Hall for his tremendous support and encouragement; Jim Lane of Telco Systems Inc, for providing his expertise on ATM technology; Benjamin Nelson of Markem Corp. for the wealth of information and expertise he has provided me on bar code theory; Gregg Castro of Pacific Bell for his expertise on the PSTN and the numerous tours of Pacific Bell's switching facilities; and Kim Gueterman of Prentice Hall for her patience, professionalism, and absolute dedication in her job of organizing this text—you were a tremendous joy to work with. Lastly, I am deeply indebted to my wife, Andie, the secretary of this manuscript and CAD architect of supplementary manuals. Where would I be without you?

WARREN HIOKI

1

Introduction
to Telecommunications

We live in an era that is often referred to as the *Information Age*. With the advent of computer technology, this era has rapidly transformed the office and factory worker into an analyzer, manipulator, gatherer, and distributor of information. This information comes in a variety of forms, all representing some meaningful arrangement of the human thought process. In this chapter, an introduction to telecommunications, one of the most explosive technological fields of our time, is presented. There is a growing need to consider the importance of why this highly dynamic field has brought millions of computers together to share their endless capabilities for providing information. Our educational system, jobs, business, entertainment, and very often, our everyday tasks necessitate skill levels in the use of information systems.

1.1 ROLE OF THE COMMUNICATION SYSTEM

The distribution of information from one location to another is the role of the communication system. Figure 1-1 illustrates the essential components of the communication system. Although this may seem simplified, the process of sending and receiving information in an orderly and successful manner can be extremely complex for today's modern communication system. Consider some of the factors involved:

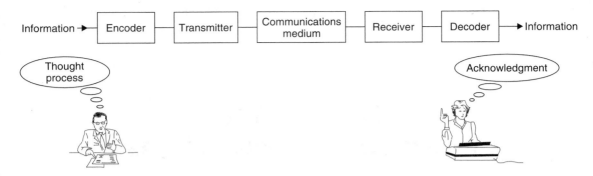

Figure 1-1   Basic components of a communication system.

— Nature of the information
— Format of the information
— Transmission speed
— Transmission medium
— Transmission distance
— Modulation technique
— Error control

1.2 WHAT IS TELECOMMUNICATIONS?

The process of distributing information can be subdivided into several categories. *Telecommunications* is perhaps the most extensive of these categories. What is telecommunications? Many definitions of this process have been given over the years, all of which are accurate to some degree, depending on the nature of the technology at that time. To begin with, the prefix *tele* is derived from the Greek meaning for "at a distance." Historically, telecommunications encompassed the telegraph system, invented by Samuel Morse in 1854. Information was transmitted (and still is) as a series of electrical impulses generated by depressing a hand key. To be able to communicate over such long distances through the use of the telegraph system was, at the time, a scientific marvel.

The invention of the telephone by Alexander Graham Bell in 1876 led to the growth of the telephone system, which further extended our ability to communicate "at a distance." Many milestones were yet to follow, each contributing to our ability to communicate over long distances. Telephone networks began to expand nationwide. Television and radio broadcasts spanning the globe via satellite communications added to the steady growth of an industry: the telecommunications industry, one of the largest and fastest-growing industries in our history. Thus telecommunications has come to be regarded as *long-distance* communications via a conglomeration of information-sharing networks all tied together.

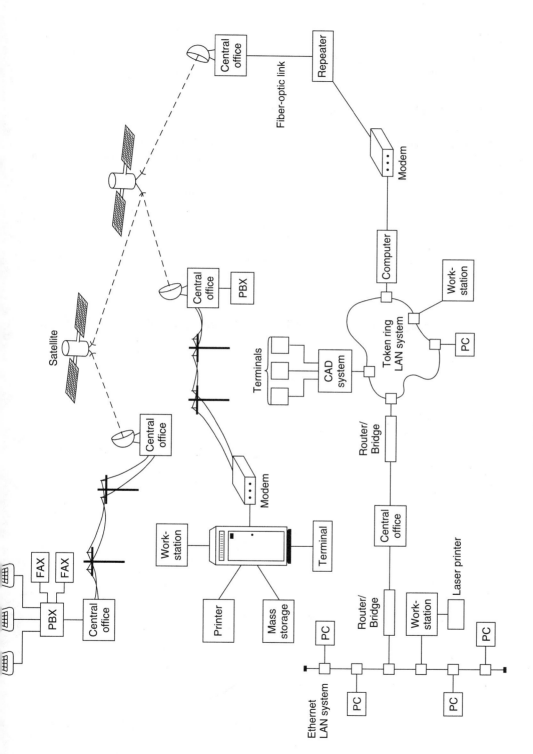

Figure 1-2 Telecommunication system.

These include the *public switched telephone network* (PSTN), radio and television networks, and lately, the emergence of worldwide computer networks. Figure 1-2 illustrates the magnitude of the telecommunication system.

1.3 THE DATA COMMUNICATION SYSTEM

In the past, most telecommunication systems transmitted voice information in *analog* form. An analog signal is one that has a continuous range of values as a function of time. Figure 1-3 illustrates the analog waveform. The PSTN, for example, was designed to accommodate voice transmission in analog form. Due to the enormous capabilities of the computer, however, the trend in telecommunications has been a gradual conversion from analog to digital transmission.

A digital transmission system transmits signals that are in *digital* form. The digital signal has discrete sets of values as a function of time, such as a binary 1 or 0. When we think of a digital waveform, we think of a pulse stream comprised of two discrete voltage levels: one for a logic high and one for a logic low. The *digital signal* can also be an analog waveform representing two discrete values. For example, a logic 1 can be represented by a 1200-Hz tone and a logic 0

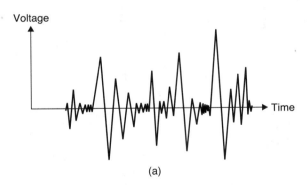

(a)

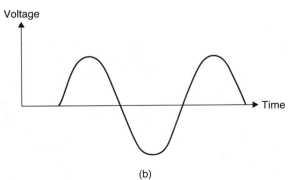

(b)

Figure 1-3 Analog waveforms: (a) analog voice signal; (b) sine wave.

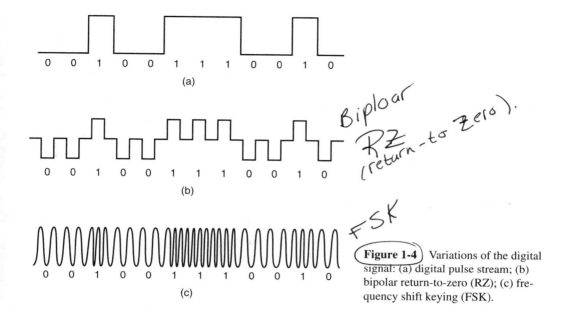

Biploar
RZ
(return-to-zero).

FSK

Figure 1-4 Variations of the digital signal: (a) digital pulse stream; (b) bipolar return-to-zero (RZ); (c) frequency shift keying (FSK).

can be represented by a 2200-Hz tone. The two discrete analog waveforms are frequency shifted from one to the other in accordance with the binary bit stream. The digital signal can also be phase shifted or amplitude shifted in the same manner. Figure 1-4 illustrates variations of the digital signal.

Data communications involves the transmission and reception of digital signals from one location to another. The digital signals typically represent information that is alphabetical, numerical, or symbolic in nature. The organization of this information is referred to as *data*. Virtually every human thought process can be represented as data and can therefore be transmitted by the data communication system. The characters that you read here are considered data. They can be transmitted by the data communication system to a printer, a disk drive, or even a remotely located terminal or computer that is a part of a *local area network* (LAN) in another city. The data communication system is actually a part of the telecommunication system that links together the capabilities of the computer.

1.4 FUTURE TRENDS IN TELECOMMUNICATIONS

Imagine living in a house where you can have access to all of the recorded music and video programs in the world without a single cassette tape or disc on the premises. Imagine a 6-foot big-screen high-definition television set (HDTV) in which you can talk to and interact with various people around the world. This vision may seem far-fetched, but the future of telecommunications is headed this way. Equipment manufacturers and service providers alike have recognized this extraordinary communications era in which we live as an opportunity for technological growth. The heated competition among them is expected to continually provide new products, new jobs, and new solutions to our ever-increasing demands.

Technological advancements in the telecommunications industry will continue to escalate in an exponential fashion. The LAN system and its dramatic growth in recent years is perhaps the most profound and visible example of this. Expansion of LAN-based systems in the workforce and into remote locations has created *metropolitan area networks* (MANs) and *wide area networks* (WANs) that span the globe. The interconnection of these networks via the PSTN have created a new generation of switching technologies, each of which promises to provide improved services to the home and workplace. Some of these technologies, such as *Asynchronous Transfer Mode* (ATM), have risen from virtual obscurity to global competition for the marketplace. Other technologies, such as *Integrated Services Digital Network* (ISDN), *Switched Multi-megabit Data Service* (SMDS), *Synchronous Optical Network* (SONET), *Fiber Distributed Data Interface* (FDDI), and more, have made considerable progress with standards organizations. These technologies combined will soon deliver integrated multimedia voice, data, text, and full-motion video to virtually any terminal throughout the world.

The socioeconomic impact resulting from the telecommunications era we are transpiring into has caused an increase in worldwide productivity through the automation of engineering, manufacturing, and office activities. The effects on modern society are more apparent than ever before. If ever there has been a time when people must be able to function and succeed in a "high-tech" world, it is now. Clearly, the advances in telecommunications in the years to come will warrant considerable effort on the part of the technologist to keep abreast of the enormous number of changes that will take place. In this book we evaluate the technical aspects of telecommunications, so vital to our educational growth and understanding.

PROBLEMS

1. Explain why it is vitally important for us to learn about telecommunications.
2. Name at least five factors that must be accounted for by today's modern communication systems.
3. Define *telecommunications*.
4. Define *data communications*.

5. Draw an example of an analog signal in the time domain.
6. Describe the difference between an *analog* signal and a *digital* signal.
7. Draw an example of a digital signal in the time domain using FSK as a modulation technique.
8. What does *ATM* stand for?
9. What do *LAN, MAN,* and *WAN* stand for?

2

Transmission Codes

Transmitted data are typically prearranged in accordance with various codes developed over the years. Many of these codes are universally recognized and have become standards, whereas others are highly cryptic and applicable to unique and limited situations. Transmission codes, such as the International Morse Code and Baudot, were developed over a century ago. They are still used in the field of digital communications. ASCII and EBCDIC are more recent codes that have gained popularity with the advent of the computer. Codes normally overlooked in the realm of data communications are those encompassed by bar code technology, a technology that has substantially improved productivity and materials management in the workforce. In each case, the intent of the code is to provide an alternative method of representing numbers, letters, and symbols.

It is often desirable to have codes readily available for development and troubleshooting purposes. Often, these codes are provided on reference cards or manuals. The intent of this chapter is to present those codes that are most widely used in today's communication systems and to serve as a reference guide for future use.

2.1 BINARY-CODED DECIMAL

Since computers function on binary information, binary bits are often grouped or encoded into a value representing a decimal character. One way to express a

TABLE 2-1 Binary-Coded Decimal

Decimal	BCD		
0	0000		
1	0001		
2	0010		
3	0011		
4	0100	1010	
5	0101	1011	
6	0110	1100	Not
7	0111	1101	valid in
8	1000	1110	BCD
9	1001	1111	

decimal character is through the use of *binary-coded decimal* (BCD). In BCD, *four* bits are used to encode one decimal character. As we know, four bits can give us a possibility of 16 binary combinations. Since there are 10 decimal characters, 0 through 9, only 10 of the 16 possible combinations are necessary for encoding in BCD. The remaining six combinations are said to be invalid numbers.

There are several BCD codes. When a decimal character is represented with straight binary bits, the BCD code is referred to more specifically as *8421 code.* As its name implies, the weighting of the four binary bits representing the decimal character are in the order of 8421. Because the 8421 code is the most widely used of the BCD codes, it is generally referred to as *BCD*. Table 2-1 depicts the BCD code.

Example 1

Convert 367_{10} to BCD.
Solution: $367_{10} = 0011\ 0110\ 0111$

Example 2

Convert 1249_{10} to BCD.
Solution: $1249_{10} = 0001\ 0010\ 0100\ 1001$

Example 3

Convert 58_{10} to BCD.
Solution: $58_{10} = 0101\ 1000$

2.1.1 BCD Addition

The addition of BCD numbers can be performed using straight binary addition as long as the result does not exceed a decimal value of 9.

Example 4

Add the decimal numbers 3 and 4 in BCD.

Solution: 3   0011
              +4   +0100
              ‾‾‾   ‾‾‾‾‾
               7   0111

Example 5

Add the decimal numbers 63 and 24 in BCD.

Solution: 63   0110 0011
              +24   +0010 0100
              ‾‾‾   ‾‾‾‾‾‾‾‾‾
               87   1000 0111

When the sum of two numbers exceeds 9, an invalid BCD number is produced. As shown in Examples 6 and 7, the invalid number is made valid by adding 0110 (6) to it.

Example 6

Add the decimal numbers 9 and 6 in BCD.

Solution: 9   1001
              +6   +0110
              ‾‾‾   ‾‾‾‾‾
               15   1111   Not a valid BCD number
                        +0110   Add 6 for correction
                        ‾‾‾‾‾‾‾‾‾
                      0001 0101 Correct BCD number

Example 7

Add the decimal numbers 46 and 79 in BCD.

Solution: 46   0100 0110
              +79   +0111 1001
              ‾‾‾   ‾‾‾‾‾‾‾‾‾‾‾
               125   1011   1111   Not valid BCD numbers
                          0110   0110   Add 6 for correction
                       ‾‾‾‾‾‾‾‾‾‾‾‾‾‾‾
                   0001 0010   0101
                     1   2   5

2.2 EXCESS-3 CODE

Another BCD code used to represent decimal numbers is the *excess-3 code*. Excess-3 code is very similar to 8421 BCD code. The only difference is that 3 is added to the decimal character before it is encoded into a four-bit word. Table 2-2 depicts the excess-3 code. A comparison is made with BCD. Notice that the excess-3 code also has six invalid characters.

TABLE 2-2 Excess-3 Code

Decimal	BCD	Excess-3
0	0000	0011
1	0001	0100
2	0010	0101
3	0011	0110
4	0100	0111
5	0101	1000
6	0110	1001
7	0111	1010
8	1000	1011
9	1001	1100

0000
0001
0010
1101 Not valid in excess-3
1110
1111

The excess-3 code is useful in some mathematical operations where 8421 BCD is too cumbersome. For example, where it is desirable to perform an addition of two 8421 BCD numbers whose sum exceeds a decimal 9, excess-3 BCD code eliminates the necessity of adding a correction of 6 to the result. Example 8 illustrates this. Notice that the sum of the two excess-3 numbers results in an answer in 8421 BCD code. This occurs due to the excess value of 3 originally encoded into each number; that is, the correction of 6 (3 in each number) is already accounted for. Another advantage of the excess-3 code is that at least one 1-bit is present in all states, thus providing an error-detection advantage.

Example 8

Add the decimal numbers 9 and 7 in excess-3.

Solution:

$$
\begin{array}{r@{\quad}r}
9 & 1100 \\
+7 & 1010 \\
\hline
16 & 1\,0110
\end{array}
$$

2.3 GRAY CODE

A disadvantage with the codes discussed thus far is that several bits can change state simultaneously between adjacent counts. The *Gray code* is unique in that successive counts result in only a single bit change. For example, when changing from a count of 7 (0111) to 8 (1000) in binary, or BCD, all four bits change state. The switching noise generated by the associated circuits may be intolerable in some environments. The same change with Gray code undergoes only a single bit change (0100 to 1100); consequently, less noise is generated. Shaft encoders used for receiver tuning dials often use Gray code for encoding the position of the rotary shaft. Gray code is also less likely to bring about errors or erroneous results due to settling times and propagation delays associated with the given hardware. Table 2-3 lists the Gray code for the decimal numbers 0 through 15.

TABLE 2-3 Gray Code

Decimal	Binary	Gray code[a]
0	0000	0000
1	0001	0001
2	0010	0011
3	0011	0010
4	0100	0110
5	0101	0111
6	0110	0101
7	0111	0100
8	1000	1100
9	1001	1101
10	1010	1111
11	1011	1110
12	1100	1010
13	1101	1011
14	1110	1001
15	1111	1000

[a] Gray code is not restricted to four bits.

From Table 2-3 it can be seen that Gray code is an *unweighted* code. That is, the bits associated with each four-bit word do not have any numerical weighting as does the binary code shown. Gray code is therefore not suitable for arithmetic operations. There is, however, a relationship between the binary code shown and Gray code. For any given binary number, the equivalent Gray code can be established by the process described below.

Binary-to-Gray Conversion

1. The *first* bit, starting from the MSB of the given binary code, becomes the leftmost bit of the Gray code.
2. EXCLUSIVE-ORing the *first* and *second* bits of the given binary code yields the second bit of the Gray code.
3. The third bit of the Gray code is found by EXCLUSIVE-ORing the *second* and *third* bits of the given binary code.
4. The process continues until the last two bits of the binary code have been EXCLUSIVE-ORed together to produce the last bit of the equivalent Gray code.

Example 9

Compute the Gray code for the binary number 11010.
Solution:

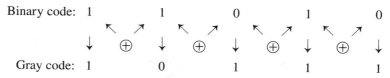

Binary code: 1   1   0   1   0

Gray code: 1   0   1   1   1

Example 10

Compute the Gray code for the binary number 10001101.
Solution:

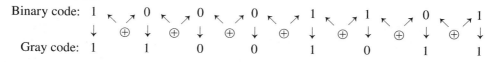

Binary code: 1   0   0   0   1   1   0   1

Gray code: 1   1   0   0   1   0   1   1

Converting from Gray code to binary is performed by the procedure described below.

Gray-to-Binary Conversion

1. The *first* bit, starting with the *left*most bit of the given Gray code, becomes the MSB of the binary code.
2. EXCLUSIVE-ORing the *second* Gray code bit with the MSB of the binary code yields the second binary bit.
3. EXCLUSIVE-ORing the *third* Gray code bit with the *second* binary bit yields the third binary bit.
4. EXCLUSIVE-ORing the *fourth* Gray code with the *third* binary bit yields the fourth binary bit, and so forth.

Example 11

Compute the binary code for the Gray code 101101.
Solution:

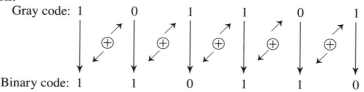

Gray code: 1   0   1   1   0   1

Binary code: 1   1   0   1   1   0

Example 12

Compute the binary code for the Gray code 11011101.

Solution:

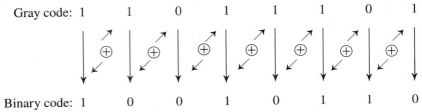

Gray code: 1   1   0   1   1   1   0   1

Binary code: 1   0   0   1   0   1   1   0

2.4 MORSE CODE

One of the oldest electrical transmission codes, still in use, was developed and patented in 1840 by Samuel F. B. Morse. The code was first used with the telegraph system. It was soon revised and has become known as *International Morse Code*. International Morse Code is used primarily in amateur radio. The digital code system is made up of a series of *dots* and *dashes* representing the alphabet and decimal number system. Punctuation is included in the code set. A dash is three times the duration of a dot. The relative duration of a dot and dash

TABLE 2-4 The International Morse Code

A	. _	N	_ .	1	. _ _ _ _
B	_ . . .	O	_ _ _	2	. . _ _ _
C	_ . _ .	P	. _ _ .	3	. . . _ _
D	_ . .	Q	_ _ . _	4	. . . . _
E	.	R	. _ .	5	. . . . .
F	. . _ .	S	. . .	6	_ . . . .
G	_ _ .	T	_	7	_ _ . . .
H	. . . .	U	. . _	8	_ _ _ . .
I	. .	V	. . . _	9	_ _ _ _ .
J	. _ _ _	W	. _ _	0	_ _ _ _ _
K	_ . _	X	_ . . _		
L	. _ . .	Y	_ . _ _		
M	_ _	Z	_ _ . .		

Period	. _ . _ . _	Fraction bar (slash)	_ . . _ .
Comma	_ _ . . _ _	Wait	. _ . . .
Question mark	. . _ _ . .	End of message	. _ . _ .
Error	. . . .	Invitation to transmit	_ . _
Colon	_ _ _ . . .		
Semicolon	_ . _ . _ .	End of transmission	. . . _ . _
Parenthesis	_ . _ _ . _	Double dash (break)	_ . . . _

International distress (SOS) . . . _ _ _ . . . (*save our ship*)

are dependent on sending speed, typically expressed in words per minute. The higher the sending speed, the shorter the dot and dash lengths become. Sending speeds vary depending on the code proficiency of the amateur radio operator. Some amateur radio operators, often referred to as *hams,* can send and receive up to 45 words per minute.

Characters in Morse code are spaced for a duration equal to that of a dash and words are spaced approximately the duration of two dashes. Table 2-4 lists the International Morse Code set.

TABLE 2-5 Baudot Code

Binary bit pattern, 54321	Letters characters, LS (↑)	Figures characters, FS (↓)
00011	A	—
11001	B	?
01110	C	:
01001	D	$
00001	E	3
01101	F	!
11010	G	&
10100	H	#
00110	I	8
01011	J	Bell
01111	K	(
10010	L	)
11100	M	.
01100	N	,
11000	O	9
10110	P	0
10111	Q	1
01010	R	4
00101	S	'
10000	T	5
00111	U	7
11110	V	;
10011	W	2
11101	X	/
10101	Y	6
10001	Z	"
00100	Space (SP)	
01000	Carriage return	
00010	Line feed	
00000	Blank	
11111	Letters shift (↑)	
11011	Figures shift (↓)	

2.5 BAUDOT

In 1874, Emile Baudot, a French engineer, developed the alphanumeric code known as *Baudot*. The Baudot code is used in the field of telegraphy and RTTY (radio teletype). CCITT (Consultative Committee for International Telephony and Telegraphy) has standardized the Baudot code into a version that has become known as CCITT Alphabet No. 2. The code is used worldwide within the international *Telex* network. The Telex network is a network of teletypes interconnected by the *public switched telephone network* (PSTN). Over a million teletype systems are still in operation throughout the world despite its slow operating speeds of 50, 75, and 110 *baud* (bits per second).

Baudot (Table 2-5) is a five-bit alphanumeric code. This allows the possibility of 32 (2^5) combinations of characters. Since there are more than 32 alphabetical and numerical characters, two of the 32 binary combinations are used for extending the character set. These two characters are the *letters shift* (LS) and *figures shift* (FS) characters. The letters shift character is a binary 11111. It is represented graphically by the upward arrow symbol (↑). When a Baudot receiver receives the letters shift character, subsequent characters are interpreted as those listed in the letters characters column of Table 2-5. The figures shift character is a binary 11011. It is represented by the downward arrow symbol (↓). Upon receiving this character, the Baudot receiver interprets characters thereafter as those listed in the figures characters column in Table 2-5. Both of these characters are nonprintable.

Baudot characters are transmitted as a series of electrical impulses, as shown in Figure 2-1. There are two possible states: a *mark,* equal to a logic 1, and a *space,* a logic 0. Each character is framed by a *start* and *stop bit.* A mark

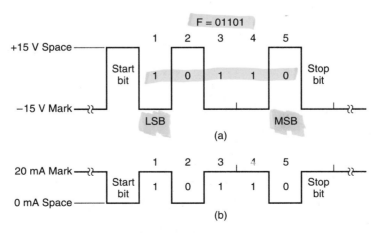

Figure 2-1 Electrical impulses for the Baudot character F, transmitted by a Teletype terminal: (a) voltage impulses; (b) current impulses.

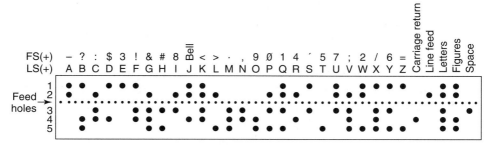

Figure 2-2 Paper tape perforated with Baudot characters.

is usually represented by a negative voltage, and a space by a positive voltage. Current pulses of 20 mA are also used, whereby 20 mA is a mark and 0 mA is a space. Note that the least significant bit is transmitted first.

Teletype machines are often equipped with a *reperforator*, a device used to punch tape with the Baudot code (Figure 2-2). The black dots represent a perforation, or mark. The absence of a perforation is a space. A complete message can be typed out in advance and sent at a later time. A special tape reader, called a *transmitter–distributor*, is used for this process. In the past, tapes were often gummed on the reverse side to allow for cutting and pasting onto a sheet of paper for documentation purposes.

Baudot code has several deficiencies compared to modern codes. The five-bit code does not allow for an error-detection mechanism such as *parity* (the parity bit will be defined later). Lowercase letters are not available, nor are the control functions offered by today's communications equipment. Another serious shortcoming of Baudot code is the time required to generate LS and FS in order to shift back and forth between letters and figures characters. Also, if a false detection of either of these characters is received, subsequent characters will be misinterpreted. These deficiencies have prevented the code from remaining popular in data and telecommunication systems.

2.6 EBCDIC

In 1962, IBM Corporation developed an eight-bit code called *Extended Binary-Coded Decimal Interchange Code* (EBCDIC). EBCDIC, often pronounced "EB-SA-DIC," is used extensively in IBM's large-scale computers and peripheral equipment. The code offers a considerable improvement over the old electromagnetic teletype code, Baudot. Lowercase alphabetical characters and control functions are included. The code (Table 2-6) is a matrix that has been designed to group characters by function. Each row and column has been identified by a hexadecimal character. The most significant hexadecimal character represents the upper four bits of the intersecting EBCDIC character. The least significant hexadecimal character represents the lower four bits of the character.

Each EBCDIC character is represented by a corresponding two-character hexadecimal quantity.

The first four rows of the matrix (0 through 3) contain *control characters*. Control characters are used to control the format and transmission of data. A description of these control characters has been provided with Table 2-6. They are listed in alphabetical order along with their corresponding hexadecimal code. Many of these control characters are used in the ASCII code described in the next section. Rows 4 through 7 contain punctuation characters, and the remaining eight rows, 8 through F, include upper and lowercase alphabetical characters and decimal numbers.

The logical placement of characters within the matrix has been designed to offer programming as well as hardware development and test advantages. The decimal digits 0 through 9, for example, are represented by F0H through F9H, respectively. The lower four bits of the code, or least significant hexadecimal characters equal the decimal number. Upper and lowercase letters have a difference of 40H; that is, only one bit, bit 6, distinguishes between the two cases (a = 81H, A = C1H).

A close look at the matrix will reveal that there are 117 blank spaces. This means that almost one-half of the matrix is not being used, or only 139 (256 − 117 = 139) of the possible 256 binary combinations are used. Since parity is not included as part of the eight-bit code, one might consider eliminating 11 of the 139 characters, thus making a total of 128 EBCDIC characters. The code could then be reduced to seven bits instead of eight ($2^7 = 128$). This would result in a reduction in transmission time of over 12%. The matrix, however, is designed to accommodate additional characters for more recently developed equipment.

2.7 ASCII

The American Standard Code for Information Interchange (ASCII) is the most widely used alphanumeric code for data transmission and data processing. ASCII was developed in 1962 specifically for computer communications systems. The code was updated by American manufacturers and published in 1967.

ASCII, pronounced "ASK-EE," is a seven-bit code that can be represented by two hexadecimal characters for simplicity. The most significant hexadecimal character in this case never exceeds 7. The ASCII code is shown in Table 2-7. Notice that there are no empty spaces. The placement of characters within the matrix, like EBCDIC, offers many of the same development and test advantages. The decimal digits 0 through 9, for example, are represented in the least significant digit of the hexadecimal characters 30H through 39H. Control characters have a value of less than 20H. They are grouped in the first two columns, 0 and 1, and are defined in numerical order next to Table 2-7. Columns 2 and 3 include punctuation and decimal numbers. The uppercase alphabet is contained in

TABLE 2-6 EBCDIC

Least significant (hex) digit

Most significant (hex) digit	0	1	2	3	4	5	6	7	8	9	A	B	C	D	E	F
0	NUL	SOH	STX	ETX	PF	HT	LC	DEL			SMM	VT	FF	CR	SO	SI
1	DLE	DC1	DC2	DC3	RES	NL	BS	IL	CAN	EM	CC		IFS	IGS	IRS	IUS
2	DS	SOS	FS		BYP	LF	EOB	PRE			SM			ENQ	ACK	BEL
3			SYN		PN	RS	UC	EOT					DC4	NAK		SUB
4	SP										¢	.	<	(	+		
5	&										!	$	*	)	;	¬
6	-	/										,	%	—	>	?
7											:	#	@	`	=	"
8		a	b	c	d	e	f	g	h	i						
9		j	k	l	m	n	o	p	q	r						
A			s	t	u	v	w	x	y	z						
B																
C		A	B	C	D	E	F	G	H	I						
D		J	K	L	M	N	O	P	Q	R						
E			S	T	U	V	W	X	Y	Z						
F	0	1	2	3	4	5	6	7	8	9						

TABLE 2-6 (continued)

Code	Abbr	Description		Code	Abbr	Description
2E	ACK	Acknowledge		1E	IRS	Interchange record separator
2F	BEL	Bell		1F	IUS	Interchange unit separator
16	BS	Backspace		06	LC	Lowercase
24	BYP	Bypass		25	LF	Line feed
18	CAN	Cancel		3D	NAK	Negative acknowledge
1A	CC	Unit backspace		15	NL	New line (LF and CR)
0D	CR	Carriage return		00	NUL	Null (all zeros)
11	DC1	Device control 1		04	PF	Punch off
12	DC2	Device control 2		34	PN	Punch on
13	DC3	Device control 3		27	PRE	Prefix
3C	DC4	Device control 4		14	RES	Restore
07	DEL	Delete		35	RS	Record separator
10	DLE	Data link escape		0F	SI	Shift in
20	DS	Digit select		2A	SM	Start message
19	EM	End of medium		0A	SMM	Repeat
2D	ENQ	Enquiry		0E	SO	Shift out
26	EOB	End of block		01	SOH	Start of header
37	EOT	End of transmission		21	SOS	Start of significance
03	ETX	End of text		40	SP	Space
0C	FF	Form feed		02	STX	Start of text
22	FS	File separator		3F	SUB	Substitute
05	HT	Horizontal tab		32	SYN	Synchronous idle
1C	IFS	Interchange file separator		36	UC	Uppercase
1D	IGS	Interchange group separator		0B	VT	Vertical tab
17	IL	Idle				

columns 4 and 5, and the lowercase alphabet is contained in columns 6 and 7. Upper and lowercase alphabetical characters can be distinguished by a difference of 20H. An uppercase A, for example, is a 41H, whereas a lowercase a is 61H. Various other nonalphabetical characters are distributed throughout columns 4 through 7.

A *parity* bit is generally provided with the seven-bit ASCII character, thus making the overall character length equal to a byte. The parity bit is an optional bit that takes the position of the MSB of the byte. Parity is used for error-detec-

TABLE 2-7 ASCII

Least significant (hex) digit	Most significant (hex) digit							
	0	1	2	3	4	5	6	7
0	NUL	DLE	SP	0	@	P	`	p
1	SOH	DC1	!	1	A	Q	a	q
2	STX	DC2	"	2	B	R	b	r
3	ETX	DC3	#	3	C	S	c	s
4	EOT	DC4	$	4	D	T	d	t
5	ENQ	NAK	%	5	E	U	e	u
6	ACK	SYN	&	6	F	V	f	v
7	BEL	ETB	'	7	G	W	g	w
8	BS	CAN	(	8	H	X	h	x
9	HT	EM	)	9	I	Y	i	y
A	LF	SUB	*	:	J	Z	j	z
B	VT	ESC	+	;	K	[	k	{
C	FF	FS	,	<	L	\	l		
D	CR	GS	-	=	M	]	m	}
E	SO	RS	.	>	N	^	n	~
F	SI	US	/	?	O	___	o	DEL

00	NUL	Null
01	SOH	Start of header
02	STX	Start of text
03	ETX	End of text
04	EOT	End of transmission
05	ENQ	Enquiry
06	ACK	Acknowledge
07	BEL	Bell
08	BS	Backspace
09	HT	Horizontal tab
0A	LF	Line feed
0B	VT	Vertical tab
0C	FF	Form feed
0D	CR	Carriage return
0E	SO	Shift out
0F	SI	Shift in
10	DLE	Data link escape
11	DC1	Device control 1
12	DC2	Device control 2
13	DC3	Device control 3
14	DC4	Device control 4
15	NAK	Negative acknowledge
16	SYN	Synchronous idle
17	ETB	End of transmission block
18	CAN	Cancel
19	EM	End of medium
1A	SUB	Substitute
1B	ESC	Escape
1C	FS	File separator
1D	GS	Group separator
1E	RS	Record separator
1F	US	Unit separator
7F	DEL	Delete

tion purposes. When parity is not used, the MSB position of the byte representing the ASCII character is normally set to a logic 0. Some terminals and computers have the feature of disabling parity and maintaining a set or reset condition of this bit position.

2.7.1 ASCII Control Characters

ASCII has a total of 32 control characters (00H to 1FH), most of which are included in the EBCDIC character set. As stated earlier, control characters are used to control the format and transmission of data. Most control characters are classified into one of four categories:

1. Device control
2. Format effectors
3. Information separators
4. Transmission control

2.7.1.1 Device control. The ASCII code includes four *device control* characters: DC1, DC2, DC3, and DC4. There are no specific functions for these characters. However, the original intent was to accommodate the control of computers and related equipment.

The control characters DC1 (XON) and DC3 (XOFF) have become a de facto standard for use in what is referred to as *flow control*: the process of regulating the flow of data from one point to another. Consider a computer sending a long text file to a printer. Because the printer operates at a relatively slow speed in comparison to the transmission rate of the computer, a buffer is used within the printer to temporarily store characters as they are being printed. To prevent the buffer from eventually overflowing, device control characters DC1 and DC3 are typically used by the printer to turn the transmitting source, in this case the computer, on and off. This type of flow control process is commonly referred to as XON–XOFF, whereby

$$DC1 = XON \text{ (transmit on)}$$
$$DC3 = XOFF \text{ (transmit off)}$$

2.7.1.2 Format effectors. *Format effectors* are used to control a video display terminal's (VDT) cursor movement or the positioning of a printing head on a printer. This allows an operator to control the physical layout of printed material. There are six ASCII format effectors, most of which can be found on an ordinary typewriter. Their values in hexadecimal range from 08H to 0DH.

08	BS	Backspace
09	HT	Horizontal tabulation
0A	LF	Line feed
0B	VT	Vertical tabulation
0C	FF	Form feed
0D	CR	Carriage return

2.7.1.3 Information separators. Five *information separators* are listed in Table 2-7:

1C	FS	File separator
1D	GS	Group separator
1E	RS	Record separator
1F	US	Unit separator
19	EM	End of medium

Information separators are used as delimiters in the organization of file transfers. A hierarchy can be formed in the order listed above. The *unit separator,* US, delimits the smallest unit of information within the file. The *record separator*, RS, delimits a record of information that contains so many units. The *group separator,* GS, delimits a number of records contained in the group, and the *file separator,* FS, contains the groups of data. *End of medium*, EM, is used to indicate the end of a file transfer in cases where different storage mediums are used.

2.7.1.4 Transmission control. The 10 *transmission control* characters listed next are used for flow control and for framing blocks of data. In Chapter 10 we consider the use of these characters for synchronous serial transmission.

01	SOH	Start of header
02	STX	Start of text
03	ETX	End of text
04	EOT	End of transmission
05	ENQ	Enquiry
06	ACK	Acknowledge
10	DLE	Data link escape
15	NAK	Negative acknowledge
16	SYN	Synchronous idle
17	ETB	End of transmission block

2.7.1.5 Other ASCII control characters. There are seven remaining ASCII control characters:

00	NUL	Null
07	BEL	Bell
0E	SO	Shift out
0F	SI	Shift in
18	CAN	Cancel
1A	SUB	Substitute
1B	ESC	Escape

NUL: The null byte is a string of zeros that is used as a programming and test aid. It is often used to identify the end of a string of text. It is also used as a redundant byte to fill memory.

BEL: When the BEL character is received by a terminal or computer, an audible response is given. Its purpose is to gain the attention of the operator.

SO: Shift out is used to shift out of the normal ASCII character set and into an extended character set. The extended character set may include graphic symbols, mathematical symbols, and so on.

SI: Shift in is used to return to the normal ASCII character set.

CAN: Cancel is used to suspend or terminate an escape or control sequence. An error character is typically displayed on the screen.

SUB: Substitute is interpreted the same as CAN.

ESC: Escape is an ASCII character used for sending special command and control functions. ESC, followed by a single character or string of characters, is called an *escape sequence.* Escape sequences are used to perform special functions, such as cursor positioning, clearing the screen, and editing of text. Various commands can be sent with escape sequences. A modem or terminal, for example, can be commanded to initiate a self-test.

2.8 BAR CODES

Those ubiquitous black-and-white stripes that we regularly encounter in our supermarkets are called *bar codes.* They are marked on everything from boxes of cereal and cartons of milk to cans of shaving cream and bottles of aspirin. Most of us have observed how our grocery items are pulled across a viewing window at the cashier's checkout stand. Through this viewing window, a laser beam scans the bar code to identify the item being purchased and its corresponding price. The key to this automated system is the bar code, a data communications technology that applies many of the latest digital signal-processing techniques and optoelectronic devices toward the identification of commercial, industrial, and consumer products.

2.8.1 What Is Bar Code?

The bar code (Figure 2-3) is a series of consistently sized white and black bars that are wide and narrow in width. The white bars, separating the black bars, are referred to as *spaces*. Wide and narrow bars and spaces are used because they ultimately find themselves being deciphered by a digital computer and therefore their widths and reflective abilities represent binary 1's and 0's. Bar codes are

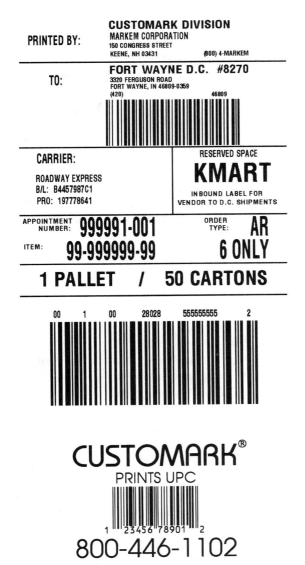

Figure 2-3 Typical bar code labels. (Courtesy of Benjamin Nelson, Markem Corp., Keene, N.H.)

TABLE 2-8 Advantages and Disadvantages of Bar Code

Advantages	Disadvantages
High throughput	Extra equipment required
Increased productivity	Computer system and special application
Low cost	software may be required
Little operator training	Primary use is for fixed, repetitive data
High data accuracy	input, not variable data input
Readable by contact and	
noncontact scanners	

typically painted on, pasted on, or burned into the item that is being tracked—a container or box, an automobile engine, a library book, and so on.

Bar codes have been around since the early 1970s. The technology has advanced into a multibillion-dollar industry that has automated the keyboard entry task far beyond the local grocery store. More and more applications requiring an accurate, easy, and inexpensive method of data storage and data entry for computerized information management systems are utilizing bar code technology. Table 2-8 lists the advantages and disadvantages of bar code. Let us now consider some of the most common applications of bar code technology today:

— Inventory management and control
— Security access
— Shipping and receiving
— Production counting
— Document and order processing

— Work progress management
— Asset management
— Functional programming
— Data entry
— Automatic billing

The basic bar code structure is shown in Figure 2-4 and a block diagram illustrating the key elements of the bar code system is shown in Figure 2-5. The *data characters* field of the bar code represents a character corresponding to the bar code symbology or format used. Serial data stored in the data characters field are extracted from this field with an optical scanner. The optical scanner develops a logic signal corresponding to the difference in reflectivity of the printed bars and underlying spaces (white bars). Scanning over the printed bars and spaces with a smooth, continuous motion is required to retrieve data correctly. This motion can be provided by an operator in basically one of three ways: moving a hand-held wand, by a rotating mirror moving a collimated beam of light, or by an operator or conveyor system moving the symbol past a fixed beam of light. The scanner is typically a laser, light-emitting diode (LED), or an incandescent light bulb. A photodetector is used by the scanner to sense the reflected light and convert it to electrical signals for decoding.

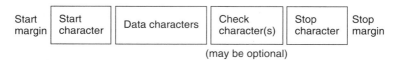

| Start margin | Start character | Data characters | Check character(s) | Stop character | Stop margin |

(may be optional)

Figure 2-4 Basic bar code structure.

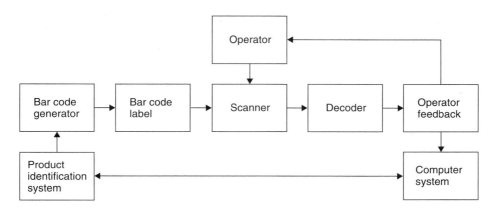

Figure 2-5 Block diagram of a bar code system.

TABLE 2-9 Common Industrial Bar Codes

Year introduced	Code	Type of code[a]	Industry used in
1972	Interleaved 2 of 5	Numeric, continuous	Industrial, retail
1972	Codabar	Numeric with special characters, discrete	Blood banks, libraries, postal services
1973	UPC	Numeric, continuous	Food and general retail, magazine and book publishing, liquor stores
1974	Code 39	Alphanumeric, discrete	Military, industrial, health care, automotive
1976	EAN	Numeric, continuous	Food retail, product marking
1977	Code 11	Numeric, discrete	Telecom equipment tracking
1981	Code 128	Full ASCII, continuous	Industrial, retail case codes
1982	Code 93	Full ASCII, continuous	Industrial

[a] *Discrete code:* a bar code that has intercharacter spaces or gaps between characters. Consequently, each character within the bar code symbol is independent of every other character. (See Figure 2-6(c) for an illustration of intercharacter space.)

Continuous code: a bar code that does not have intercharacter spaces as part of its structure.

2.8.2 Bar Code Formats

There are several standard bar code formats used in industry. Which format is best suited for a given application depends on what type of data are being stored, how they are being stored, system performance, and what is currently being used by the majority of similar businesses one is engaged in. Table 2-9 lists the most common industrial codes used today and the type of industry in which they are being used. The type of code and year of inception are also listed. Although each of these codes has advantages and disadvantages over the others, we limit our discussion to two bar code formats: *Code 39* and *Universal Product Code* (UPC).

2.8.3 Code 39

The most popular alphanumeric bar code is *Code 39,* also known as *Code 3 of 9* or *3 of 9 Code.*[†] The code consists of 36 defined numeric and uppercase alphabetic characters, seven special characters, and a special start/stop character decoded as an asterisk(*). The Code 39 character set along with its structure and label density is shown in Figure 2-6.

Code 39 characters can be deciphered by noting that each character is comprised of nine vertical *elements,* which are defined as bars and spaces (a space is a white bar). Each element is "width modulated" to encode the logic value of the nine binary bits of data. Elements can be either a logic 1 or 0. A wide element, bar or space, is encoded as a logic 1, and a narrow element, bar or space, is encoded as a logic 0. A 3:1 ratio in widths is used to distinguish between a logic 1 and 0 (i.e., a bar or space representing a logic 1 is three times the width of a bar or space representing a logic 0). For higher resolutions or greater information density, a 2.2:1 ratio is used.

Three of the nine elements in Code 39 are always a logic 1, and the remaining six are always a logic 0, hence the name Code 3 of 9. This can be seen in the Code 39 character set shown in Figure 2-6(a). Furthermore, of the three wide elements, two are bars and one is a space. Each Code 39 character begins and ends with a bar (black) with alternating spaces (white) in between. Individual characters within the label are separated by an *intercharacter space* (CS), which classifies Code 39 as a *discrete code.* Discrete codes are codes whose characters are separated by gaps. The CS, or "gap" between characters, is nominally one element (wide or narrow) in width.

[†] Code 39 is a registered trademark of Interface Mechanisms Inc. (now Intermec). The symbology is public domain.

ASCII Character	Binary Word	Bars	Spaces	Check Character Value
0	000110100	00110	0100	0
1	100100001	10001	0100	1
2	001100001	01001	0100	2
3	101100000	11000	0100	3
4	000110001	00101	0100	4
5	100110000	10100	0100	5
6	001110000	01100	0100	6
7	000100101	00011	0100	7
8	100100100	10010	0100	8
9	001100100	01010	0100	9
A	100001001	10001	0010	10
B	001001001	01001	0010	11
C	101001000	11000	0010	12
D	000011001	00101	0010	13
E	100011000	10100	0010	14
F	001011000	01100	0010	15
G	000001101	00011	0010	16
H	100001100	10010	0010	17
I	001001100	01010	0010	18
J	000011100	00110	0010	19
K	100000011	10001	0001	20
L	001000011	01001	0001	21
M	101000010	11000	0001	22
N	000010011	00101	0001	23
O	100010010	10100	0001	24
P	001010010	01100	0001	25
Q	000000111	00011	0001	26
R	100000110	10010	0001	27
S	001000110	01010	0001	28
T	000010110	00110	0001	29
U	110000001	10001	1000	30
V	011000001	01001	1000	31
W	111000000	11000	1000	32
X	010010001	00101	1000	33
Y	110010000	10100	1000	34
Z	011010000	01100	1000	35
–	010000101	00011	1000	36
.	110000100	10010	1000	37
SPACE	011000100	01010	1000	38
*	010010100	00110	1000	–
$	010101000	00000	1110	39
/	010100010	00000	1101	40
+	010001010	00000	1011	41
%	000101010	00000	0111	42

Check character example:

Message: CODE 39

Characters:	C	O	D	E		3	9
Value:	12	24	13	14	38	3	9

Sum of character values: 113

$$113 \div 43 = 2 \text{ remainder } 27$$
$$27 = \text{R check character}$$

Final message: CODE 39R

Figure 2-6(a) Code 39 character set.

CODE CONFIGURATION

CODE 39 derives its name from its structure which is 3 out of 9. Each character is represented by 9 elements (5 bars and 4 spaces between the bars): 3 of the 9 elements are wide (binary value 1) and 6 elements are narrow (binary value 0). Spaces between characters have no code value. The specific structure of each character is given in the table below.

At the standard printing density of 9.4 characters per inch, the narrow bars and spaces are .0075 inches in width and the wide bars and spaces are .0168 inch. In other printing densities, the nominal ratio of wide to narrow elements should exceed 2.2:1 but be no greater than 3:1. Figure 2-6(c) shows dimensions (enlarged) for standard 9.4 character-per-inch bar code.

The start/stop (∗) code precedes and follows all encoded data, defining the beginning and end of the code. These symbols typically are not transmitted by the bar code scanner.

A white margin must be present at both ends of the printed code. The minimum white margin is one-half of a nominal character. If practical, the margin should exceed 1/4 inch so that an operator can easily begin scanning in the margin.

CODE 39 Code Configuration

CHAR.	PATTERN	BARS	SPACES	CHAR.	PATTERN	BARS	SPACES
1		10001	0100	M		11000	0001
2		01001	0100	N		00101	0001
3		11000	0100	O		10100	0001
4		00101	0100	P		01100	0001
5		10100	0100	Q		00011	0001
6		01100	0100	R		10010	0001
7		00011	0100	S		01010	0001
8		10010	0100	T		00110	0001
9		01010	0100	U		10001	1000
0		00110	0100	V		01001	1000
A		10001	0010	W		11000	1000
B		01001	0010	X		00101	1000
C		11000	0010	Y		10100	1000
D		00101	0010	Z		01100	1000
E		10100	0010	-		00011	1000
F		01100	0010	.		10010	1000
G		00011	0010	SPACE		01010	1000
H		10010	0010	∗		00110	1000
I		01010	0010	$		00000	1110
J		00110	0010	/		00000	1101
K		10001	0001	+		00000	1011
L		01001	0001	%		00000	0111

∗ Denotes a start/stop code which must precede and follow every bar code message.
Note that ∗ is used only for the start/stop code.

Figure 2-6(b) Code 39 code configuration (courtesy of Benjamin Nelson, Markem Corp., Keene, N.H.).

CS = intercharacter space

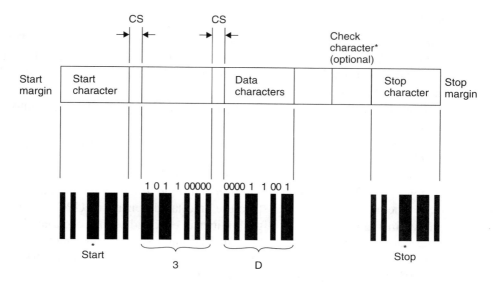

*An optional checksum character may be added at the end of a Code 39 message. This character is used to check that the correct number and type of data is present, thus providing data security.

Figure 2-6(c) Code 39 label structure (courtesy of Hewlett-Packard Co.).

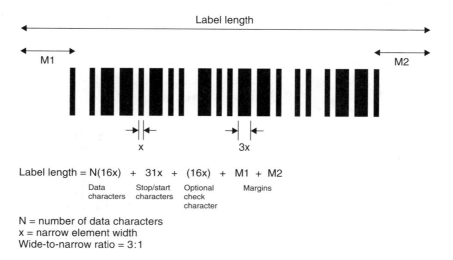

Label length = N(16x) + 31x + (16x) + M1 + M2

Data	Stop/start	Optional	Margins
characters	characters	check	
		character	

N = number of data characters
x = narrow element width
Wide-to-narrow ratio = 3:1

Figure 2-6(d) Code 39 label density (courtesy of Hewlett-Packard Co.).

2.8.4 Universal Product Code

Universal Product Code (UPC) was developed by the grocery industry in the early 1970s for product identification. It was officially adopted in 1974 by the National Association of Food Chains. A UPC bar code label can be found on your can of soup or box of cereal or crackers. There are three versions: A, D, and E. Version A, the regular version, is used for encoding a 12-digit number. Version E (zero-suppressed version) is used for encoding 12 digits into 6. It is used primarily for labeling small packages. Version D is a variable-length version limited to special applications such as credit cards. International interest in UPC led to the adoption of JAN (Japanese Article Numbering) and EAN (European Article Numbering). Both JAN and EAN codes have several features in common with UPC. Our discussion is limited to UPC, Version A.

The character set, label structure, and label density of UPC, Version A, are shown in Figure 2-7. A 12-digit number is encoded within each label. The two long bars on the outermost left- and right-hand sides of the label are called the *start guard pattern* and *stop guard pattern,* respectively. They form a binary pattern of 101 (bar space bar) and are used to frame the 12-digit number. Six of the 12 numbers (five data characters and one number system character) are encoded on the left half of the label. These digits are called the *left-hand characters.* The remaining six numbers (five data characters and one check character) are encoded on the right-hand side of the label. These are called the *right-hand characters.* The two halves of the label are separated by two long bars in the center of the label that are called the *center guard pattern.* The UPC guard pattern is 01010, and therefore the two long bars are separated by a space in between and to either side of the bars, as illustrated in Figure 2-7(b) and (c).

Left-hand character	Decimal number	Right-hand character
0001101	0	1110010
0011001	1	1100110
0010011	2	1101100
0111101	3	1000010
0100011	4	1011100
0110001	5	1001110
0101111	6	1010000
0111011	7	1000100
0110111	8	1001000
0001011	9	1110100

Figure 2-7(a) UPC A character set.

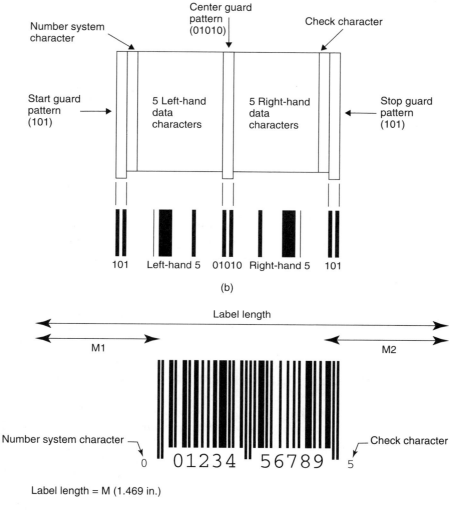

Figure 2-7(b) UPC A label structure; **(c)** UPC A label density. (Courtesy of Hewlett-Packard Co.)

The first of the 12 numbers encoded is referred to as the *UPC number system character.* The number system character identifies how the UPC symbol is used. For example, a number system character of 5 implies that the item is for use with a coupon. Table 2-10 lists the 10 possible usage characters. The next 10 digits, five on each side, are data characters. Data characters specify the de-

TABLE 2-10 UPC Number System Characters

Character	Usage
0	Regular UPC Codes
1	Reserved
2	Random weight items which are symbol marked at the store level
3	National Drug Code and National Health Related Items Code
4	For use without code format restrictions and with check digit protection for in-store marking of non food items
5	For use on coupons
6	Regular UPC Codes
7	Regular UPC Codes
8	Reserved
9	Reserved

tails of the product. The last digit, or twelfth digit, called the *check character,* is used for error detection. Notice that the number system character is always printed in decimal to the left of the UPC label. On many UPC labels, the check character is printed in decimal to the right of the label.

Unlike Code 39, the width of the bars and spaces for UPC does not correspond to binary 1's and 0's. Instead, individual characters, 0 through 9, are encoded into a combination of two variable-width bars and two variable-width spaces, which occupy a total of seven *modules,* as shown in Figure 2-8. A left-hand 5 and a right-hand 5 are shown here. Bars and spaces making up the seven modules correspond to the seven 1's and 0's representing the decimal character. A single bar represents a binary 1, or up to four consecutive 1's (of the seven-bit code), as in the left-hand characters representing the decimal numbers 3 and 6. A single space represents a binary 0, or up to four consecutive 0's, as in the right-hand characters representing the decimal numbers 3 and 6 (see the UPC character

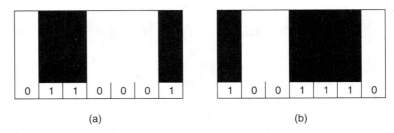

(a)   (b)

Figure 2-8 (a) Left and (b) right hand UPC A characters representing the digit 5.

set in Figure 2-7a). A close look at the UPC character set will reveal that the seven-bit binary code representing each of the 10 decimal characters does in fact make up two bars and two spaces for each UPC character. In addition, the left-hand characters always have odd parity, so that the number of modules that are bars are always odd, while the parity for right-hand characters is always even.

Another distinction between Code 39 and UPC is that UPC is a *continuous code,* meaning that there are no intercharacter spaces between characters. Instead, characters are directly adjacent to each other.

PROBLEMS

1. Specify the following decimal numbers in BCD.
    (a) 7
    (b) 123
    (c) 96
2. Convert the following decimal numbers to BCD and perform an arithmetic addition in BCD.
    (a) 3 and 11
    (b) 789 and 165
    (c) 14 and 92
3. Add the decimal numbers 9 and 8 in excess-3 code and explain why the result is a valid BCD number.
4. Determine the Gray code for the following binary numbers.
    (a) 1010111
    (b) 1100
    (c) 11101100
    (d) 10001
5. Determine the binary code for the following Gray code numbers.
    (a) 1101
    (b) 10000001
    (c) 10011110
    (d) 110011
6. Write the sequence of Morse code characters for the message "TESTING 1 2 3 4. DO YOU READ ME?" Use adequate spacing to separate characters and words.
7. Draw the voltage waveform for the Baudot character "W." Include start and stop bits.
8. Refer to Figure 2-2. Draw the paper tape perforations that would be produced by a Teletype terminal for the same message as in Problem 6.
9. How are upper and lowercase letters distinguished from each other in EBCDIC?
10. What unique bit pattern is associated with EBCDIC decimal characters?

11. How many control characters are there in ASCII, and what can be said about their codes?
12. How many device control characters are there in ASCII, and what are their names?
13. Which ASCII control character is represented by a binary string of all zeros?
14. What is the ASCII code for "ACK" in binary, hexadecimal, and octal?
15. What are format effectors, and how are they used?
16. Name at least three applications of bar code.
17. Name two advantages and two disadvantages of bar code.
18. What is a *continuous* bar code?
19. What is a *discrete* bar code?
20. What is the binary word for the ASCII character "$" in Code 39?
21. What is the binary word for the ASCII character "R" in Code 39?
22. Draw the Code 39 bar code pattern for the ASCII character "R."
23. Draw the Code 39 bar code symbol for the ASCII character "$."
24. Draw the UPC A bar code symbol for a left-hand 7.
25. Draw the UPC A bar code symbol for a right-hand 8.

3

Terminals

Over the years, the *terminal* has clearly become one of the most visible and readily identifiable pieces of computer-related equipment. The terminal, simply stated, is an input/output device used by an operator to communicate with a host computer. It includes a keyboard that can generate an alphanumeric character set, a display (or printer) to monitor alphanumeric characters, and a communications interface. Terminals can be electromechanical, a Teletype terminal for example (TTY), or a more sophisticated microprocessor-controlled video display terminal (VDT).

Over the past few decades, technology has brought a variety of new features to the terminal. These new features, coupled with technological advances in other segments of the computer market, have resulted in a great deal of confusion over what is best suited for a given application. Classifications of *dumb, smart,* and *intelligent* terminals have surfaced in order to help define a terminal type. In this chapter, an overview of some of the basic features of today's terminals, including their classifications, is presented.

3.1 TERMINAL CLASSIFICATIONS

In the midst of tough competition among manufacturers of personal computers and graphics workstations, the terminal has taken on a new meaning. Terminal manufacturers have succumbed to the attitude "if you can't beat 'em, join 'em," with virtually all terminals manufactured today being microprocessor controlled. Computer-intensive applications for CAD/CAM (computer-aided design/computer-aided manufacturing), medical, research, and business continue to pressure manufacturers into pumping more power and capabilities into their terminals. In spite of the continued growth in the terminal market, many of the resulting improvements have been subtle. Often these improvements are needed only in specific installations. High-resolution graphics terminals and *ergonomically* improved displays and keyboards are often justified, but not always necessary. The end result is an extensive range of terminals with varying degrees of features. It has therefore become necessary to classify terminals into three general categories that suit the end user's needs. These categories are the *dumb terminal,* the *smart terminal,* and the *intelligent terminal* or *workstation.*

3.1.1 Dumb Terminals

There are literally millions of dumb terminals on the market, still being used today and still adequately serving the needs of their users. In the past, a dumb terminal was simply one that was not microprocessor controlled. Transmission parameters were manually set through the use of DIP (dual-in-line package) switches. Virtually all of today's terminals, including dumb terminals, are microprocessor controlled and sophisticated enough to offer programmable communications features, including self-test capabilities. Thus what distinguishes the dumb terminal from other classes is not clear. A precise definition is subject to change over the years. Generally, the dumb terminal offers a limited number of features, yet provides the essential input/output capability necessary to send and receive data from a host computer. A block diagram of Zentec Corporation's ADM-3A terminal is shown in Figure 3-1.

3.1.2 Smart Terminals

Smart terminals offer programmable features that require extensive memory, both ROM and RAM. In addition to the normal features that a dumb terminal offers, the distinguishing feature of the smart terminal is its capability to transmit *blocks* of data, that is, a *block mode* of operation. Other features include the following:

— Programmable communications setup
— Screen editing
— Display enhancement
— Alternative character sets
— Protocol emulation

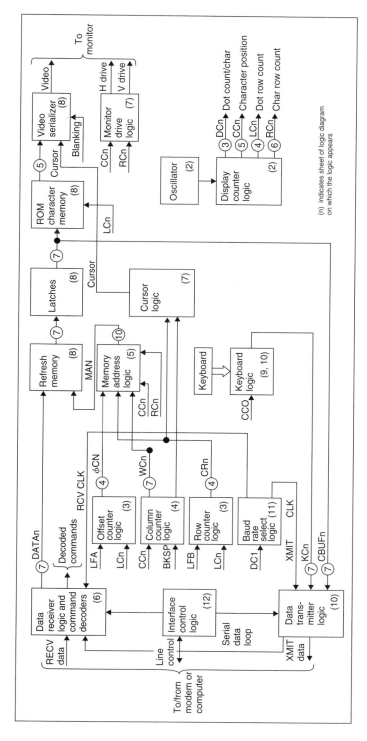

Figure 3-1 Block diagram of Zentec Corporation's ADM-3A terminal. (Reprinted with permission from Zentec Corporation.)

3.1.3 Intelligent Terminals

Intelligent terminals or *workstations* are the third category of terminals. These types of terminals promise to transform the way in which we work. The workstation features a complete operating system with a high-resolution graphics display. Built-in communications protocols allow the interactive use of shared resources, including those of other workstations on a network. Microcomputing power with 32-bit architecture is necessary to obtain the highest level of workstation performance. High-resolution graphic displays, managed independently by a coprocessor, allow users to create visual aids that would otherwise be difficult, if not impossible, to understand. Bar graphs, pie charts, and line graphs, for example, are simple ways to make information readily understandable. Three-dimensional models can be created on the workstation for engineering, manufacturing, government, scientific, and educational applications. More complex applications performed by the workstation may involve *modeling* and *simulation*. An example could be the modeling and simulation of atmospheric and oceanic conditions resulting from temperature changes. Other applications of the workstation include artificial intelligence (AI), computer-aided publishing, finance, and earth resources.

Many manufacturers of workstations have embraced the use of *X Windows*. X Windows, developed at MIT (Massachusetts Institute of Technology), provides high-performance, high-level, device-independent graphics. Using X Windows, the workstation can run a program on another computer, interacting with that computer as if it were the user's own workstation. Several "windows" can be opened up simultaneously on the user's display, allowing the user to view and work with several different programs simultaneously from different systems. Software provided within the workstation permits the emulation of various other types of terminals, including their communications protocols, to allow the presentation of windows across the network. Intelligent terminals are rapidly becoming more and more prevalent in business offices, laboratories, and educational institutes

3.2 ERGONOMICS

Recently, considerable attention has been given to the science of *ergonomics:* the study of people in adjusting to their working environments. This environment includes everything from equipment, layout, temperature, lighting, and sound levels, to the atmosphere and social context of the task. Ergonomics seeks ways in which we may improve the working conditions within this environment.

For the millions of people who work day in and day out on VDTs, terminal manufacturers are considering more seriously the ergonomics of terminal design. Consider the following:

— Keyboard layout
— Thickness and size of keys
— Sculptured key caps
— Variable phosphor color displays
— Mouse/joystick/data tablet entry
— Color graphic displays
— Slope of keyboard
— Keyboard detachability
— Key click intensity
— Independent numeric keypads
— Swivel/tilt display mounts
— Touch and light pen data entry
— Reflection free screens
— Clock/calendar function

Numerous studies have been conducted by ergonomic specialists on the human factors involved in the use of terminals. These include keyboard and display discomforts, lighting and glare control, eye strain, and radiation effects. While some studies indicate that keyboards should be flat, others contend that they should be pitched from front to back and from center to side. Studies on phosphor colors have indicated that some colors may strain the eyes more than others. Amber, for example, may be better for your eyes than green. However, some reports indicate that green produces fewer headaches. Questions raised about whether VDTs possibly emitting unsafe levels of radiation have led to several investigations on radiation emissions. One such study performed by the American Council on Science and Health reports that there is no scientific evidence indicating that VDTs cause birth defects or miscarriages.[†]

Ultimately, it has been decided by terminal manufacturers that the ergonomic design of terminals is largely a matter of personal preference. For this reason, many terminals are equipped with multicolor modes, swivel mounts for the display, detachable keyboards, and a number of keyboard layout designs to choose from. Wyse Technology's popular WY-60, shown in Figure 3-2, includes many of the latest ergonomic design features.

3.3 THE ASCII TERMINAL

Most terminals manufactured today are alphanumeric or ASCII character terminals used for data entry. Although the number of features built into these terminals continues to grow, most manufacturers adhere to the standard functional guidelines set forth in ANSI (American National Standards Institute) standards.

[†] *Health and Safety Aspects of VDTs,* 2nd ed., a report by the American Council on Science and Health, 1985.

Figure 3-2 Wyse Technology's WY-60 ASCII terminal. (Courtesy of Wyse Technology.)

3.3.1 Alphanumeric Display

The display most often used for the ASCII terminal is the cathode-ray tube (CRT). High-voltage and sweep circuits are contained within the CRT display section. A raster scan deflection system is used to control an electron beam that is swept horizontally across the screen from top to bottom. The procedure is similar to that used by a television set. Characters are formed by a series of line segments and dots that are produced by turning the CRT's beam on and off at the appropriate time as it traverses the screen. For many terminals, characters are represented by a 5 × 7 or 7 × 9 dot matrix pattern on a display of typically 12 to 15 in. (diagonal measurement). Included in the matrix is two-dot spacing between characters, both horizontally and vertically. Figure 3-3 depicts the formation of uppercase letters for a 5 × 7 dot matrix pattern.

3.3.2 Screen Capacity

A terminal's screen capacity is the maximum number of characters that can be displayed. Characters are typically arranged in a 24-row by 80-column format for a screen capacity of 1920 characters (24 × 80 = 1920). More recent terminals offer 24 rows by 132 columns for a screen capacity of 3168 characters. Figure 3-4 illustrates the standard 1920-character format.

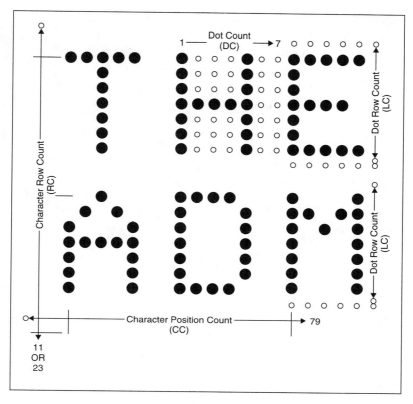

Figure 3-3 CRT display formation of a 5 X 7 dot matrix pattern. (Reprinted with permission from Zentec Corporation.)

3.3.3 Keyboard Layout

The keyboard is the user's direct link to the computer. It is by far the most common input device to a computer. The placement of keys and their respective spacing are of major concern to the operator, who spends a great part of the day entering data. Keys are typically sculptured to fit the shape of the user's fingertips. Alphabetic and numeric keys have been standardized to the layout of a typewriter. Also, the popular numeric keypad has been designed into most terminal keyboards to offer the same rapid data entry found in adding machines and calculators.

Digital Equipment Corporation's VT300 series keyboard layout is shown in Figure 3-5. Introduced in 1986, the VT320 almost instantly assumed its role as a standard. We will use the VT320 keyboard layout to describe the functions of each key.

Figure 3-4 Screen capacity arranged in the standard 24-row by 80-column (1920 characters) format. The 24-row by 132-column (3168 characters) format is also becoming popular. (Courtesy of Digital Equipment Corporation.)

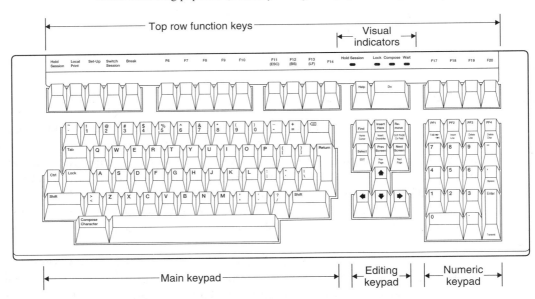

Figure 3-5 Keyboard layout for Digital Equipment Corporation's VT300 series video display terminal (North American/United Kingdom version). (Reprinted with permission from Digital Equipment Corporation.)

TABLE 3-1 Definition of Function Keys

Key	Description
Tab	Generates a horizontal tab that normally moves the cursor and the text following it to the next tab stop setting.
CTRL	When pressed in combination with another key, the CTRL key causes the terminal to transmit a code that has a special meaning to your system.
Lock	When pressed, the LOCK key makes the alphabetic keys generate uppercase characters. When the LOCK key is pressed again, the alphabetic keys generate lowercase characters.
Shift (two keys)	When either the right or left side SHIFT key is pressed, the uppercase function of all keys is enabled. If a key does not have an uppercase function, the SHIFT key will be disregarded. In some cases this key is used in combination with another key to generate a predefined control function.
Return	Transmits either a carriage return (CR) code or a carriage return (CR) and linefeed (LF) code. In some cases it moves the cursor to the next line when editing text. If NEW LINE was selected in Set-Up mode, RETURN can be a signal to the applications program that a particular operation is finished.
Delete	Pressing this key generates a DE character. Normally, this erases one character to the left of the cursor.
Compose Character	This key is used to create special characters that do not exist as standard keys on your keyboard. Use of this key and compose character sequences are described in detail in the *VT220* and *VT240 Owner's Manual*.

Source: Digital Equipment Corporation.

3.3.4 Function Keys

Function keys and their definitions are listed in Table 3-1. They are used to command the terminal's microprocessor to perform special functions based on programs stored in firmware. Figure 3-6 illustrates their relative locations on the main keypad.

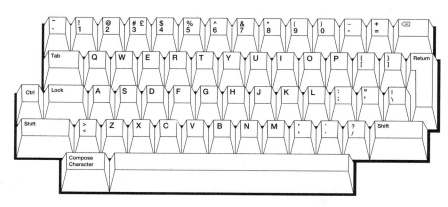

Figure 3-6 Main keypad for the VT300 series terminal. (Reprinted with permission from Digital Equipment Corporation.)

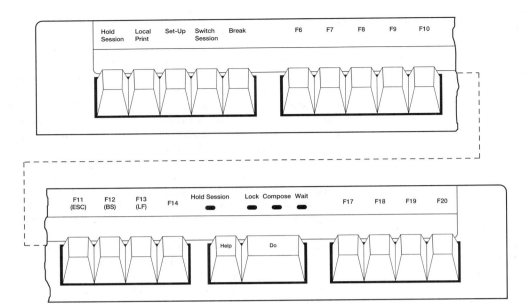

Figure 3-7 VT320 special function keys. (Reprinted with permission from Digital Equipment Corporation.)

3.3.5 Special Function Keys

The *special function keys* are located at the top row of the keyboard, as depicted in Figure 3-7. Most of these keys are defined by the software in use. They may also be programmed by the user. Each key is therefore capable of performing multiple functions, ranging from communications control setup to editing commands. These keys are tremendous time-savers for the user because a series of instructions or keystrokes can be executed with the touch of a single key.

There are 15 special function keys for the VT300 series terminals (**F6** through **F20**). They may be programmed by the user to store and recall text and commands applicable to a given program. Up to 256 characters per function key can be programmed and saved in the terminal's nonvolatile memory. When defined by the user, these keys are referred to as *user-defined keys* (UDKs). They are also referred to as *soft keys*.

3.3.6 Editing Keypad

The editing keypad is illustrated in Figure 3-8. This dedicated keypad is used exclusively for editing text. It includes four arrow keys that may be used for cursor movement in the direction indicated by the arrow. Six additional editing keys are defined by the user's software.

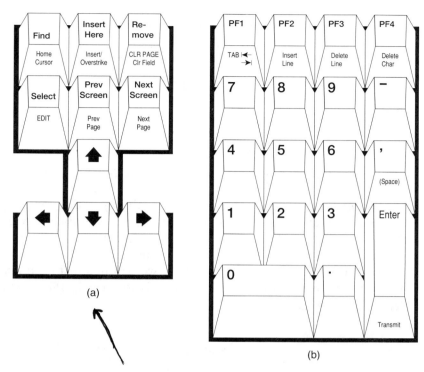

(a)

(b)

Figure 3-8 (a) Editing keypad for the VT300 series terminal; (b) numeric keypad for the VT300 series terminal. (Reprinted with permission from Digital Equipment Corporation.)

3.3.7 Numeric Keypad

The numeric keypad is shown in Figure 3-8b. The numbers on the keypad are arranged in standard calculator format for ease of numerical data input. The numeric keypad can also have special functions assigned by an applications program. For example, the keys **PF1** through **PF4** (programmed function 1–4) may be used for performing commands related to a spreadsheet or word-processor program. The applications manual must be referred to in this case. Often, a rubber or plastic template that fits over the numeric keypad is provided. The template defines the newly assigned functions for each key.

The **ENTER** key normally acts as a carriage return. It can also serve as a **TRANSMIT** key when the terminal is set up for block mode of operation.

3.3.8 Baud Rate Selection

A terminal's *baud rate* is the rate at which its characters are transmitted in bits per second. This form is illustrated in Figure 3-9. The character shown includes a *start bit, seven-bit data word, parity bit,* and a *stop bit.* The duration of each

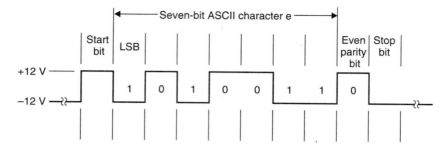

Figure 3-9 Asynchronously transmitted character from a terminal.

bit, the *bit time*, may be found mathematically by taking the reciprocal of the baud rate. A baud rate of 1200, for example, has a bit time of 833 μs (1/1200). Baud rate is selectable either through the terminal's soft keys or through hardware switches provided by the terminal manufacturer. Most terminals offer standardized baud rate settings. In Chapter 9 we will find that baud rate and bit rate have different meanings in terms of transmission speed. Table 3-2 lists the standard baud rates used in data communications.

TABLE 3-2 Standard Terminal Baud Rates

75	1800
110	2400
150	4800
300	9600
600	19,200
1200	

3.3.9 Parity

Parity is a simplified method of error detection. When an operator types a key, an asynchronous character is transmitted by the terminal to a remote device, such as a computer. A single bit, called the *parity bit*, is added to the transmitted seven-bit ASCII character.

The logic level of the parity bit depends on the number of 1 bits set within the transmitted character. When parity is enabled on a terminal, it can be set to either *even* or *odd*. For even parity, a terminal's transmitter section sets the parity bit to a level making the total number of 1 bits (including the parity bit itself) an even number. The opposite is true for odd parity. Consider the following examples of even and odd parity.

Even parity generation:

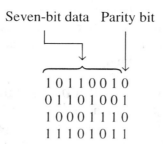

Seven-bit data   Parity bit

```
1 0 1 1 0 0 1 0
0 1 1 0 1 0 0 1
1 0 0 0 1 1 1 0
1 1 1 0 1 0 1 1
```

Odd parity generation:

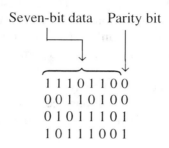

Seven-bit data   Parity bit

```
1 1 1 0 1 1 0 0
0 0 1 1 0 1 0 0
0 1 0 1 1 1 0 1
1 0 1 1 1 0 0 1
```

Enabling parity on a terminal also enables the receiver section of the terminal to perform a parity check on the received character transmitted from a remote device. The terminal's parity setting must be equivalent to that of the remote device. An agreement is decided on beforehand. If the number of 1 bits for the received character does not match with the parity setting, a transmission error has occurred. A special character is displayed in place of the received character. Each manufacturer has its own unique character. Digital Equipment Corporation's terminals use the checkerboard character: (▒).

Parity is an optional feature on a terminal. It can be enabled or disabled by an operator via the special function keys on the terminal or by hardware switches provided by the terminal. If parity is disabled, the bit following the seven-bit data word will be the (first) stop bit. If the terminal has been configured for eight-bit data with no parity, a logic 0 is typically transmitted in place of the parity bit. Some terminals allow the feature of forcing this bit, the eighth bit, to a logic 1 or a 0.

3.3.10 Transmission Modes

There are essentially two modes of terminal operation: *character mode* and *block mode*. Most terminals are set up to operate in character mode.

3.3.10.1 Character mode. When a terminal is in its character mode, characters are sent to the host computer only when a key is typed by the operator. The character is said to be transmitted *asynchronously;* that is, there is no synchronism with the operator's keystrokes. When the operator is not typing, the terminal is said to be in its *idle* state. Conversely, when the computer sends a character to the terminal, the character is displayed at the current position of the cursor. If the character is a nonprintable character (CR, LF, BEL, and the like), the terminal will act accordingly.

Half-Duplex (HDX) versus Full-Duplex (FDX). A terminal operating in its character mode must be configured for either *half-duplex* or *full-duplex* operation. In half-duplex (Figure 3-10) characters typed are transmitted and internally *echoed* back to the receiver section. The echoed character is displayed for the operator to see what has been typed. This echoed character is referred to as a *local echo*. In *full-duplex* operation, the characters typed are transmitted only. The characters are not displayed unless they are received by a remote device and echoed back. An echo generated from a remote location is referred to as a *remote echo*.

When remote echoes are displayed as a result of typing, an operator can be assured that the transmission link has been properly connected. Much of the associated hardware and software of the sending and receiving stations must function successfully in order for echoed characters to appear on the display. Echoed characters are therefore a test measure for the system beyond just the terminal itself.

Occasionally, a terminal will display duplicate characters as they are typed. In this case, the terminal must be reconfigured from half-duplex to full-duplex. What has happened is that a local echo and a remote echo have been received by the terminal. The first of the two characters displayed is the local echo, and the second character is the remote echo generated from the remote computer.

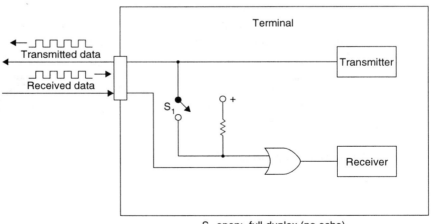

S₁ open: full-duplex (no echo)
S₁ closed: half-duplex (echo)

Figure 3-10   Half-duplex versus full-duplex operation.

3.3.10.2 Block mode. Smart terminals can be configured to operate in either character mode or block mode. In block mode, characters are *not* transmitted as they are typed. Instead, the operator works locally with the information displayed on the screen. Characters are stored in a buffer, rather than transmitted as they are typed. In the meantime, the host computer can be servicing one of several other terminals that are part of a *multidrop* communications system. When the operator is satisfied with the information displayed on the screen, the **ENTER** (or **TRANSMIT**) key of the terminal is depressed and the buffer contents are transmitted in accordance with the given communications protocol.

When a terminal is operating in block mode, every character position on the display is assigned into one of two *fields:* a *protected field* or an *unprotected field.* Protected fields are areas of the display that cannot be written to. Unprotected fields can be written to in the normal manner. These fields are defined by software through the use of *escape sequences.* Escape sequences are discussed in Section 3.5. Figure 3-11 depicts the two fields. Using the editing features that the terminal has been programmed for, characters, words, and lines are inserted and deleted throughout the unprotected fields of the display. When the end of an unprotected field is written to, the cursor moves to the beginning of the next unprotected field. The terminal's tab function can be used to move the cursor from one unprotected field to the next, skipping over the protected fields.

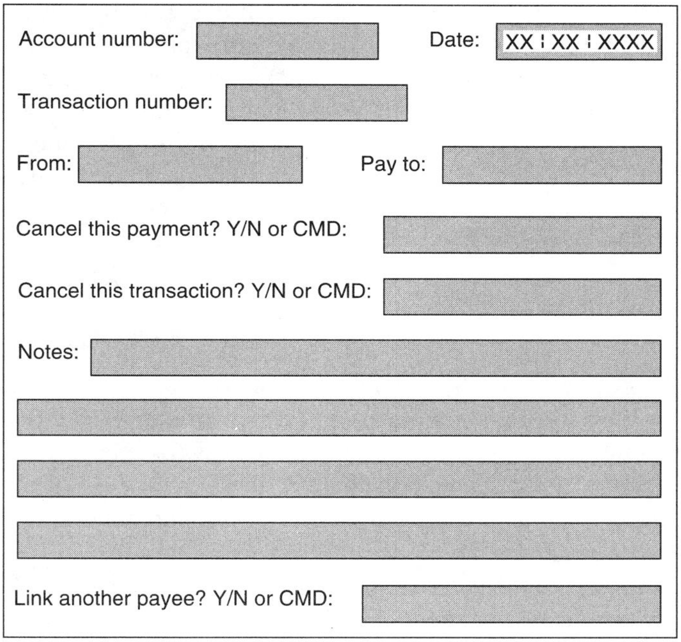

Figure 3-11 Block mode operation showing protected and unprotected fields of the display.

If a character is typed within a protected field, the character will be displayed in the first position of the next unprotected field.

The maximum block size usually depends on the requirements of the computer. Unlike character mode, whereby characters are transmitted asynchronously, in block mode the entire block is sent synchronously; that is, each character is sent in contiguous order. The block of data may include several hundred contiguous bytes, including framing, control, and error-detection information. Start and stop bits are eliminated from each character, in addition to any idle time between characters. Hence, transmission efficiency is greatly improved in block mode operation. Block mode terminals are much more expensive than character mode terminals due to the supporting hardware and software that are necessary.

3.4 USING THE CONTROL (CTRL) KEY

Most ASCII characters are represented by dedicated terminal keys. Control characters, excluding the format effectors CR, LF, HT, and the like, are an exception. These characters are listed in the first two columns of Table 2-7. To transmit control characters, ASCII terminals provide the operator with a *control* (CTRL) key. The CTRL key allows an operator to send control characters, similar to the manner in which the SHIFT key is used to send an uppercase letter. The CTRL key is simply depressed, followed by one of the noncontrol characters listed in column 4 or 5 of Table 2-7. The transmitted ASCII control character is equal to 40H subtracted from this character (in software, 40H is masked with the character). For example, if an operator wanted to transmit the ASCII code for the BEL character, the CTRL key, followed by G, would be depressed. Since 40H subtracted from 47H; a G, is equal to 07H, the ASCII code for BEL is generated. Another example that perhaps many personal computer owners are familiar with is CTRL Q and CTRL S, which generate the ASCII control characters XON and XOFF, respectively.

$$\text{CTRL Q} = 51\text{H} - 40\text{H} = 11\text{H DC1 (XON)}$$

$$\text{CTRL S} = 53\text{H} - 40\text{H} = 13\text{H DC3 (XOFF)}$$

CTRL Q and CTRL S, as many of us know, are used to control the *flow* of data transmitted from a computer to a VDT or printer. A large directory file is often transmitted to a VDT that is either too fast for one to read, or, the directory is larger than the screen capacity. As the screen fills, typing a CTRL S: DC3 (XOFF) turns the transmitter section of the computer off. Printing of the directory file on the display halts, thus allowing one to read the text before it scrolls off the screen. Once it is read by the operator, a CTRL Q: DC1 (XON) turns the transmitter section of the computer back on, and the directory file continues to be printed on the display from where it left off. This sequence of key strokes is repeated until the entire directory file has been read.

3.5 ESCAPE SEQUENCES

Most ASCII terminals today offer an entire set of commands and control functions that can be performed through the use of the ASCII character *escape* (ESC). As defined in Chapter 2, ESC, followed by a single character or a string of characters, is called an *escape sequence.* Escape sequences are used to perform special command and control functions, most of which are not immediately apparent to the operator. Clearing the screen for example or positioning the cursor to a specific location on the screen can be performed through an escape sequence.

The escape sequence begins with ESC ($1B_{16}$ or 033_8) and ends with the final character for the sequence. Any subsequent characters after the escape sequence are interpreted as normal ASCII text. Table 3-3 lists common escape sequences used by various manufacturers.

Escape sequences are normally sent to a terminal from a remote computer. An operator, however, can test an escape sequence for its effect, independent of the computer, by simply configuring the terminal to its half-duplex mode. Recall that in half-duplex a local echo is produced within the terminal. The ESC key can be typed, followed by the appropriate character(s) to perform the escape sequence. For example, some terminals (DEC VT52, HP2648) use **ESC A** and **ESC B** to position the cursor up or down, respectively. **ESC H** followed by **ESC J** (clear the screen from cursor on) will clear the entire screen.

Cursor up:

ESC	A	
1B	41	(hex)
033	101	(octal)

Cursor down:

ESC	B	
1B	42	(hex)
033	102	(octal)

Clear screen:

ESC	H	
1B	48	(hex)
033	110	(octal)
ESC	J	
1B	4A	(hex)
033	112	(octal)

TABLE 3-3 Common Escape Sequences Used by Various Manufacturers

Command function	ANSI/ AT&T4410	DEC VT52	DEC VT100	Hazeltine 1420	HP 2648	IBM 3101	Lear-Siegler ADM3/5	Televideo 910
READ CURSOR POSITION	ESC [6 n	N/A	ESC[6n	ESC CTRL E	ESC · DC1	ESC 5	ESC ?	ESC ?
CURSOR UP	ESC [pn A	ESC A	ESC[Pn A	ESC CTRL L	ESC A	ESC A	CTRL K	CTRL K
CURSOR DOWN	ESC [pn B	ESC B	ESC[Pn B	ESC CTRL K	ESC B	ESC B	CTRL J	CTRL V
CURSOR RIGHT	ESC [pn C	ESC C	ESC[Pn C	CTRL P	ESC C	ESC C	CTRL L	CTRL L
CURSOR LEFT	ESC [pn D	ESC D	ESC[Pn D	CTRL H	ESC D	ESC D	CTRL H	CTRL H
HOME CURSOR	ESC [H	ESC H	ESC[H	ESC CTRL R	ESC H	ESC H	CTRL	CTRL ^
CLEAR TO END OF PAGE	ESC J	ESC J	ESC[O J	ESC y	ESC J	ESC J	ESC y	ESC y
CLEAR TO END OF LINE	ESC K	ESC K	ESC[O K	ESC t	ESC K	ESC I	ESC t	ESC T
LINE INSERT	ESC L	N/A	N/A	ESC CTRL Z	ESC L	ESC N	ESC E	ESC E
LINE DELETE	ESC M	N/A	N/A	ESC CTRL S	ESC M	ESC O	ESC R	ESC R
CHARACTER INSERT	ESC [@	N/A	N/A	ESC Q	ESC Q	ESC P	ESC Q	ESC Q
CHARACTER DELETE	ESC [P	N/A	N/A	ESC W	ESC P	ESC Q	ESC W	ESC W
POSITION CURSOR	ESC [:n;pnH	ESCY r + 31 c + 31	ESC[Pn; Pn H	N/A	ESC&a #r#c	ESC Y xy	ESC = rc	ESC = rc
CLEAR ALL	ESC [2J	N/A	ESC[2 J	ESC CTRL L	ESC g	ESC ;	ESC *	ESC *
UNDERSCORE	ESC [4m	N/A	ESC[4m	N/A	ESC &d D	N/A	N/A	ESC G8
BLINK	ESC [5m	N/A	ESC[5m	N/A	ESC &d A	ESC 3 I	N/A	ESC G2
REVERSE VIDEO	ESC [7m	N/A	ESC[7m	N/A	ESC &d B	ESC 3 E	N/A	ESC G4
BLANK VIDEO	ESC [7m	N/A	N/A	N/A	ESC &d S	ESC 3 M	N/A	ESC G1
132 COLUMN MODE	ESC [?3h	N/A	ESC[?3h	N/A	N/A	N/A	N/A	N/A
80 COLUMN MODE	ESC [?31	N/A	ESC[?31	N/A	N/A	N/A	N/A	N/A
HORIZONTAL TAB SET	N/A	N/A	ESC H	ESC 1	N/A	ESC 0	ESC 1	ESC 1
LOCK KEYBOARD	N/A	N/A	N/A	ESC CTRL U	N/A	ESC :	ESC #	ESC #
UNLOCK KEYBOARD	N/A	N/A	N/A	ESC CTRL F	N/A	ESC ;	ESC "	ESC "
NEXT PAGE	N/A	N/A	N/A	N/A	ESC V	N/A	N/A	N/A
PREVIOUS PAGE	N/A	N/A	N/A	N/A	ESC U	N/A	N/A	N/A
PROTECT ON	N/A	N/A	N/A	ESC CTRL Y	ESC &dJ	ESC 3C	ESC)	ESC& ESC)
PROTECT OFF	N/A	N/A	N/A	ESC CTRL _	ESC &d@	ESC 3B	ESC (	ESC' ESC(
RESET DEVICE	N/A	N/A	ESC c	N/A	ESC E	N/A	N/A	N/A

Notes: The letters n, r, c, pn, Pn, and xy represent a number used to specify a row, column, or the number of times to perform a sequence.

Numerous other escape sequences can be performed beyond just cursor positioning. Consider the following.

— Local echo on/off
— Interrogate current cursor position
— Set/clear tabs
— Initialization of data communications test
— Inverse video
— Lock keyboard
— Caps mode
— Click on/off
— Block mode on/off
— Define block size
— Display data communications menus
— Display soft key labels
— Define character sets
— Define datacomm/printer ports
— Definition of protected and unprotected fields
— Definition of function (soft) keys
— Initialization of terminal self-test

Since escape sequences vary among terminal manufacturers, application programs must be tailored to suit a given terminal type. It is essential that the programmer refer to the manufacturer's operating manual in order to make use of the terminal's available escape sequences.

Escape sequences can also be sent by a terminal to command and control computers and peripheral devices such as printers, plotters, and modems. Here, again, the operating manual of each device must be referred to.

3.6 TERMINAL INTERFACES

Many of today's terminals support more than one interface standard. Aside from the usual interface to a host computer, modems, printers, plotters, and even local area networks (LANs) are being directly connected to. A brief presentation of some of the standard terminal interfaces used today will be given here.

3.6.1 RS-232-C Interface

Virtually all terminals offer the standard EIA (Electronics Industries Association) RS-232-C interface. RS-232-C is a serial interface standard that includes the electrical, mechanical, and functional specifications for interconnecting data communications equipment. The voltage specification for a logic 1 is −3 to −15 V, whereas a logic 0 is +3 to +15 V; and ±12V is typically used for terminals. The maximum data transfer rate is 20 kbps for a 15-m cable. A 25-pin D-type connector is used to connect the interfacing cable. In Chapter 4 the RS-232-C interface standard is discussed in greater depth.

3.6.2 RS-422-A and RS-423-A Interface

In 1978, EIA derived two new and improved electrical interface standards: RS-422-A and RS-423-A. The intent was to improve on the transmission characteristics (speed, noise immunity, and so on) of the existing RS-232-C standard. RS-422-A defines the electrical characteristics for balanced line drivers and receivers. RS-423-A is for unbalanced line drivers and receivers. Valid logic levels are +200 mV to +6 V for a logic 0 and −200 mV to −6 V for a logic 1. Higher transmission speeds are attainable through these two interfaces. For a 10-m cable, 10 Mbps can be achieved. RS-422-A and RS-423-A are discussed in greater depth in Chapter 4.

3.6.3 Current Loop Interface

Some terminal manufacturers continue to offer the 20-mA current loop interface. This standard has evolved from the old electromechanical teletype, which used current to activate its relays. A logic 1 is represented by the presence of 20 mA in the interface loop. A logic 0 is the absence of current in the loop. The same serial bit stream used to drive RS-232-C circuits is used to drive the current loop circuit.

3.6.4 IEEE-488 Interface Bus

The IEEE-488 Interface Standard was published in 1975. This standard specifies a method of interconnecting *digital programmable measurement instruments* (DPI) via an eight-bit, bidirectional, asynchronous parallel bus. Hewlett-Packard Corporation developed the original standard, *HP-IB* (Hewlett-Packard interface bus), on which the IEEE-488 standard is based. The bus has also become known as the *general-purpose interface bus* (GPIB). Many terminals have adopted the IEEE-488 interface for interconnecting to printing devices and other peripherals.

Four basic functional elements are used to organize and manage information among IEEE-488 interconnecting devices. These elements are:

Controller: A device that can address other devices to control their transmission or reception of data and their operating functions. The controller is typically the host computer.

Talker: A device that can be addressed by an interface mesage to transmit data to other devices on the bus. Examples include frequency counters, voltmeters, spectrum analyzers, and terminals.

Listener: A device that can be addressed by an interface message to receive data from other devices on the bus. Examples include terminals, printers, and plotters.

Talker/listener: A device that can be addressed by an interface message to act as a talker or a listener.

The *controller* assigns which device (or devices) on the bus will be a *listener,* that is, receiver, or a *talker,* the transmitting device. Only one device may transmit on the bus at any one time. The transmitting device can be the controller itself or an assigned talker.

Sixteen lines make up the IEEE-488 bus. Their functions are divided into three groups, as shown in Table 3-4. Figure 3-12 depicts an IEEE-488 interface

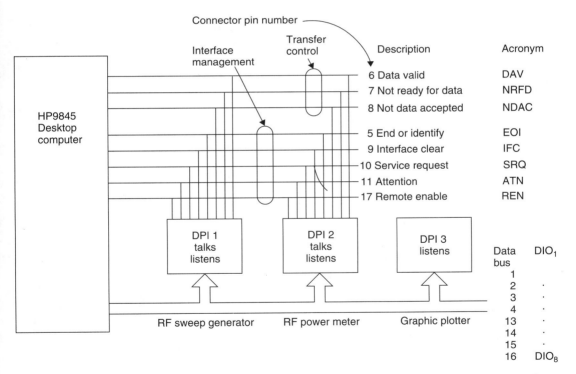

Figure 3-12 IEEE-488 bus interface setup. (From *PET and the IEEE 488 Bus [GPIB]* by Eugene Fisher and C. W. Jensen. Copyright © 1980 McGraw-Hill, Inc.)

TABLE 3-4 IEEE-488 Signal Definition

Signal name	Group	Description	Pin number
DIO1-DIO8	Data bus	Bidirectional data transfer lines	1–4 and 13–16
DAV	Handshake lines	Data Valid indicates that data on DIO lines are valid. DAV is issued by a talker.	6
NRFD		Not Ready For Data indicates the readiness of a listener to accept data.	7
NDAC		Not Data Accepted indicates whether or not a listener has accepted data from the bus.	8
ATN	Interface management	Attention is issued by a controller to indicate whether information on the bus is data or an interface control message.	11
IFC		Interface Clear is used by a controller to reset the complete interface system into its quiescent state.	9
SRQ		Service Request is used by a device to indicate to the bus contoller that it requires attention.	10
REN		Remote Enable is used by a controller to provide selection between two sources of device programming data.	17
EOI		End or Identify is issued by a talker to indicate the end of a multiple-byte transfer. EOI is also used by a controller with ATN to initiate a polling sequence.	5

between a system controller and several digital programmable instruments (DPIs).

Data transactions between devices occur via the eight-bit bidirectional data bus lines **DIO1** through **DIO8.** A *master/slave* relationship is used to transfer data bytes between two devices. Three lines are used for handshaking the data: **NRFD** (Not Ready For Data), **NDAC** (Not Data Accepted), and **DAV** (Data Valid). The remaining five control lines are **ATN** (Attention), **IFC** (Interface Clear), **SRQ** (Service Request), **REN** (Remote Enable), and **EOI** (End Or Identify). Figure 3-13 illustrates the handshaking sequence for an eight-bit data transaction between a talker and a listener.

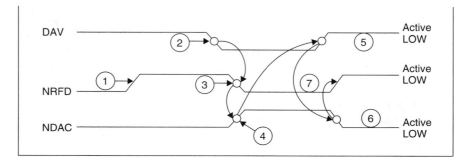

Figure 3-13 IEEE-488 handshaking sequence between talker and listener. (From *PET and the IEEE 488 Bus [GPIB]* by Eugene Fisher and C. W. Jensen. Copyright © 1980 McGraw-Hill, Inc.)

PROBLEMS {DUE FRIDAY}

1. What primary feature distinguishes a smart terminal from a dumb terminal?

2. Define *ergonomics*.

3. Name three applications for a workstation.

4. Name five ergonomic features of a terminal.

5. What is the typical screen capacity for a standard ASCII terminal?

6. Name the function keys of the main keypad for the VT300 series terminal.

7. What are special function keys used for?

8. A terminal is programmed for a baud rate of 4800. What is the bit time associated with the transmitted character?

9. Determine the parity bit for the following data words:

Odd parity
1 1 0 1 0 1 0 ____
0 0 1 1 0 1 0 ____
1 1 1 0 0 0 0 ____
1 0 1 1 0 1 0 ____

Even parity
0 0 1 1 0 1 0 ____
1 0 1 0 1 1 0 ____
1 1 1 1 0 0 0 ____
0 0 1 1 0 1 0 ____

10. Explain the difference between block mode and character mode.

11. Is a local echo produced in half-duplex or full-duplex mode?

12. Explain how a character is displayed when typed at a terminal that is set for full-duplex operation.

13. What happens when a terminal is set for half-duplex and a remote echo is returned to the terminal as a result of typing a key?

Problems 59

14. How may the CTRL key be used to generate the ASCII character BEL?
15. What is an escape sequence?
16. Explain how a terminal's cursor may be moved up by a remote device.
17. Name three types of terminal interfaces.
18. What four functions may IEEE-488 devices be programmed for?
19. What is the term for a DPI that can receive data only?
20. What are the names of the IEEE-488 handshake lines?

4

Serial Interfaces

On May 24, 1844, Samuel F. B. Morse transmitted the famous message *"What hath God wrought."* The message was transmitted from the U.S. Supreme Court in Washington, D.C., to the city of Baltimore. It was translated by a receiving party and returned. A short conversation followed. Morse, inventor of the telegraph, had successfully demonstrated to the U.S. Congress that messages could be sent in excess of 10 miles by transmitting electrical impulses down a wire. The apparatus used for the demonstration included a telegraph key, conductive wire, a battery, and an electromagnetic relay. By manually opening and closing the telegraph key, Morse was able to activate the remote electromagnetic relay acting as the receiving device. The relay opened and closed accordingly. A series of *clicks* was transmitted and received in accordance with his specially devised code.

Morse's historical event paid tribute to many pioneers of electricity and marked perhaps the first electrical transmission of serial data. His serial communications link was capable of transmitting a few words per minute. Little did he realize that serial transmission of data would evolve into today's enormous transmission rates over *the public switched telephone network* (PSTN).

The PSTN still remains the largest and most common facility for the transmission of serial data. Due to its vastness in size and complexity, data transfer rates are still limited to the bandwidth constraints of existing facilities. Although the original intent of the PSTN was for analog voice communications, it has evolved today as a standard medium for data communications. Binary serial data are converted from parallel format to a serial bit stream used to modulate carrier frequencies that are compatible with the characteristics of the PSTN. Although the transmission of the original information from one point to another is less expedient in this manner, there is an enormous savings in hardware, and existing communications facilities of the PSTN can be utilized for long-distance communications. In this chapter we look at some of the fundamental concepts, rules, and regulations regarding the format of serial data and how data are transferred between devices.

4.1 SYNCHRONOUS VERSUS ASYNCHRONOUS SERIAL TRANSMISSION

The transmission of serial data between two devices needs to be coordinated in the manner in which it is sent. The receiver must synchronize with the incoming data in order to interpret the orderly sequence in which it was transmitted. The set of rules that govern the manner in which the data are both transmitted and received is called a *protocol.*

Transmitted serial data can be classified into two categories: *synchronous* and *asynchronous.* We begin our discussion with asynchronous serial transmission. The term *asynchronous* literally means *lack of synchronism.* An example of an asynchronous event is a person typing characters at a video display terminal (VDT). Each time a character is typed, a serial bit stream representing the character typed is transmitted by the terminal. The person can proceed to type at any speed and pause for any length of time between characters. There is no synchronism to the manner in which the characters are typed. The transmitted characters therefore lack synchronism. Since characters are typed in a more-or-less random fashion, VDTs are well suited for asynchronous serial transmission. How widespread is asynchronous serial transmission? Think of the millions of people who work on a daily basis typing at a terminal.

4.1.1 Asynchronous Serial Link Protocol

The lack of synchronism in asynchronous transmission is solved by having the transmitting device insert *framing* bits at the beginning and end of each transmitted character. These framing bits are referred to as the character's *start bit* and *stop bit* (Figure 4-1). The start bit is always a logic 0 and the stop bit is always a logic 1. A character's start and stop bits mark the beginning and end of the character, thereby allowing the receiving device a means of achieving frame sync with each character. The character can then be sent at any time. The time interval between characters is referred to as the *idle time.* This time varies between characters depending on the speed of the typist. Since there will always be at least one stop bit between characters, the receiver simply searches for the occur-

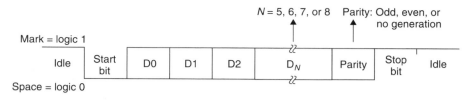

Figure 4-1 Format of an asynchronous serial character.

rence of the next start bit to synchronize and detect additional serial characters. The stop bit is verified to ensure frame sync. Framing is said to occur on a *character-by-character* basis for asynchronously transmitted serial data.

4.1.2 Synchronous Serial Link Protocol

Synchronous transmission of serial data involves the high-speed transmission of data in the form of *blocks*. A block of data can represent a contiguous series of data bytes. In contrast to asynchronously transmitted characters, the idle time between characters as well as the start and stop bits are eliminated, making it possible to send data at higher rates. Synchronization of data is performed on a *block-by-block* basis. The transmitting device often provides a separate clock pulse that is in sync with the center of each transmitted data bit. The synchronizing clock is carried on a line separate from the transmitted data line. (See the discussion of RS-232-C circuits DA, DB, and DD in Section 4.4.6.) For long-distance communications via the telephone system, the synchronizing clock on a separate circuit becomes impractical. In this case the clock is encoded with the data by a device called a *modem*. The modem modulates a carrier frequency with this information and sends it down the telephone lines. The receiving modem demodulates the synchronous clock along with the corresponding data. The clock is separated from the data and used to sample the data at the center of each bit, thus determining the state of each bit. Figure 4-2 illustrates these two methods of synchronous transmission.

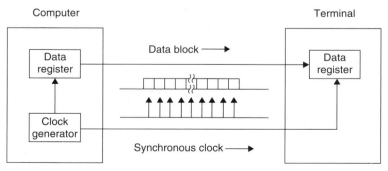

Figure 4-2 (a) Synchronous data transmitted with a separate clock; (b) synchronizing clock encoded with data and transmitted together.

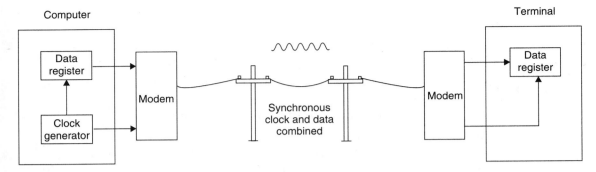

Figure 4-2 *(continued)*

Another method of achieving synchronization with the transmitted data is to insert a unique bit pattern at the beginning and end of each block. The receiver can synchronize with the data block by recognizing the occurrence of the unique bit pattern. This eliminates the need for a separate clock. This unique bit pattern, and the number of bits contained in the block, are a function of the synchronous serial link protocol. In some protocols, this special synchronizing bit pattern is referred to as the *sync character*. In others, it is referred to as the *opening flag* and *closing flag*. The synchronous receiver *hunts* for the sync character or opening flag until a recognition has been made. When the receiver has recognized this unique bit pattern, it goes into *sync-lock* with the data block that has been transmitted. More than one sync character can be sent in succession as an added measure of ensuring sync-lock. Figure 4-3 illustrates two examples of synchronous protocols that are commonly used: *IBM's Synchronous Data Link Control*

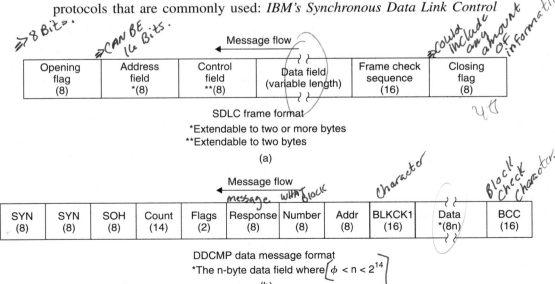

Figure 4-3   (a) SDLC frame format; (b) DDCMP message format.

(SDLC) and Digital Equipment Corporation's *Digital Data Communications Message Protocol* (DDCMP).

Many different synchronous serial link protocols are used today. They have been developed by several manufacturers and standards organizations. A detailed discussion of these protocols will be presented in Chapter 10. Some of the more common synchronous serial link protocols used today are as follows:

> **BISYNC or BSC:** Binary Synchronous Communications
> **DDCMP:** Digital Data Communications Message Protocol
> **HDLC:** High-level Data Link Control
> **SDLC:** Synchronous Data Link Control

4.1.3 Protocol Efficiency

One often considers the relative advantages and disadvantages of serial data transmission protocols in terms of *efficiency.* The efficiency of a protocol can be thought of as the maximum transfer of data in a minimal amount of time, that is, with the least likelihood of getting garbled during the interim. In asynchronous serial link protocols, a significant amount of time is expended by including the start, stop, and parity bits with the actual data. A seven-bit ASCII character transmitted in asynchronous format may include one start bit, as many as two stop bits, and a parity bit, as illustrated earlier in Figure 4-1. If only one stop bit is used, the total composite character is ten bits. The *protocol overhead* in this case is 30%, three overhead bits and seven data bits. It is now apparent why asynchronous serial transmission is often thought of as *inefficient:* the overhead is high. Protocol overhead for serial transmission of data can be computed by the following equation:

$$\text{Protocol overhead} = \left(1 - \frac{N}{N + C}\right) \times 100\%$$

where N represents the number of data bits transmitted and C represents the number of control bits transmitted.

Example 1

Compute the protocol overhead for a DDCMP message block. The message block in this example will include the maximum allowable number of data bytes: 16,384 (2^{14}).

Solution: Refer to Figure 4-3.

$$\text{Protocol overhead} = \left(1 - \frac{131,072}{131,072 + 96}\right) \times 100 = 0.073\%$$

where $N = 16,384$ bytes $\times 8$ bits/byte $= 131,072$ bits and $C = 96$ bits.

The overhead for a DDCMP synchronous data block in this case is less than one-tenth of 1%! It is apparent that synchronous transmission of data is much more efficient than asynchronous transmission. The example here, however, uses the maximum allowable bytes in the DDCMP data field. The data field within the DDCMP message format can be anywhere between 0 and 16,384 bytes. What would be the case if a DDCMP message block were sent with only two bytes?

Example 2

Compute the protocol overhead for a DDCMP message block containing two data bytes in the data field.

Solution:

$$\text{Protocol overhead} = \left(1.0 - \frac{16}{16 + 96}\right) \times 100 = 85.7\%$$

where $N = 2$ bytes $\times$ 8 bits/byte $= 16$ bits and $C = 96$ bits.

The overhead for synchronous transmission is very high for small messages. The number of small DDCMP message blocks should therefore be kept at a minimum. Likewise, if an error occurs in synchronous transmissions, the entire message block may have to be retransmitted. This can be very time consuming for large DDCMP message blocks. With asynchronous transmissions, if an error occurs, only one character may be lost, since each character is individually synchronized with its own start and stop bits.

As we can see, there are advantages and disadvantages to both synchronous and asynchronous transmission of data. They are listed in Table 4-1.

TABLE 4-1 Characterizing Synchronous and Asynchronous Serial Transmission

	Asynchronous	Synchronous
Advantages	Suitable for ASCII terminals and data entry	High-speed transmission
	Minimal hardware to implement	Maximum throughput
	Bit errors can readily be seen as erroneous characters displayed	Low overhead
	Low-speed transmission means fewer errors	Error-detection methods are extremely reliable
	Suitable for electromechanical teletype	
Disadvantages	Slow and inefficient	Expensive to implement
	High overhead due to start and stop bits	Communication protocols must be compatible
		Entire blocks may need to be retransmitted if a single bit error occurs
		Cannot be used with electromechanical teletype

4.2 DEVELOPMENT OF THE MODEM

With the advent of the computer and the increasing need to render its services from a remote location, a device called a *modem* was developed in the early 1960s. The term *modem* is a contraction of the words *modulator–demodulator.* Telephone companies refer to a modem as a *data set.* The modem, which has become a household word, would make it possible to communicate serial binary data between computers and terminals via the telephone lines. Data could be sent in either synchronous or asynchronous form.

A modem simply utilizes the computer's digital pulse streams to *modulate* an analog carrier frequency compatible with the communications facilities of the telephone system. The modem also performs the reverse task of *demodulating* an analog carrier frequency (generated from a remote modem via the telephone system) into digital pulse trains for the computer; hence the name *modulator–demodulator* or modem was derived (Figure 4-4). A detailed discussion of the modem will be presented in Chapter 8.

It is necessary at this time to define two new very important data communication terms: DTE and DCE.

DTE: *Data terminal equipment* corresponds to a device that transmits or receives binary digital data. DTE acts as the primary source or destination of transmitted or received data. Examples of DTE include VDTs, printers, and computers, in cases where the computer is interfaced to a modem.

DCE: *Data communications equipment* corresponds to devices that *transfer* binary digital data between two transmission mediums. The binary digital data can be represented as an analog or digital signal. In either case, signal processing (modulation, demodulation, encoding, or decoding) is necessary for DCE to transfer the data between the two mediums. An example of DCE is a modem or data set. The two transmission mediums in this case would be the telephone lines and an RS-232-C interface. A computer can also be DCE. Take, for example, when it is connected to the printer. The printer is DTE. It is the primary destination of the transmitted data. The computer, DCE in this case, transfers the data from disk drive or memory to the terminal.

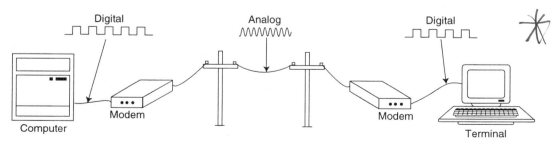

Figure 4-4 Diagram depicting the operation of a modem.

4.3 A NEED FOR STANDARDIZATION

Within a short period of time after the invention of the modem, computers and business machines within the same room, different buildings, and different cities were exchanging data over the phone lines. Eminent problems surfaced as the need to communicate continued to grow in orders of magnitude. Signal names and functions were incompatible between computers, modems, and other related equipment. Transmit and receive voltages varied between modems. When one business machine was in the process of sending data, the remote business machine was not ready to receive it and, subsequently, data were lost. In addition, connectors between devices varied. Special interfacing cables were necessary. These cables were adaptable only to a given situation. A standard was necessary to ensure compatibility between DTE and DCE. The standard would ensure that interfacing devices met the following criteria:

— The transmitted and received voltage levels must fall within a specified region.
— Electrical characteristics of the transmission line, including source and terminating loads, must meet specification, thus setting limitations on maximum data transfer rates.
— Interfacing cables between DTE and DCE must include cable connectors at each end that are compatible in size and pin number for all DTE and DCE.
— The electrical function of signals between DTE and DCE must be compatible.
— The function of each electrical signal must have common names and pin numbers.

4.4 EIA RS-232-C INTERFACE STANDARD

A serial interface standard was developed that encompassed the preceding criteria. This standard, known as RS-232-C, was developed and published in 1969 by the Electronics Industries Association (EIA), a U.S. organization of electronics manufacturers that establishes and recommends industrial standards. RS-232-C conforms to the V.24 standard set within the European countries by the Consultative Committee on Telegraphy and Telephony (CCITT).

RS-232-C defines the electrical, mechanical, and functional interface between *data terminal equipment,* **DTE**, typically a computer terminal or computer, and *data communications equipment,* **DCE**, typically a modem or data set, employing serial binary data interchange. The standard has been universally observed throughout the world and has evolved to this date as a standard, not only for connecting computers and terminals to modems, but also for connecting com-

Figure 4-5 RS-232-C communications interface.

puters to printers, plotters, PROM programmers, and many other peripherals. Virtually all mainframe computers, personal computers, and microprocessor-based communication systems have provisions for an RS-232-C interface to associated peripherals. Figure 4-5 depicts a typical RS-232-C interface.

The RS-232-C specification is comprised of three parts:

1. Electrical signal specification
2. Mechanical specification
3. Functional specification

4.4.1 Electrical Signal Specification

The RS-232-C electrical specification defines the complete electrical characteristics that DTE and DCE shall adhere to. Figure 4-6 illustrates a model of the interchange equivalent electrical circuit.

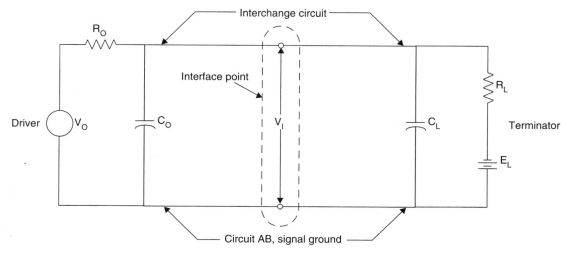

Figure 4-6 RS-232-C equivalent circuit.

— V_o is the open-circuit voltage presented by the driver.

— R_o is the internal resistance of the driver.

— C_o represents the equivalent capacitance associated with the driver and cable measured at the interface point.

— V_i is the voltage measured at the interface point.

— C_L is the equivalent capacitance associated with the terminating load and any cable capacitance to the interface point.

— R_L is the terminating dc load resistance.

— E_L is the open-circuit terminator voltage presented by any dc bias associated with the terminating circuit.

The interfacing parameters associated with the electrical specification apply to both synchronous and asynchronous serial binary data communication systems. These parameters are intended to set limits on the impedances and voltages that each interchange circuit must adhere to. Some of the most useful parameters are summarized as follows:

1. V_o and V_i may not exceed ±25 V with respect to signal ground.
2. The short-circuit current between any two or more pins or conductors within the interface may not exceed 0.5 A.
3. The total effective capacitance of the interchange circuit may not exceed 2500 pF.
4. $3000\ \Omega < R_L < 7000\ \Omega$.
5. $5\ \text{V} < V_i < 15\ \text{V}$ when $E_L = 0\ \text{V}$.
6. $-15\ \text{V} < V_i < -5\ \text{V}$ when $E_L = 0\ \text{V}$.
7. Data signaling rates must be in the range from zero to a nominal upper limit of 20,000 bps.
8. The driver slew rate must be less than 30 V/μs.
9. Cable length should be kept under 50 ft.

The RS-232-C specification includes limits that define the voltage range of a logic 1, a MARK, and a logic 0, a SPACE. A logic 1 at the driver output must be between −5 and −15 V. A logic 1 at the terminating load must be between −3 and −15 V. A logic 0 at the driver output must be between +5 and +15 V. A logic 0 at the terminating load must be between +3 and +15 V. Figure 4-7 illustrates these ranges. Notice that the voltage levels for a logic 1 and logic 0 are inverted from what we normally think of in terms of HIGH and LOW. The driver and receiver circuits for the interchange will have to be buffered and level translated in order to be compatible with TTL, ECL, and other logic families.

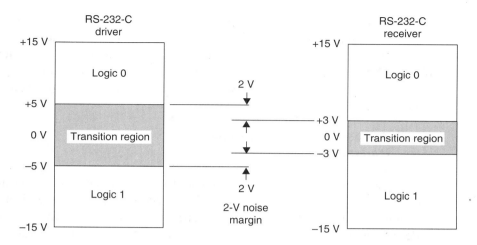

Figure 4-7 RS-232-C logic levels.

4.4.2 Noise Margin and Noise Immunity

An implied *noise margin* of 2 V (the difference between 3 and 5 V or −3 and −5 V) exists between the driver and the terminator (Figure 4-7). Noise margin is a quantitative measure of a circuit's ability to tolerate noise transients inherent within the driving and receiving circuit or induced by external sources. If a circuit's noise margin is high, it is said to have a high *noise immunity.* Noise immunity is the general term associated with the same concept. If an RS-232-C signal is received with a voltage swing of ±5 V, a noise transient in excess of 2 V_P will fall into the transition region, as depicted in Figure 4-8. There is no guarantee that

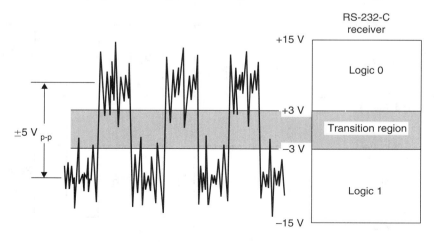

Figure 4-8 Excessive noise on a signal causes data to fall into the transition region.

the voltages in this region are interpreted as the original logic level. If an RS-232-C signal were to be transmitted by the driving circuit at ±5 V, the noise margin would be less than 2 V at the receiving end due to the voltage (IR) dropped across the interfacing cable. The noise margin can be increased simply by increasing the output voltage swing of the driver up to the limits of the specification.

4.4.3 An Asynchronous RS-232-C Transmitted Character

The asynchronous protocol used to transmit serial data has long been established since the days of the electromechanical Teletype terminals. These electrical signals must conform to the RS-232-C electrical specifications outlined above. The order in which the transmitted data is sent under this convention is reversed from what is customarily thought of as correct. Normally we think of a structured group of characters, numbers, or letters as being read from left to right. ASCII characters coded in binary are transmitted in reverse order, so that the *least* significant bit is transmitted *first*. The remaining bits are transmitted in the order of their increasing significance. Figure 4-9 illustrates examples of asynchronous RS-232-C characters transmitted at the output of a VDT, pin 2, *Transmitted Data*. The waveforms appear as they would on an oscilloscope if the given ASCII characters e or $ were typed.

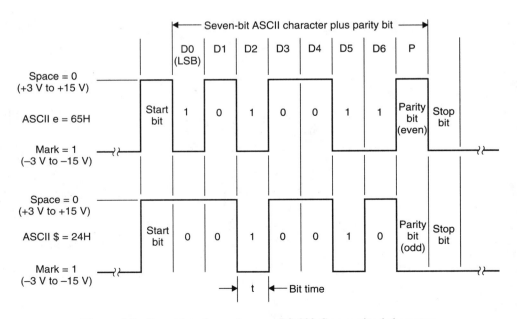

Figure 4-9 Examples of asynchronous RS-232-C transmitted characters.

Notice that other bits have been inserted as part of the total character length. Prior to sending the first bit of the character, or LSB, an additional bit has been inserted. This bit is referred to as the *start bit*. The start bit indicates to the receiving device that it should initialize its circuitry to receive a data character. This occurs the moment the start bit is asserted. The start bit is always a logic 0. In effect, it is acting as a control bit to the receiving device. The last bit sent is referred to as the character's *stop bit*, which acts as a controlling indicator to the receiving device that it is the end of a character. The stop bit is always a logic 1, *opposite* to that of the start bit. The time durations of the start and stop bits are each equivalent to that of a data bit. This discrete amount of time for each bit is referred to as the character's *bit time*. Once the stop bit has been transmitted, the logic level of the transmitting device remains in an *idle* state, which has the same logic level as the stop bit, a logic 1. The advantage to this technique is that it allows characters to be sent in an asynchronous fashion; that is, the time difference between characters can vary. A typical case would be a person typing characters at a terminal. The typist can proceed to type at any speed and can pause for any length of time between the characters being typed. The (odd or even) *parity* bit is inserted just prior to the stop bit. Its logic level depends of the number of binary 1-bits set within the data character. The parity bit can be set to a logic 1 or a logic 0 by the transmitting device depending on the character itself. When a character is received, the receiving device checks the number of 1-bits set in the received character against the level of the corresponding parity bit. If they do not agree, the receiving device can exercise its option of flagging an error to the operator. Parity is perhaps the most fundamental of all error-checking schemes.

4.4.4 Mechanical Specification

The RS-232-C mechanical specification states that the interface circuit must consist of a *plug* on one end to connect to DTE and a *receptacle* on the opposite end to connect to DCE. This implies that, in order to connect the interfacing cable, DTE must have a receptacle and DCE must have a plug. Figure 4-10 illustrates the connector interface.

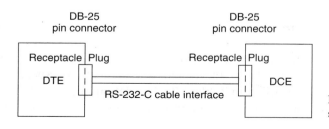

Figure 4-10 RS-232-C connector assignment.

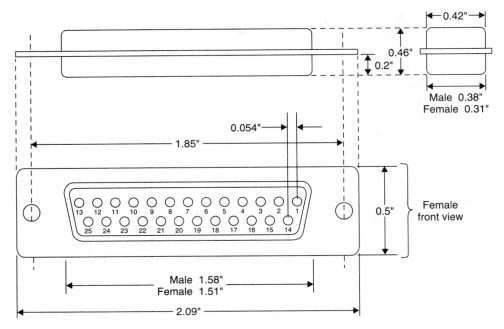

Figure 4-11   DB-25 pin connector.

The RS-232-C standard does not specify the type of mechanical connector to use, however, the DB-25 pin connector illustrated in Figure 4-11 has become almost universally accepted and associated with the standard itself. Detailed mechanical dimensions have been shown here for both plug and receptacle.

4.4.5 Functional Specification

The complete connector pin assignment for the RS-232-C interface standard is listed in Table 4-2, with an unofficial abbreviation for each circuit that has been adopted as part of the standard. Each pin has a reference designator (AA, BA, and so on) that identifies its *circuit* name. The first letter of each circuit name designates the functional category to which that circuit belongs. For example, circuit CA, Request to Send, pin 4, is a control circuit. Its circuit name begins with the letter C. There are five functional categories, as shown in the table.

4.4.6 Functional Definition

The following material contains extracts from the Electronics Industries Association RS-232-C Standard. The extracts include the functional specification for each interchange circuit listed in Table 4-2. The definitions of some circuits have been summarized, whereas others are written in their entirety. A com-

TABLE 4-2 RS-232-C Interchange Circuits

Interchange circuit	C.C.I.T.T. equivalent	Description	Gnd	Data		Control		Timing	
				From DCE	To DCE	From DCE	To DCE	From DCE	To DCE
AA	101	Protective Ground	X						
AB	102	Signal Ground/Common Return	X						
BA	103	Transmitted Data			X				
BB	104	Received Data		X					
CA	105	Request to Send					X		
CB	106	Clear to Send				X			
CC	107	Data Set Ready				X			
CD	108.2	Data Terminal Ready					X		
CE	125	Ring Indicator				X			
CF	109	Received Line Signal Detector				X			
CG	110	Signal Quality Detector				X			
CH	111	Data Signal Rate Selector (DTE)					X		
CI	112	Data Signal Rate Selector (DCE)				X			
DA	113	Transmitter Signal Element Timing (DTE)							X
DB	114	Transmitter Signal Element Timing (DCE)						X	
DD	115	Receiver Signal Element Timing (DCE)						X	
SBA	118	Secondary Transmitted Data			X				
SBB	119	Secondary Received Data		X					
SCA	120	Secondary Request to Send					X		
SCB	121	Secondary Clear to Send				X			
SCF	122	Secondary Rec'd Line Signal Detector				X			

plete listing of the specification can be obtained from EIA at the following address:

Electronic Industries Association
Engineering Department
2001 Eye Street, N.W.
Washington, D.C. 20006

Pin 1, Circuit AA: Protective Ground, PG.   This conductor shall be electrically bonded to the machine or equipment frame. It may be further connected to external grounds as required by applicable regulations.

Pin 7, Circuit AB: Signal Ground or Common Return, SG.   This conductor establishes the common ground reference potential for all interchange circuits except pin 1, Protective Ground.

Pin 2, Circuit BA: Transmitted Data, TD (or TxD).   Signals on this circuit are generated by DTE and transmitted to DCE. DTE shall hold this line in the MARK condition between characters or words, and at all times when no data are being transmitted. In all systems, DTE shall not transmit data unless an ON condition is present on all of the following four circuits:

1. Pin 4, Circuit CA: Request to Send, RTS
2. Pin 5, Circuit CB: Clear to Send, CTS
3. Pin 6, Circuit CC: Data Set Ready, DSR
4. Pin 20, Circuit CD: Data Terminal Ready, DTR

Pin 3, Circuit BB: Received Data, RD (or Rxd).   The signals on this line are transmitted from DCE to DTE. DCE shall hold this line in the MARK condition at all times when pin 8, Circuit CF, Received Line Signal Detector, is in the OFF state. In half-duplex operation, this line shall be held in the MARK condition when pin 4, RTS, is in the ON condition and for a brief interval following the ON to OFF transition of RTS to allow for the completion of transmission and the decay of line reflections.

Pin 4, Circuit CA: Request to Send, RTS.   This circuit is turned ON by DTE and is used to indicate to DCE that it is ready to transmit data. In the half-duplex mode, this line is used to control the direction of data transmission of DCE. RTS is used as a handshake line with Clear to Send, CTS, pin 5. When RTS is asserted by DTE, DCE responds when it is ready to accept the transmitted data from DTE by asserting CTS. When DTE turns RTS from ON to OFF, it is an instruction to DCE that it has completed the transmission of its data on pin 2, Transmitted Data. DCE responds (after a brief period of time to allow all the transmitted data to be received) by turning OFF its CTS line. DCE is then prepared to respond to any subsequent RTS signals from DTE. When RTS is turned OFF, it shall not turn ON again until CTS is turned OFF by DCE. It is permissible for DTE to turn on RTS at any time DCE's CTS is OFF regardless of the condition of any other interchange circuit.

Pin 5, Circuit CB: Clear to Send, CTS. Signals on this circuit are generated by DCE to indicate whether or not DCE is ready to transmit the data from DTE onto the communications channel (the telephone system). The ON condition of CTS is a handshake to the ON condition of Request to Send, RTS, pin 4, and Data Set Ready, DSR, pin 6. The OFF condition of CTS is an indicator to DTE that it should not send data across the interchange circuit on Transmitted Data, TD, pin 2.

Pin 6, Circuit CC: Data Set Ready, DSR. The signal on this circuit is sent from DCE to DTE as a status line indicating the condition of the local data set (the local modem). Data Terminal Ready, DTR, pin 20 of DTE, is assumed to be in the ON condition indicating its readiness. The ON condition of DSR is presented to indicate the following:

1. The local DCE equipment is connected to the communications channel, for example, the *off-hook* condition of the telephone network.
2. The local DCE equipment is not in test (local or remote) mode, talk (alternate voice), or dial mode.
3. The local DCE has completed where applicable:
   (a) Any timing functions required by the switching system (telephone system) to complete call establishment.
   (b) The transmission of any discrete answer tone, the duration of which is controlled solely by the local data set.

The OFF condition shall appear at all other times and should serve as an indicator to DTE to disregard signals appearing on any other interchange circuit with the exception of circuit CE, Ring Indicator, RI, pin 22. If DSR is turned OFF during which time DTR is still ON, DTE shall interpret this as a lost or aborted connection to the communications channel and take action to terminate the call.

Pin 20, Circuit CD: Data Terminal Ready, DTR. This line is turned ON by DTE, indicating that it is ready to transmit or receive data from DCE. DSR is a response to DTR. Both DTR and DSR work in conjunction with each other and indicate equipment readiness.

Pin 22, Circuit CE: Ring Indicator, RI.   The Ring Indicator, RI, is sent from DCE to DTE. This circuit indicates that a ringing signal is being received on the communications channel. Its ON–OFF duration should be approximately coincidental with the ON–OFF duration of the ringing signal received from the communications channel.

Pin 8, Circuit CF: Received Line Signal Detector, RLSD (or Carrier Detect, CD).   The ON condition of this circuit is presented by DCE when it is receiving a carrier signal from the remote DCE that meets its suitability criteria. These criteria are established by the DCE manufacturer. The OFF condition of this circuit indicates to DTE that no signal is being received that is suitable for demodulation. DCE shall clamp its Received Data, RD, pin 3, signal to the MARK condition. On half-duplex channels, CD is held in the OFF condition whenever Request to Send, RTS, pin 4, is in the ON condition and for a brief interval of time following the ON-to-OFF transition of RTS.

Pin 21, Circuit CG: Signal Quality Detector, SQ.   Signals on this circuit are sent by DCE to DTE. They are used to indicate whether or not there is a high probability of an error in the received data. An ON condition indicates there is no reason to believe that an error has occurred. An OFF condition indicates a high probability of an error. It may be used in some instances to call automatically for the retransmission of the previously transmitted data.

Pin 23, Circuit CH (DTE) or Circuit CI (DCE): Data Signal Rate Selector, SS

Circuit CH.   This signal is sent from DTE to DCE and is used to select between one of two data signaling rates in cases where two are offered. The ON condition selects the higher of the two rates.

Circuit CI.   This circuit is similar to Circuit CH, except that it is sent from DCE to DTE.

Pin 24, Circuit DA: Transmitter Signal Element Timing (DTE), TSET.
Signals on this circuit are sent from DTE to DCE to provide the transmitting signal converter with signal element timing information. The ON to OFF transition shall nominally indicate the center of each signal element on Transmitted Data, TD, pin 2. When implemented in DTE, DTE shall normally provide this timing information if in the power ON condition.

Pin 15, Circuit DB: Transmitter Signal Element Timing (DCE), TSET.
Signals on this circuit are sent from DCE to DTE to provide signal element timing information. DTE shall provide a data signal on Transmitted Data, TD, pin 2,

in which the transition between signal elements nominally occurs at the time of the transitions from the OFF to ON condition of the signal on this circuit. When implemented in DCE, DCE shall normally provide this timing information if in the power ON condition.

Pin 17, Circuit DD: Receiver Signal Element Timing (DCE), RSET. Signals on this circuit are sent from DCE to DTE to provide received signal element timing information. The transition from the ON to OFF condition shall nominally indicate the center of each signal element on pin 3, Received Data. Timing information on this circuit shall be provided at all times when Received Line Signal Detector (Carrier Detect), CD, pin 8, is in the ON condition. It may, but need not, be present following the ON to OFF transition of pin 8, Received Line Signal Detector.

Pin 14, Circuit SBA: Secondary Transmitted Data, (S)TD. This circuit is equivalent to Transmitted Data, TD, pin 2, except that it is used to transmit data via the secondary channel. When the secondary channel is usable only for circuit assurance or to interrupt the flow of data in the primary channel, this circuit is normally not provided, and the channel carrier is turned ON or OFF by means of Secondary Request to Send, (S)RTS, pin 19. Carrier OFF is interpreted as an interrupt condition.

Pin 16, Circuit SBB: Secondary Received Data, (S)RD. This circuit is equivalent to Received Data, RD, pin 3, except that it is used to receive data on the secondary channel from DCE. When the secondary channel is usable only for circuit assurance or to interrupt the flow of data in the primary channel, this circuit is normally not provided.

Pin 19, Circuit SCA: Secondary Request to Send, (S)RTS. This circuit is equivalent to Request to Send, RTS, pin 4, except that it requests the establishment of the secondary channel instead of requesting the establishment of the primary data channel. Where the secondary channel is used as a backward channel, the ON condition of RTS shall disable this circuit and it shall not be possible to condition the secondary channel transmitting signal converter to transmit during any time interval when the primary channel transmitting signal converter is so conditioned.

When the secondary channel is usable only for the assurance or to interrupt the flow of data in the primary data channel, this circuit shall serve to turn ON the secondary channel unmodulated carrier. The OFF condition of this circuit shall turn OFF the secondary channel carrier and thereby signal an interrupt condition at the remote end of the communication channel.

Pin 13, Circuit SCB: Secondary Clear to Send, (S)CTS. This circuit is equivalent to Clear to Send, CTS, pin 5, except that it indicates the availability of the secondary channel instead of indicating the availability of the primary channel. This circuit is not provided where the secondary channel is usable only as a circuit assurance or an interrupt channel.

Pin 12, Circuit SCF: Secondary Received Line Signal Detector, (S)RLSD or (S)CD (Secondary Carrier Detect). This circuit is equivalent to Carrier Detect, CD, pin 8, except that it indicates the proper reception of the secondary channel line signal. Where the secondary channel is usable only as a circuit assurance or an interrupt channel, this circuit shall be used to indicate the circuit assurance status or to signal the interrupt. The ON condition shall indicate circuit assurance or a noninterrupt condition. The OFF condition shall indicate circuit failure (no assurance) or the interrupt condition.

Recall that the RS-232-C interface was written with the intention of interfacing DTE (a terminal or computer) to DCE (a modem) for the purpose of long-distance data communications over the telephone network. The intention at that time was certainly not to satisfy the interface requirements of today's communications between microcomputers, printers, ROM blasters, and the like, all of which did not exist when the standard was originally published! Nevertheless, the standard has adapted itself to this date as perhaps the most widely recognized data communication interface standard.

To improve our understanding of some of the circuits most commonly used, attention will be focused on building a typical RS-232-C interface. For purposes of clarification, we will part from the two-letter (or three) designator assigned for each circuit in the standard. Instead, the circuit descriptive name listed in Table 4-2 will be used.

4.5 INTERFACING DTE TO DCE

The minimal amount of wires necessary to establish a two-way communications link between DTE and DCE is depicted in Figure 4-12. The interfacing cable includes only three interconnecting lines: pin 2, pin 3, and pin 7. For many applications today, this is desirable since it greatly simplifies the RS-232-C cable alone.

The RS-232-C specification states that pin 2 of DTE will be used for **Transmitted Data,** TD (or TxD), to DCE. From our diagram in Figure 4-12, this implies that pin 2 of DCE *receives* the transmitted data from DTE. Pin 3 of DTE is **Received Data,** RD (or RxD), from DCE. Pin 3 of DCE is therefore transmitted data to DTE. Pin 7 of both DTE and DCE are **Signal Ground,** SG. It all seems so simple, it is no wonder so many of today's RS-232-C interfaces

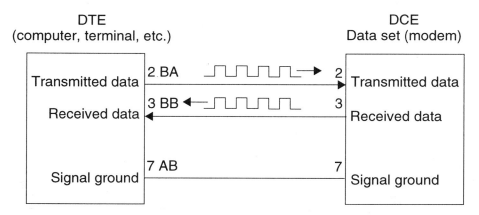

Figure 4-12 Minimal interface between DTE and DCE.

include only three interconnect lines. The setup described is certainly acceptable since it is functional; that is, it works and does so with a minimal amount of hardware. It does, however, leave much to be desired in terms of control. The following assumptions are made: Interconnecting devices are ready to transmit and receive data; the communications channel has been established through the telephone company; carrier frequencies of both modems are idling at their correct frequencies; interrupting of the transmitted and received data will not occur; and so on. A much more intelligent interface is often needed. The resulting interface would include the RS-232-C *control circuits* (circuits that begin with C) for regulating and controlling the flow of data across pins 2 and 3.

4.5.1 A Complete RS-232-C Interface

Figure 4-13 illustrates a complete RS-232-C interface between DTE and DCE. The diagram depicts the most commonly used circuits compared to those that are not often used. Notice that some of the control circuits that are included in the most commonly used group have nothing to do with an RS-232-C interface between a computer and printer. Circuits such as **Ring Indicator** and **Received Line Signal Detector** or **Carrier Detect** are superfluous as far as a printer is concerned. These circuits, however, must be considered in *any* RS-232-C interface. Even though some devices, such as printers, may not implement many of the circuits depicted in Figure 4-13, other devices may. Any attempt to interface two devices employing a different number of circuits will inevitably result in *floating* inputs and signals with a lost cause. What may seem a dilemma at first will soon be resolved.

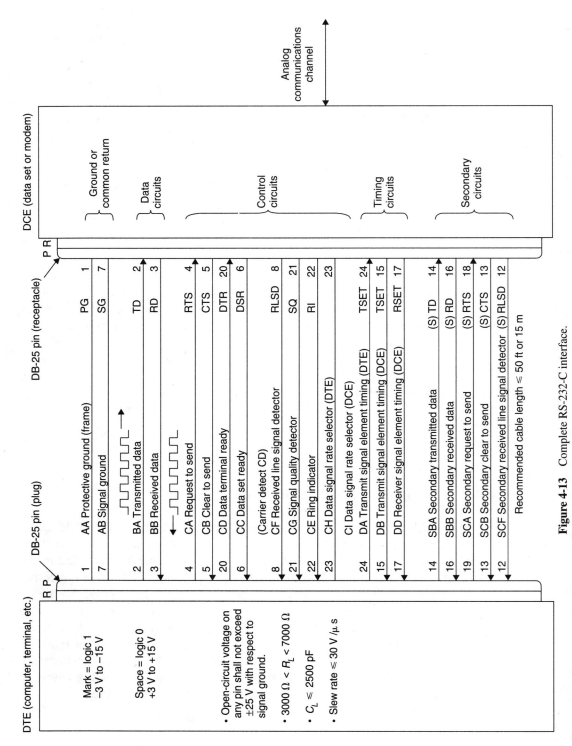

Figure 4-13 Complete RS-232-C interface.

4.6 HANDSHAKING

The regulation and control of data between DTE and DCE is performed by a process called *handshaking*. The process stems from the literal meaning itself. When two people shake hands, they are communicating. The communication can range from a simple acknowledgment of one's presence to an agreement. The process is much the same in the interactive control of data between two devices, DTE and DCE. To illustrate the concept of handshaking, refer to Figure 4-14. Handshaking is said to occur between DTE's **RTS** and DCE's **CTS.** The entire process is a series of inquiries and acknowledgments between DTE and DCE, ultimately resulting in the successful transmission of data between the two devices. Let us now consider a complete RS-232-C control sequence. A half-duplex communications link will be used for simplicity (Figure 4-15).

Suppose that DTE and DCE are both powered up with **Data Terminal Ready** and **Data Set Ready** both ON (Figure 4-15). The ON condition of both of these circuits indicates that DCE has dialed a number and completed call establishment to a remote data set. When DTE is ready to transmit data, it turns ON its **Request to Send** signal. When the modem receives **Request to Send** from DTE, it will turn its carrier frequency (MARK condition) ON. The modem is now ready to transmit data from DTE. It informs DTE of its readiness by as-

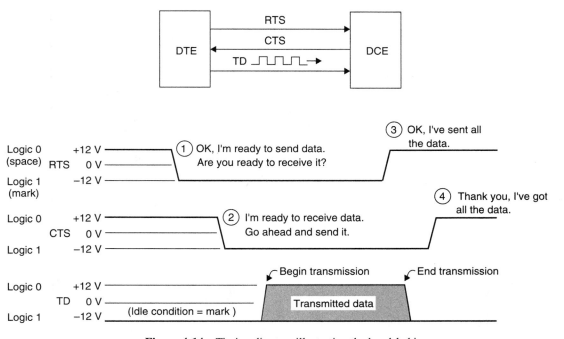

Figure 4-14   Timing diagram illustrating the handshaking process.

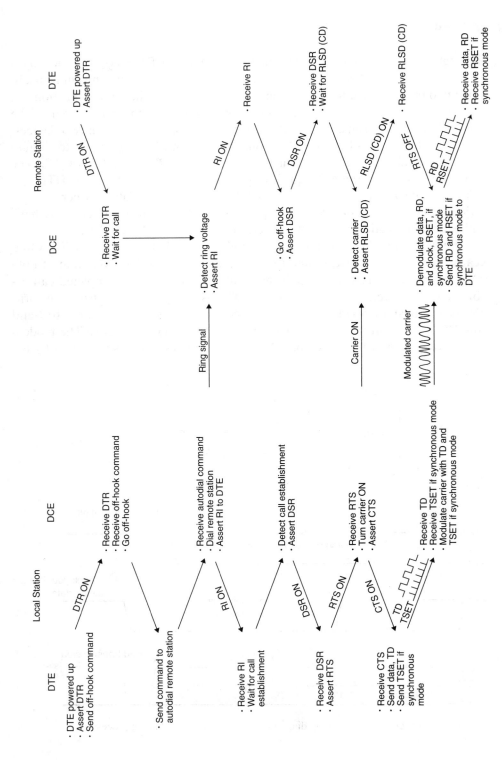

Figure 4-15 RS-232-C half-duplex control sequence.

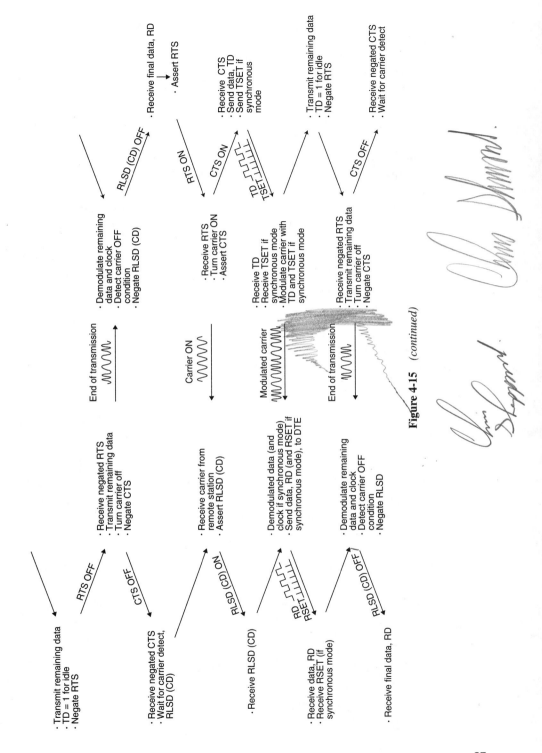

Figure 4-15 *(continued)*

serting its **Clear to Send** signal. During the course of this action, the remote modem has sensed the local modem's carrier frequency and consequently activates its **Carrier Detect** line. The local DTE can now begin to transmit its data on its **Transmitted Data** line. DCE in turn modulates its carrier frequency with the transmitted data. The resulting signal is transmitted over the communications channel and demodulated at the remote data set. The demodulated data are sent on the **Received Data** circuit of the remote DTE. Once the local DTE has completed its transmission of data, it turns its **Request to Send** line OFF and **Transmitted Data** is left in its quiescent state (MARK condition). The transition of **Request to Send** from ON to OFF is an instruction to the local DCE to complete the transmission (onto the communications channel) of the last data sent from DTE. The local DCE responds by completing the transmission and turning OFF its carrier frequency and **Clear to Send** line. Since the control sequence is half-duplex, the carrier frequency is turned OFF to allow the remote station to transmit data in return on the same frequency. The process is now repeated in the reverse order, as depicted in Figure 4-15. Any subsequent data to be transmitted will follow the same procedure.

4.7 RS-232-C INTERFACE TO DEVICES OTHER THAN A MODEM

Although the procedure just outlined may seem comprehensive and systematic, it has in effect caused a great deal of confusion among those of us who have attempted to connect DTE to DCE. The underlying problem is that the RS-232-C standard was never really intended to satisfy the interfacing requirements of today's microcomputers and peripheral devices. Computers, for example, may be interfaced in one situation as DTE and in another as DCE. For instance, in the previous example the computer is DTE, interfaced to a modem, DCE. However, what would be the case if the same computer were interfaced to a printer? The printer, acting as the *terminating* device, or DTE in this case, would be interfacing with the computer as DCE. Computer manufacturers today are left with the dilemma of having to decide whether to configure their RS-232-C ports as DTE or DCE or to provide the option of both. Figure 4-16a shows the seemingly inevitable situation that arises, that is, two devices that transmit and receive data on the *same* pin numbers. The output drivers for control and data are in conflict with each other. Communications obviously cannot exist in this situation. One solution to this problem is to simply reverse the wires on our RS-232-C interface cable. A cable configured as such is referred to as a *null modem*. Null modems are also used to test two DTEs.

Obviously, we have deviated from the standard. As mentioned earlier, signal names like **Carrier Detect** and **Ring Indicator** have nothing to do with

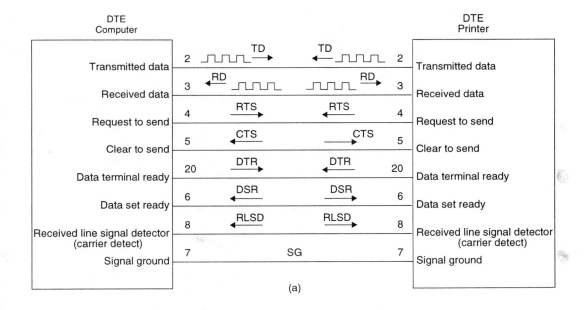

(a)

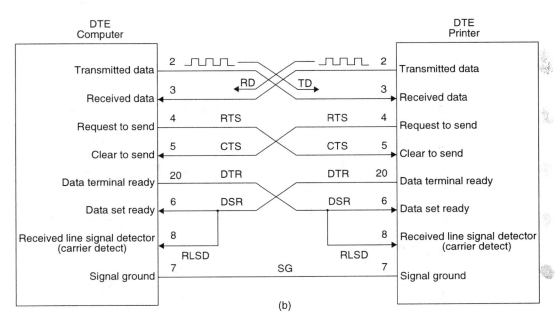

(b)

Figure 4-16 (a) Two devices that have been configured as DTE cannot communicate; (b) null modem.

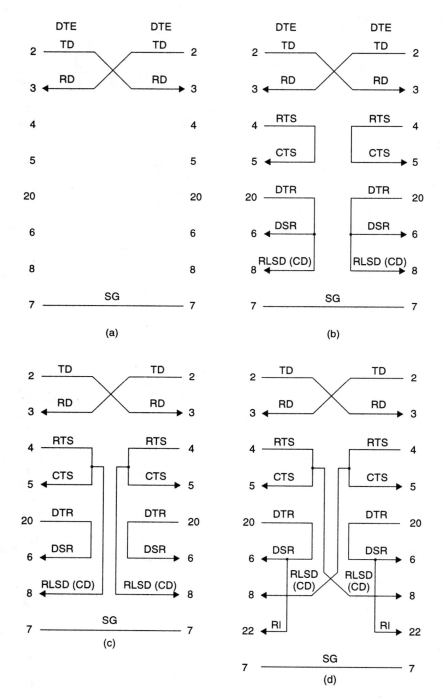

Figure 4-17 Nonstandard interface cable techniques.

printers, PROM programmers, and many other devices on the market that boast RS-232-C compatibility. The fact of the matter is that many of the RS-232-C circuits discussed so far are either emulated, cross-connected, or, in some cases, not even used. Figure 4-17 illustrates several tricks-of-the-trade nonstandard connector configurations that are commonly employed to successfully interface the so-called RS-232-C compatible devices of today.

4.8 EIA RS-449, RS-422-A, AND RS-423-A

Despite all its shortcomings, RS-232-C has allowed most computer manufacturers to consolidate their efforts toward standardization of serial interfaces. In the absence of such formalities, the computer industry could very well face a deluge of incompatible products. The RS-232-C standard is here and is likely to be around for awhile. Several related standards have since evolved for which RS-232-C has served as a predecessor and a basis for those standards to improve on.

4.8.1 Improving the Old Standard

In November 1977, the Electronics Industries Association published a new and improved serial interface standard: EIA RS-449. EIA RS-449 addresses the mechanical and functional specification of an enhanced version of RS-232-C. The new standard is stated as follows:

EIA Standard RS-449: General Purpose 37-Position and 9-Position Interface for Data Terminal Equipment and Data Circuit-Terminating Equipment Employing Serial Binary Interchange.

New semiconductor interfacing products have given more reason to improve on the old standard's electrical specification. Two additional standards immediately followed the publication of RS-449. *RS-423-A* and *RS-422-A* directly address the enhancement of the electrical characteristics of RS-232-C. These two additional standards are stated as follows:

EIA Standard RS-423-A: Electrical Characteristics of Unbalanced Voltage Digital Interface Circuits, September 1978.
EIA Standard RS-422-A: Electrical Characteristics of Balanced Voltage Digital Interface Circuits, December 1978.

4.8.2 EIA Objectives in Setting the New Standards

RS-449, together with RS-422-A and RS-423-A, is intended to gradually replace RS-232-C as the specification for interface between DTE and DCE. The new specifications meet the following objectives:

— Maintain compatibility with existing RS-232-C equipment.

— Derive a new and more meaningful set of circuit names and mnemonics to differentiate between RS-232-C and RS-449.

— Increase the data transmission rate beyond 20 kbps.

— Increase the transmission distance over twisted-pair wires.

— Specify a standard connector type that can be used without the use of a special tool.

— Provide a separate connector to carry secondary interchange circuits.

— Improve on the cross-talk characteristics of the signal lines.

— Provide for differential signal lines for transmitted and received data.

— Provide for loopback testing.

— Utilize the latest advances in semiconductor technology to improve on performance and reliability.

4.8.3 RS-499 Interchange Circuits

Table 4-3 is a complete listing of the RS-449 interchange circuits. The table includes RS-232-C equivalent circuits where applicable. Notice that RS-449 circuit names and mnemonics are more readily identifiable with their associated functions.

4.8.3.1 Category I and category II circuits.

A main objective set forth by EIA in the establishment of RS-449 is to maintain compatibility with the existing RS-232-C standard. To fulfill this objective, EIA has divided the new interchange circuits into two categories: category I and category II. Category I includes those signals that are functionally compatible with RS-232-C. The remaining signals are category II signals. Category I and II circuits and their equivalent RS-232-C circuits are shown in Tables 4-4 and 4-5.

Category I circuits can be implemented with either RS-422-A balanced drivers and receivers or RS-423-A unbalanced drivers and receivers (see Section 4.8.5 for balanced versus unbalanced circuits). The transmission data rates should be less than 20 kbps (the upper limit of RS-232-C) if RS-423-A is used. Above 20 kbps, category I circuits should always be implemented under RS-422-A. Notice in Table 4-4 that category I circuits have been allotted two adjacent pins per signal function to allow for the option of either balanced or unbalanced circuits. Category II circuits are always implemented under RS-423-A since these circuits provide the interconnection of only unbalanced drivers.

TABLE 4-3 RS-449 and RS-232-C Equivalent Interchange Circuits

EIA RS-449 37-pin connector			EIA RS-232-C DB-25 pin connector		
SG	Signal ground	19	AB	Signal ground	7
SC	Send common	37			
RC	Receive common	20			
IS	Terminal in service	28			
IC	Incoming call	15	CE	Ring indicator	22
TR	Terminal ready	12, 30	CD	Data terminal ready	20
DM	Data mode	11, 29	CC	Data set ready	6
SD	Send data	4, 22	BA	Transmitted data	2
RD	Receive data	6, 24	BB	Received data	3
TT	Terminal timing	17, 35	DA	Transmitter signal element timing (DTE source)	24
ST	Send timing	5, 23	DB	Transmitter signal element timing (DCE source)	15
RT	Receive timing	8, 26	DD	Receiver signal element timing	17
RS	Request to send	7, 25	CA	Request to send	4
CS	Clear to send	9, 27	CB	Clear to send	5
RR	Receiver ready	13, 31	CF	Received line signal detector	8
SQ	Signal quality	33	CG	Signal quality detector	21
NS	New signal	34			
SF	Select frequency	16			
SR	Signal rate selector	16	CH	Data signal rate selector (DTE source)	23
SI	Signal rate indicator	2	CI	Data signal rate selector (DCE source)	23
LL	Local loopback	10			
RL	Remote loopback	14			
TM	Test mode	18			
SS	Select standby	32			
SB	Standby indicator	36			

9-pin connector					
	Shield	1	AA	Protective ground	1
SG	Signal ground	5	AB	Signal ground	7
SC	Send common	9			
RC	Receive common	2			
SSD	Secondary send data	3	SBA	Secondary transmitted data	14
SRD	Secondary receive data	4	SBB	Secondary received data	16
SRS	Secondary request to send	7	SCA	Secondary request to send	19
SCS	Secondary clear to send	8	SCB	Secondary clear to send	13
SRR	Secondary receiver ready	6	SCF	Secondary received line signal detector	12

TABLE 4-4 Category I Circuits

RS-449	RS-232-C equivalent circuit
SD Send data (4, 22)	BA Transmitted data (2)
RD Receive data (6, 24)	BB Received data (3)
TT Terminal timing (17, 35)	DA Transmitter signal element timing (24)
ST Send timing (5, 23)	DB Transmitter signal element timing (15)
RT Receive timing (8, 26)	DD Receiver signal element timing (17)
RS Request to send (7, 25)	CA Request to send (4)
CS Clear to send (9, 27)	CB Clear to send (5)
RR Receiver ready (13, 31)	CF Received line signal detector (8)
TR Terminal ready (12 , 30)	CD Data terminal ready (20)
DM Data mode (11, 29)	CC Data set ready (6)

TABLE 4-5 Category II Circuits

RS-449
SC Send common (37)
RC Receive common (20)
IS Terminal in service (28)
NS New signal (34)
SF Select frequency (16)
LL Local loopback (10)
RL Remote loopback (14)
TM Test mode (18)
SS Select standby (32)
SB Standby indicator (36)

4.8.3.2 RS-449 functional specification for new circuits.   There are 10 functionally new interchange circuits in RS-449 that differ from its predecessor, RS-232-C. A brief definition of these new functions are as follows:

Pin 37, Circuit SC, Send Common: This circuit acts as a common signal return line for unbalanced signals transmitted from DTE to DCE.

Pin 20, Circuit RC, Receive Common: This circuit acts as a common signal return line for unbalanced signals transmitted from DCE to DTE.

Pin 28, Circuit IS, Terminal in Service: This circuit is used to indicate to DCE whether or not DTE is operational.

Pin 34, Circuit NS, New Signal: This circuit is used primarily for multipoint applications where two or more terminals or computers share a common communications channel. In this mode, DCEs are operating in a switched carrier mode. A control DTE polls other DTEs for messages to be sent to DCE. The

ON state of this signal is an indicator to DCE from the control DTE that a message from a remote DTE is completed and a new one is about to begin.

Pin 16, Circuit SF, Select Frequency: This circuit is used primarily for multipoint applications where DTE selects the transmit and receive frequencies used by its connecting DCE.

Pin 10, Circuit LL, Local Loopback: This circuit is used by DTE to request from DCE a local loopback test. When asserted by DTE, DCE loops data and control signals back to DTE to verify the functionality of the local DTE and DCE.

Pin 14, Circuit RL, Remote Loopback: This circuit is used by DTE to request from DCE a remote loopback test. When asserted by DTE, data and control signals generated by the local DTE are routed through the local DCE to the remote DCE and back to the local DTE. This, in effect, verifies the functionality of the local DTE, DCE, the communications channel, and the remote DCE.

Pin 18, Circuit TM, Test Mode: This circuit is used by DCE to inform its DTE that a test condition has been established.

Pin 32, Circuit SS, Select Standby: This circuit is used by DTE to request the use of standby equipment in the event of failure of primary equipment.

Pin 36, Circuit SB, Standby Indicator: This circuit is intended to be used as a response to circuit SS. In the event of an equipment failure, when standby equipment has replaced the failure, this circuit informs DTE of that replacement.

4.8.4 EIA RS-449 Mechanical Specification

With the addition of 10 new circuit functions, a larger connector type than the RS-232-C DB-25 pin connector was necessary. EIA decided to use a 37-pin connector to accommodate the new signals along with the basic existing signals, and a 9-pin connector to separately carry the secondary interchange circuits. Figure 4-18 shows both connector types.

The new connectors permit latching and unlatching of the interfacing connector without the use of a special tool. Anyone who has experience with connecting RS-232-C cables knows the numerous problems encountered: screws and standoffs prevent the plug and receptacle from being inserted, the absence of a proper screwdriver, incorrect match, and so on.

4.8.4.1 EIA RS-449 cable length.
Since higher data transfer rates over longer distances are supported by the new standard, it was necessary for EIA to establish guidelines pertaining to the maximum cable length. The maximum cable length is primarily a function of the data transfer rate to be used. In general, the higher the data transfer rate, the shorter the cable length should be. Figure 4-19 has been provided by EIA as a guideline toward establishing the maximum cable length versus signal rate in bits per second.

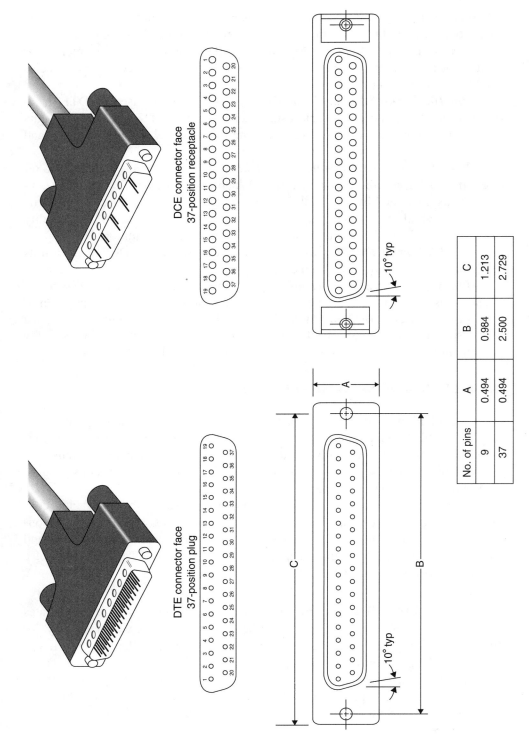

Figure 4-18 EIA RS-449 37- and 9-position connectors.

DCE connector face
37-position receptacle

DTE connector face
37-position plug

10° typ

No. of pins	A	B	C
9	0.494	0.984	1.213
37	0.494	2.500	2.729

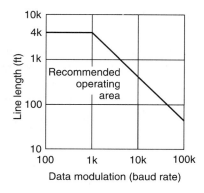

Figure 4-19 EIA graph depicting data signaling rate versus cable length.

4.8.5 EIA RS-422-A and RS-423-A: Balanced versus Unbalanced Circuits

These two electrical standards are intended to support the mechanical and functional specifications of EIA-449. RS-422-A specifies the electrical characteristics of a balanced interface, whereas RS-423-A specifies the electrical characteristics for an unbalanced interface. Let us consider the meaning of *balanced* versus *unbalanced* circuits.

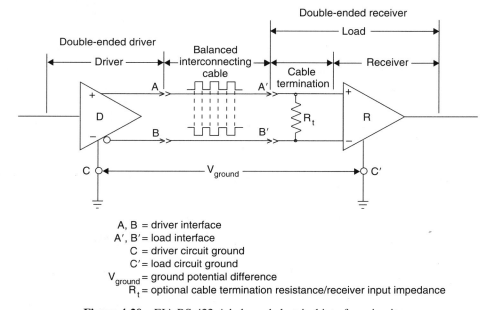

A, B = driver interface
A′, B′ = load interface
C = driver circuit ground
C′ = load circuit ground
V_{ground} = ground potential difference
R_t = optional cable termination resistance/receiver input impedance

Figure 4-20 EIA RS-422-A balanced electrical interface circuit.

Balanced Electrical Circuit.   EIA RS-422-A specifies the use of balanced electrical interface circuits. A balanced electrical circuit is one in which the transmitted signal propagates on two signal paths (Figure 4-20). Both signal paths appear to the transmitting device as having the same impedances with respect to signal ground. For this reason, they are said to be *balanced*. The transmitted signals are the *difference* in the outputs of a differential amplifier or differential line driver. If the signal to be transmitted is digital in nature, then the output waveforms for the two signal paths are *complemented* with each other, as depicted in Figure 4-20.

The receiving device at the terminating end is referred to as a differential line receiver. The line receiver is capable of amplifying the *difference* in signal levels between the two signal paths. Since the circuit is said to be balanced, both outputs of the transmitting device must have the same output impedances. Likewise, the receiving device must have the same input impedances for both signal paths. Transmitting and receiving devices operating in this mode are said to be *double-ended* amplifiers.

Unbalanced Electrical Circuit.   EIA RS-423-A specifies the use of unbalanced electrical interface circuits. An unbalanced electrical circuit is one in which the transmitted signal is propagated down a single transmission line, accompanied by its signal return or common reference line typically shared by other signals (Figure 4-21). Unlike the balanced electrical circuit, the transmitting device has only one output that is connected to a single impedance: the impedance of the interface with respect to signal ground. The receiving device re-

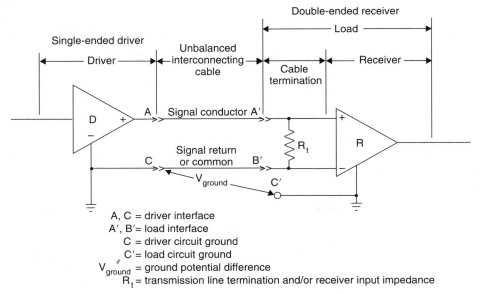

A, C = driver interface
A', B'= load interface
C = driver circuit ground
C'= load circuit ground
V_{ground} = ground potential difference
R_t = transmission line termination and/or receiver input impedance

Figure 4-21   EIA RS-423-A unbalanced electrical interface circuit.

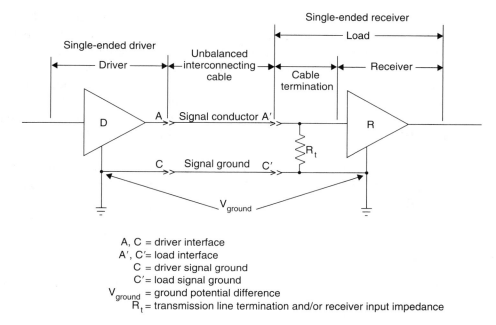

A, C = driver interface
A', C'= load interface
C = driver signal ground
C'= load signal ground
V_{ground} = ground potential difference
R_t = transmission line termination and/or receiver input impedance

Figure 4-22   RS-232-C electrical interface circuit.

ceives the transmitted signal referenced to the same signal ground. These two lines, depicted in Figure 4-21, are said to be *unbalanced* since their impedances are different. The transmitting device can still be a differential amplifier. Only one output is used in this case. The receiving device can also be a differential amplifier. One of its inputs is simply tied to signal ground. Amplifiers operating in this mode are said to be *single-ended* amplifiers.

Figure 4-22 depicts the RS-232-C electrical interface circuit. Notice that this circuit, similar to RS-423-A, is also unbalanced. The major advantage of the electrical circuits for RS-422-A and RS-423-A are their *differential receivers*. RS-232-C does not employ differential receivers, a device that greatly enhances the noise immunity characteristics of the circuit.

4.8.6 RS-422-A and RS-423-A
Common-mode Rejection

One significant feature of the differential amplifier is its ability to cancel or reject undesirable signals that are *common* to both differential inputs of the receiver. Typically, in data communications employing RS-449, a signal transmitted on a balanced or unbalanced circuit uses twisted-pair wire as the transmission medium. There are good reasons for the selection of twisted-pair wire. For one, the transmitted signal on each wire of the twisted-pair passes through the *same* electrical environment. Any noise induced by adjacent circuits will be approximately equal in amplitude and phase on both wires. This interference is *common* to both inputs of the receiving amplifier. This is where the major advantage lies over RS-232-C. The differential amplifier employed as a receiver in an RS-449 interface amplifies the *difference* between its two inputs and rejects signals that are common to both inputs. A measure of this rejection to signals common to both inputs of the differential amplifier is called the *common-mode rejection ratio* (CMRR).[†] The effect of this differential amplifier characteristic is best utilized when using both outputs of the line driver as specified in RS-422-A. The differential signal at the receiver is much greater in this case (since the signals are complemented with each other) than when the signal is transmitted unbalanced, as in RS-423-A.

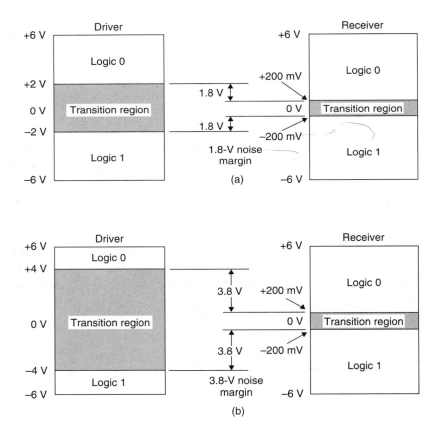

Figure 4-23   (a) RS-422-A and (b) RS-423-A logic levels.

[†] A more detailed explanation of differential amplifiers and CMRR can be found in most books covering the subject of operational amplifiers.

4.8.7 RS-422-A and RS-423-A Voltage Specification

The voltage specifications for RS-422-A and RS-423-A are shown in Figure 4-23. Notice the voltage swing for a logic 1 and 0 are considerably less, at ±6 V, than the RS-232-C specification of ±15 V. Also, the noise immunity for both specifications is different from RS-232-C. RS-422-A has a noise margin of 1.8 V, whereas RS-423-A has a noise margin of 3.8 V, over twice as much as RS-422-A. The larger noise margin has been allotted for RS-423-A to compensate for the unbalanced circuit specification.

Distances of up to 4000 ft can be achieved with RS-422-A. A maximum data transfer rate of 10 Mbps is specified. In addition, the differential line drivers are intended to support full-duplex operation with a capability of driving up to a maximum of 10 modules sharing a common receive-line pair.

4.8.8 RS-485

The newest standard is RS-485, which employs the same signal levels as RS-422. The RS-485 drivers and receivers are of better quality than their RS-422 counterparts, so you can expect to achieve maximum hardware performance over distances of up to 4000 ft at rates of 10 Mbaud. Up to 32 devices can share the same connection. RS-485 cannot generally be interfaced to RS-422, however, because RS-485 has only one differential pair signal path and is limited to half-duplex operation where only one device transmits at a time. Since there are no communications protocols, traffic on the bus must be controlled by software.[†] Table 4-6 provides a comparison of RS-232-C, RS-422-A, and RS-485.

TABLE 4-6   Summary Table for RS-232, RS-422, and RS-485

	RS-232-C	RS-422-A	RS-485
Mode of operation	Single-ended	Differential	Differential
Number of drivers and receivers allowed on line	1 driver 1 receiver	1 driver 10 receivers	32 drivers 32 receivers
Maximum cable length (ft)	50	4000	4000
Maximum data rate (bps)	20k	10M	10M

[†] Reprinted with permission from *Keithley Metrabyte Data Acquisition Catalog and Reference Guide,* Vol. 25, Taunton, Mass: (1992), p. 199. Keithley Metrabyte.

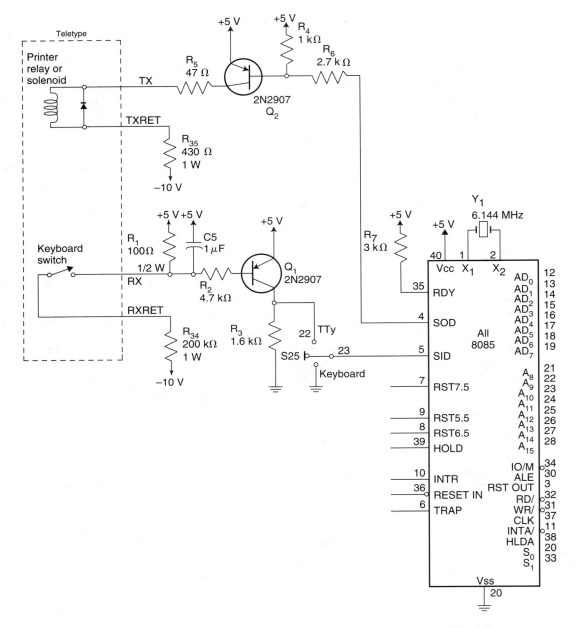

Figure 4-24 8085A current loop interface to a Teletype. (Courtesy of Intel Corporation.)

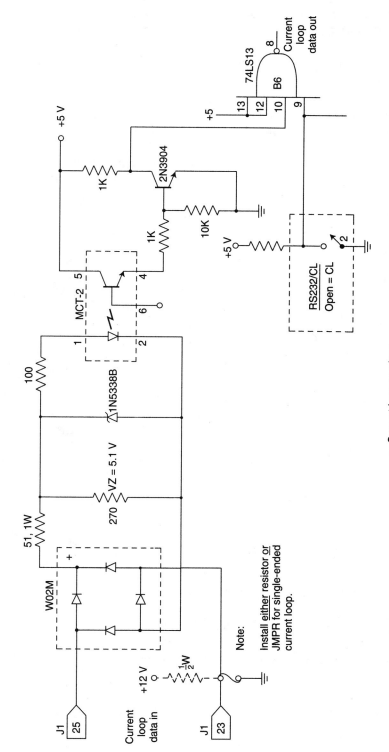

Figure 4-25 Twenty-milliampere current loop circuits for Zentec Corporation's ADM-3A employing optocouplers. (Courtesy of Zentec Corp.)

Current loop receiver

4.9 TWENTY- AND SIXTY-MILLIAMPERE CURRENT LOOP STANDARDS

In the 1960s, mechanical Teletype (TTY) printers were interfaced to computers using 20- and 60-mA current loops. The currents were used to drive electromechanical relays in conjunction with the printing action of the terminals. A typical setup between the computer and Teletype includes two current loops: one for the computer to receive characters from the Teletype and one for the computer to transmit characters to the Teletype. Figure 4-24 illustrates how the Intel 8085A microprocessor can be interfaced to a Teletype terminal using a 20-mA current loop. ASCII characters are represented by sequential combinations of current ON (MARK) and current OFF (SPACE) intervals, similar to the manner in which voltage levels are used in RS-232-C. The 8085A SID and SOD serial interface lines are used to send and receive TTL levels from the 20-mA current loop buffers.

Teletype terminals typically include built-in modems interfaced to the telephone lines for long-distance communications. Due to advances in technology, the Teletype is rapidly becoming extinct. The 20-mA current loop standard, however, has been adapted to many of today's terminals and high-speed printers. Instead of electromagnetic relays, these devices utilize *optocouplers* to interface to the current loop transmitter and receiver. Optocouplers, or *optoisolators,* are devices used to isolate high voltage and noisy circuits (found in printers and display terminals) from their respective controlling circuit. The optocoupler consists of a sealed infrared emitter and photodetector, completely isolated from each other. The 20-mA current loop driver and receiver circuit can therefore run completely isolated from high-voltage and noisy circuits. Figure 4-25 depicts how the optocoupler is used in Zentec Corporation's ADM-3A Video Display Terminal. The circuits shown are bidirectional (current can flow in either direction) and operate over the range 16 to 24 mA.

PROBLEMS

1. Define *protocol.*
2. What are the logic levels of the start and stop bits of an asynchronous serial character?
3. Compute the protocol overhead for a seven-bit ASCII character that includes a parity bit, a start bit, and two stop bits.
4. Compute the protocol overhead for a DDCMP message block that includes 512 bytes in its data field.
5. What does *modem* stand for?
6. Define *DTE* and *DCE.*

7. Is a printer considered DTE or DCE?

8. The RS-232-C specification can be divided into three categories. What are they?

9. What is the maximum recommended RS-232-C cable length?

10. Compute the noise margin for an RS-232-C transmitted signal that has an output voltage of ±9 V.

11. Draw the asynchronous serial RS-232-C waveforms for the ASCII characters y, 3, and ;. Include one stop bit and even parity generation for each character.

12. What RS-232-C signal informs DTE that it is ready to accept data?

13. What types of signals are specified for the EIA RS-449 nine-pin connector?

14. Specify whether the following EIA specifications use balanced or unbalanced electrical interface circuits.
    (a) EIA RS-232-C
    (b) EIA RS-422-A
    (c) EIA RS-423-A

15. What type of amplifier specifies CMRR?

16. What is the maximum data transfer rate specified for RS-422-A?

17. What is the maximum data transfer rate specified for RS-485?

5

The UART

Data terminal equipment (DTE) such as terminals and host computers must transmit digital information over a single transmission line. To perform this operation, DTE uses a device called a *universal asynchronous receiver/transmitter* (UART). The UART (Figure 5-1) is a peripheral device; that is, it operates external to the CPU. It is an essential component within terminals, computers, and many other serial communications devices. The UART must be programmed by the CPU in order to communicate between two devices. For *synchronous* communications, a USART, *universal synchronous asynchronous receiver/transmitter,* must be used. In this chapter, asynchronous communications is considered.

5.1 UART INTERFACE

The UART can accept a data character from the CPU in parallel format and convert it into a serial data stream. Likewise, the UART can receive a serial data stream and convert it to a parallel data character for the CPU. These two fundamental tasks of the UART can be performed simultaneously, since its internal receiver and transmitter sections operate virtually independently of each other.

Consider the case of the UART transmitting a character in serial format to a video display terminal (VDT). In the setup shown in Figure 5-1, decoding is necessary to enable the UART. The decoded output port enables the UART to receive a parallel word sent by the CPU via its data bus. The UART then begins its internal procedure to send the parallel word, bit by bit, as a serial bit stream to

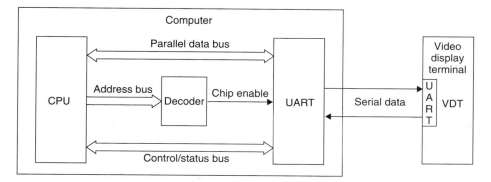

Figure 5-1 Block diagram of the UART interface.

the terminal. Conversely, the UART must be enabled by the CPU whenever a character from the terminal has been received in serial format by the UART and is ready to be input by the CPU as a parallel data word.

The CPU can establish whether or not the UART is ready to send a character, or has available a character, by addressing the device through its decoded port. This will enable the UART and allow the CPU to read a complete status of the device at any time. The internal registers that are accessed contain specific information, such as transmitter and receiver ready flags and error checking flags. External output pins are also provided by the UART for indicating its readiness to send or receive data. They may be used to interrupt the CPU for interrupt-driven programs.

Now that we have developed a fundamental idea of what the UART does, the complexities of the internal logic necessary to perform these tasks becomes increasingly apparent. As in any complex device used as a building block, it behooves the technician and design engineer to investigate and understand the internal functions of that device. On this premise, the device is said to be best selected, utilized, and maintained with its associated hardware.

5.2 UART RECEIVER SECTION

The primary task of the receiver section of the UART is to convert a serial bit stream to a parallel data word for the CPU to read. This process includes the following functions:

— Start bit detection
— Synchronization of data bit cells
— Shift register timing and control
— Serial-to-parallel conversion
— Stop bit detection

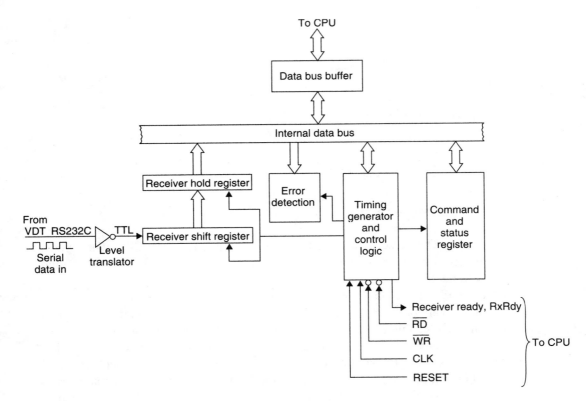

Figure 5-2   Block diagram of the UART receiver section.

— Error detection

— Status

The basic components of the receiver section are illustrated in Figure 5-2. When a character is transmitted from a terminal, it is received by the UART receiver section as Serial Data In. The signal levels at this point have been level translated from RS-232-C format to TTL. The TIMING GENERATOR and CONTROL LOGIC provide the RECEIVER SHIFT REGISTER with the necessary clock pulses and control signals to shift the serial character in. It is then loaded into the RECEIVER HOLD REGISTER, where further processing such as error detection and status operations are performed. The CPU is notified by the Receiver Ready signal, RxRdy, that a parallel formatted character is ready to be read from the UART's RECEIVER HOLD REGISTER.

5.2.1 Double Buffering the UART Receiver

Why not do away with the RECEIVER HOLD REGISTER all together and process the parallel character directly out of the RECEIVER SHIFT REGISTER? This is possible; however, the CPU would have to read the character immediately following its last bit shifted in. This would ensure that serial characters sent to the UART in direct succession to each other would not write over each other before the CPU has a chance to read them. Often, it is not convenient for the CPU to input the character as soon as it is available, as in the case when it is performing other more pertinent tasks. Holding the character in the RECEIVER HOLD REGISTER allows ample time for the CPU to input the character, while the UART can be shifting a new character in. This process is referred to as *double buffering*.

5.2.2 Composite Serial Character

If a character received from DTE by the UART were the ASCII character A, it would appear as the waveform illustrated in Figure 5-3. The composite serial data stream that the UART receives is comprised of 10 bits:

— One start bit
— Seven data bits
— One parity bit
— One stop bit

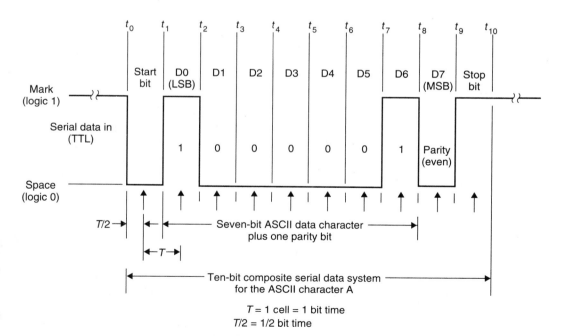

Figure 5-3 Composite serial data stream for the ASCII character A = 41H.

The number of bits per character can vary depending on the data character length, parity, and the number of stop bits that DTE has been configured for. Ten bits is commonly used in serial communication. Each of the ten bits illustrated in Figure 5-3 has a fixed interval of time. This interval of time is referred to as the *bit time.* The reciprocal of the bit time results in what is referred to as the character's *baud rate:* the number of bits transmitted per second. If the character shown in Figure 5-3 were transmitted or received from DCE, a modem, the baud rate would equal the transmission rate in bits per second *only* if the modulation rate of the DCE onto the communications channel were to vary at this same rate. In Chapter 9, we will consider cases where the baud rate does not equal the number of bits transmitted per second. For now, we assume that the UART is transmitting or receiving serial data from a terminal.

In Figure 5-3, if the bit times were measured to be 833 μs, the baud rate would equal 1200 (1/833 μs = 1200). Furthermore, if there are 10 bits per character, dividing the baud rate by 10 will establish the character rate or maximum number of characters that can be sent or received per second.

$$\text{Baud rate} = \frac{1}{\text{bit time}} \quad \text{bps}$$

$$\text{Maximum character rate} = \frac{\text{baud rate}}{\text{no. of bits per character}} \quad \text{characters per second}$$

Example

Compute the baud rate and the number of characters transmitted per second given a bit time of 104.16 μs and a character length of 10 bits.

Solution:
$$\text{Baud rate} = \frac{1}{\text{bit time}} = \frac{1}{104.16 \ \mu s} = 9600 \text{ bps}$$

$$\text{Maximum character rate} = \frac{\text{baud rate}}{\text{no. of bits per character}}$$

$$= \frac{9600}{10} = 960 \text{ characters per second}$$

5.2.3 Receiver Timing

The character shown in Figure 5-3 is represented as a seven-bit ASCII character transmitted with an even parity bit. Its level has been translated from RS232C format to TTL. Since the character has been transmitted by another device asynchronous to the UART's timing, the UART has the initial task of searching for the occurrence of the start bit. Ideally, this is precisely that instant in time, t_0, in which the signal changes from its idle condition of a MARK, a logic 1, to a SPACE, a logic 0. The terms MARK and SPACE are synonymous to a logic 1 and a logic 0 in the field of communications. A delay of $T/2$, referenced to t_0, is necessary before the first *sample* of Serial Data In is taken by the UART's RECEIVER SHIFT REGISTER. We will define a sample as that instant in time in which the RECEIVER SHIFT REGISTER is clocked. All bits within the register shift one bit position. It is at this instant in time, $T/2$, that the UART's hardware checks the level of the data bit sampled. It must be a SPACE, further confirming that the start bit of a character has been detected. If this were not the case, presumably a noise transient, otherwise known as a *glitch*, may have occurred. If sampling of the remaining bits were to continue, the data shifted into the RECEIVER SHIFT REGISTER would be erroneous. The process of seeking the start bit would be repeated in this case. If the data are in fact a SPACE at $T/2$, then the remaining bits are sampled. As depicted by the arrows shown in Figure 5-3, the sampling of each bit continues at a delay of T, the bit time, until all bits have been shifted into the register.

5.2.4 Start Bit Detection

The UART's task of shifting serial data into its RECEIVER SHIFT REGISTER is rudimentary and lends itself to further discussion. Whenever serial data into a shift register is being sampled by a clock that runs asynchronous to the serial

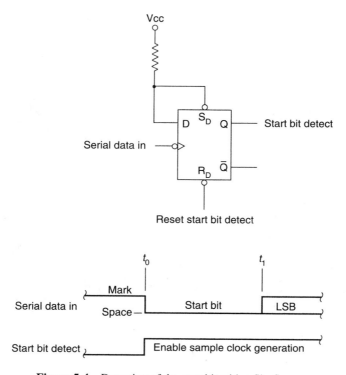

Figure 5-4 Detection of the start bit with a flip-flop.

data, the center of each bit that is being sampled can only be *approached*. Further discussion is necessary to find out why.

As stated earlier, the initial task of the UART receiver is the detection of the start bit. The timing necessary for shifting the serial data into the RECEIVER SHIFT REGISTER will rely upon the detection of the start bit at t_0. A flip-flop can be used to perform this task (Figure 5-4). When Serial Data In changes from a MARK to a SPACE, the start bit detection flip-flop will SET, indicating t_0, the beginning of the start bit. Once SET, the output of the flip-flop can be used to enable the sampling of Serial Data In.

5.2.5 Receiver Sample Clock Generation

A *sample clock* pulse must be generated at the center of each bit of the incoming serial data, including the start bit. Figure 5-5 illustrates the development of the sample clock and how it is used to shift Serial Data In into the UART's RECEIVER SHIFT REGISTER.

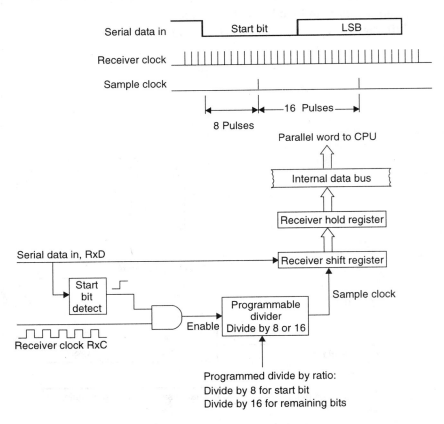

Figure 5-5 Developing the UART's sample clock.

With respect to t_0, a delay of $T/2$ is necessary before the first sample clock is generated and a delay of T for samples thereafter (see Figure 5-3). Since the baud rate of the UART must be programmed prior to its operation, a PROGRAMMABLE DIVIDER is initially preset to one-half the number of receiver clock pulses that occur within the duration of a single bit. This will result in the initial delay of $T/2$ when the PROGRAMMABLE DIVIDER is allowed to count down to zero.

For many UARTs, a typical number of receiver clock pulses per bit time is 16. This is shown in Figure 5-5 to illustrate the relative occurrence of the sample clock for the RECEIVER SHIFT REGISTER. At t_0, the counter is enabled by our start bit detection flip-flop to begin its countdown sequence from its preset value of 8, that is, one-half the full count of 16. When the counter has counted down to zero, a sample clock pulse out of the PROGRAMMABLE DIVIDER is generated. At this time, the center of what is perceived to be the start bit occurs and gets shifted into the first flip-flop internal to the RECEIVER SHIFT REGISTER. The PROGRAMMABLE DIVIDER continues its division of the receiver clock by 16, producing the sample clock pulses necessary to shift the remaining bits into the RECEIVER SHIFT REGISTER.

Since the receiver clock runs asynchronous to Serial Data In, the timing of the resulting sample clock pulse shown in Figure 5-5 may not be precisely at the center of the start bit as we would want it. By observing the timing of the sample clock pulse out of the PROGRAMMABLE DIVIDER in relation to Serial Data In, this anomaly would be apparent.

To illustrate the asynchronous nature of the situation, consider the effect of a receiver clock frequency of $4\times$, $8\times$, and $32\times$ the rate of Serial Data In. Figure 5-6 shows the resulting sample clock produced by the PROGRAMMABLE DIVIDER for these three frequencies. As mentioned earlier, the PROGRAMMABLE DIVIDER is preset initially to one-half its full count. In the three examples shown, this would be either 2, 4, or 16. When START BIT DETECT becomes active at t_0, the receiver clock is gated to the PROGRAMMABLE DIVIDER, which in turn counts its preset value down to zero. The resulting sample clock pulses shown in each example do not sample Serial Data In at the *ideal* center of the cells. The PROGRAMMABLE DIVIDER continues with its count of either 4, 8, or 32, producing the same relative *sampling error* for the remaining bits.

Notice that the sampling error is greatly reduced with higher receiver clock frequencies. A conclusion may be drawn from this resulting effect: the higher the receiver clock frequency is relative to the baud rate, the greater the resolution of actual Serial Data In. To approach the ideal center of each bit, the receiver clock frequency would have to approach an infinite rate. Unfortunately, due to the frequency limitations of TTL devices (or of all devices for that matter), the speed of the clock must be a practical value. The ratio of the UART's clock frequency to the baud rate of the incoming serial data is a factor that must be considered in the recovery of the received character.

For the three examples depicted in Figure 5-6, a *worst-case* condition exists for each in terms of accuracy in detecting the center of each bit. The maximum error for any case would be the *period* the receiver clock used. A character's start bit may become active precisely at a time when the active edge of the receiver clock occurs. A race condition exists here between Serial Data In and the

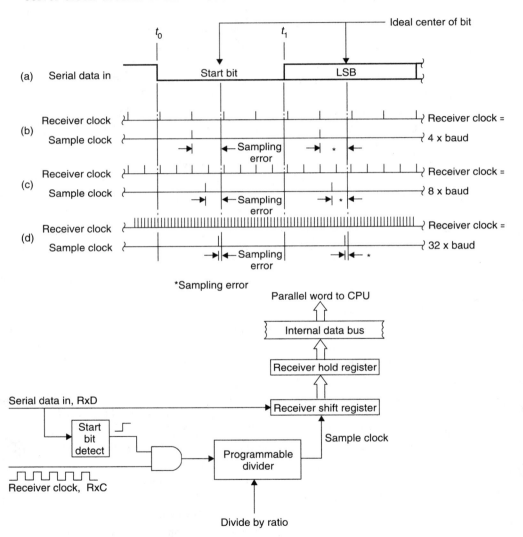

Figure 5-6 Timing diagram illustrating the effect of the receiver clock frequency on the generation of the sample clock; (a) serial data in; (b) receiver clock of 4 times the baud rate; (c) receiver clock of 8 times the baud rate; (d) receiver clock of 32 times the baud rate.

receiver clock pulse since they are running asynchronous to each other. If the PROGRAMMABLE DIVIDER (enabled through the start bit) misses this first receiver clock pulse, it will have to wait until the next receiver clock pulse to begin its count. This would be the worst-case condition in which the sample clock pulse shown in Figure 5-6 would be off by the period of the receiver clock. If, on the other hand, the race condition allows the PROGRAMMABLE DIVIDER to count this first clock pulse, then the resulting sample clock pulse would be generated at the ideal center of each bit (within a propagation delay of the PROGRAMMABLE DIVIDER). This would be the most desirable condition. It is purely random and is as likely to happen as the worst-case condition. The sample clock will shift between both extremes with each new character sent from DTE.

As a rule, when a given concept has been realized through hardware implementation, the worst-case condition must always be considered a factor in the analysis of the design. In this case, the period of the receiver clock has been considered. If the period of the receiver clock is a significant percentage of the bit time, as is the case with the example of $4\times$ baud in Figure 5-6, it is apparent that the data sent from the DTE may be sampled too close to the boundaries of each cell. The margin for error is minimal. An appreciable amount of distortion that a character picks up en route from the DTE will likely be sampled and shifted by the UART at an erroneous time.

5.2.6 False Start Bit Detection

Earlier we discussed provisions for checking the state of Serial Data In at $T/2$ to ensure that a SPACE condition was clocked into the RECEIVER SHIFT REGISTER. It is possible for a noise glitch to have occurred, falsely triggering the Start Bit Detection flip-flop. If a noise glitch falls below the maximum input LOW condition, $V_{IL\ (MAX)}$, as depicted in Figure 5-7, it will inadvertently trigger the Start Bit Detection flip-flop. Typically, this value is specified as 0.8 V for TTL. Noise transients generally have a duration in microseconds. The duration that ac-

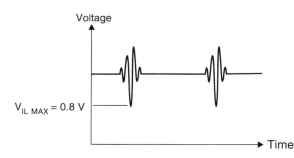

Figure 5-7 Noise transients that can cause false detection of the start bit.

tually causes the false start has elapsed by the time the first sample clock is generated at $T/2$; consequently, a MARK instead of a SPACE is shifted into the first flip-flop internal to the RECEIVER SHIFT REGISTER. This being the case, our counter is simply preset to eight and the Start Bit Detection flip-flop is reinitialized. The process of seeking the start bit of a character is repeated. If a SPACE occurs at $T/2$, the PROGRAMMABLE DIVIDER is allowed to continue its count, dividing the receiver clock by 16 and generating sample clock pulses as illustrated in Figure 5-5. Most UARTs manufactured today have a *false start bit detection* feature.

5.2.7 Flowcharting the Operation of the Receiver

The foregoing material can be presented in a flowchart, thus providing an overview of the operation of the receiver (Figure 5-8). Notice that error detection, an essential task of the receiver, has been included in the chart.

Once all the bits associated with a character have been shifted into the RECEIVER SHIFT REGISTER and stored in the RECEIVER HOLD REGISTER, error checking can be performed. The parity bit is checked in accordance with the number of data bits set in the character. If the UART's hardware detects an inconsistent result, even or odd depending on how it has been programmed by the user, the parity error bit is set in the UART's status register. A framing error is detected if the last bit sampled (see Figure 5-3) is not a MARK; that is, the end of the character, the stop bit, should always be a MARK. A framing error will occur if the baud rate is incorrect or the transmission of a character is incomplete. The overrun error flag is set if the CPU does not read a character before the next one is available. The UART will continue updating its receiver buffer with the latest character unless otherwise programmed. The previous overrun character is lost. Detailed examples of parity, framing, and overrun errors are given in Chapter 6.

The preceding error conditions do not inhibit the operation of the UART. It is up to the programmer to take the necessary steps to handle the error condition specified by the status register. The UART's error flags will remain active until the UART is either internally reset through a command or has undergone a power-up initialization sequence.

5.3 UART TRANSMITTER SECTION

The transmitter section of the UART includes the hardware to convert a parallel formatted word from the CPU into a composite serial word. The transmitter's timing is virtually independent of its receiver counterpart. There is no need to search for a start bit, nor is there a need to create a delay of $T/2$ in search of the

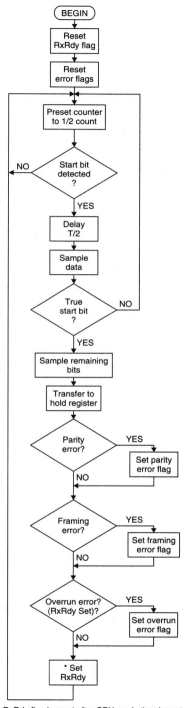

Figure 5-8 Flowchart of the UART's receiver operations.

* RxRdy flag is reset after CPU reads the character.

center of each bit. The basic functions of the UART transmitter section include the following:

— Start bit generation
— Shift register timing and control
— Parallel to serial conversion
— Stop bit generation
— Status
— Parity

Note that we have not included error detection as a part of the transmitter. Error detection is generally associated with the receiver section of a UART. Some UARTs have provisions for testing transmitter overrun and underrun errors. In Chapter 10 we look at the interaction of both the transmitter and receiver in terms of an error that has been detected followed by a retransmission request. The basic components of the transmitter section are illustrated in Figure 5-9.

The composite serial data stream is output to the Serial Data Out pin, TxD. The UART must effectively insert the start bit, parity bit, and the programmed number of stop bits as part of this data stream.

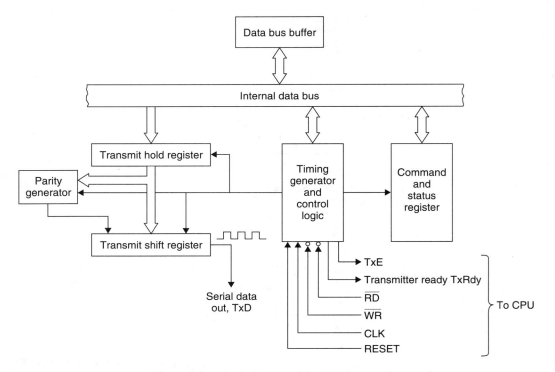

Figure 5-9 Block diagram of the UART transmitter section.

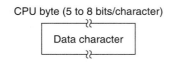

CPU byte (5 to 8 bits/character)

Data character

Assembled serial data output (TxD)

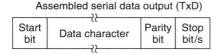

Start bit	Data character	Parity bit	Stop bit/s

Figure 5-10 Transmitted composite serial character format.

5.3.1 Transmitter Composite Character

The timing generator and control logic of the UART monitors the $\overline{WR}$ signal from the CPU. When $\overline{WR}$ becomes active, the UART loads the parallel data on the bus into its TRANSMIT HOLD REGISTER and generates the necessary timing pulses for transmitting the data in serial format. Notice in Figure 5-9 that the Timing Generator and Control Logic accepts signals and issues signals from the CPU, both internally and externally to accomplish this task.

Processing the parallel data word from the CPU is accomplished by loading the output of the TRANSMIT HOLD REGISTER into the TRANSMIT SHIFT REGISTER. It is here where the start bit, a logic 0, the stop bit, a logic 1, the programmed number of stop bits, and parity, if enabled, are inserted. The format of the entire composite serial data character is depicted in Figure 5-10. The DATA CHARACTER and the STOP BITS are shown here to be variable in length. Both are programmed by the user.

For the variable-length STOP BIT(S) illustrated in Figure 5-10, the programmer is typically provided with the option of controlling the number of stop bits, consequently increasing the overall length of time it takes to send the serial character. For example, the length of time it would take for a character to be sent with one stop bit would be one bit time less than the length of time it would take to send that same character programmed for two stop bits. Programming the variable-length Data Character to five, six, or seven bits has the same effect.

5.3.2 Transmit Shift Clock Generation

The transmitter section of the UART includes a PROGRAMMABLE DIVIDER similar to the receiver section. There is no need to preset the divider with one-half the divide-by ratio in search of the center of a bit. As illustrated in Figure 5-11a, the divider is simply programmed to a value that allows the external transmitter clock, TxC, to be divided down to a value equal to the programmed baud rate. The output of the PROGRAMMABLE DIVIDER is the Transmit Shift CLK. The Transmit Shift CLK has a period equal to the bit time of the transmitted character.

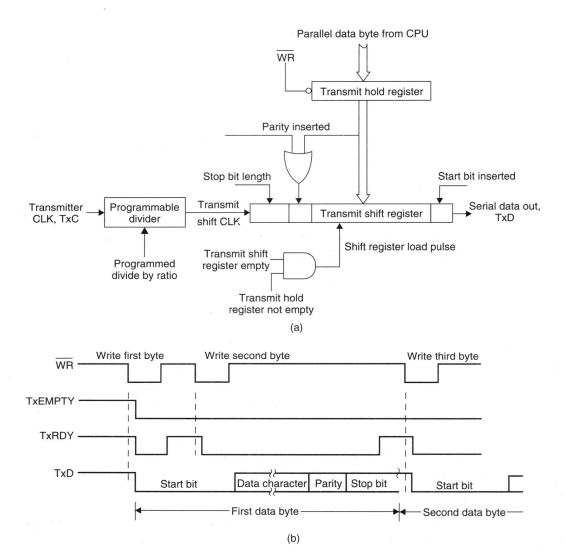

Figure 5-11 (a) Generation of transmitted serial data out, TxD; (b) transmitter timing sequence.

5.3.3 Transmitter Timing

Figure 5-11b depicts the timing sequence that the transmitter section of the UART undergoes while transmitting characters. Initially, the TIMING GENERATOR and CONTROL LOGIC's Transmit Ready line, TxRDY is active, indicating to the CPU that the UART transmitter is ready for a parallel word. TxEMPTY, when HIGH, implies that the transmitter has no further characters to send. This occurs when the TRANSMIT HOLD and TRANSMIT SHIFT REGISTERS are both empty. A $\overline{\text{WR}}$ can be issued by the CPU to output a parallel word to the UART via the data bus. In the timing diagram of Figure 5-11b, TxRDY immediately becomes inactive after the first byte has been written by the CPU to the Transmit Hold Register. The UART is now busy. The parallel word is loaded into the TRANSMIT HOLD REGISTER and TxEMPTY becomes inactive; that is, both buffers are NOT empty. The TIMING GENERATOR and CONTROL LOGIC will provide the signals for loading the TRANSMIT HOLD REGISTER into the TRANSMIT SHIFT REGISTER. This occurs when the TRANSMIT SHIFT REGISTER is empty and the TRANSMIT HOLD REGISTER is NOT empty. The start bit, parity, and stop bits are inserted in the TRANSMIT SHIFT REGISTER. The Transmitter CLK, TxC, after getting divided down by the PROGRAMMABLE DIVIDER, will begin its procedure to shift the composite serial data stream out of the Serial Data Out, TxD output pin. The foregoing sequence of events occurs immediately following the first $\overline{\text{WR}}$ from the CPU.

A closer look at the timing in Figure 5-11b will reveal an additional $\overline{\text{WR}}$ pulse immediately following the first $\overline{\text{WR}}$ pulse. It is justified by the preceding TxRDY pulse. How is this possible in the middle of a serial data stream that is being shifted out? This is the concept of *double buffering* that we have spoken of earlier with regard to the UART's receiver section. Double buffering is an essential part of the transmit section of the UART also. Its purpose here is to *hold* the current data word sent by the CPU while processing and sending the previous one. Note that after the second $\overline{\text{WR}}$ pulse occurs, both the TRANSMIT HOLD and TRANSMIT SHIFT REGISTERS are full. TxRDY is inactive until the first BYTE has been completely shifted out of the TRANSMIT SHIFT REGISTER. At this time, the second BYTE, held in the TRANSMIT HOLD REGISTER, is transferred into the TRANSMIT SHIFT REGISTER. TxRDY becomes active again, signaling the CPU that the UART is ready for a third BYTE. At this time, the second BYTE sent to the UART begins its process of getting shifted out of the TRANSMIT SHIFT REGISTER. TxEMPTY remains inactive all the while. It will remain LOW until the UART has no further characters to transmit. Both buffers in this case would be empty and TxEMPTY would become active again.

5.3.4 Implementing the Transmit Hold and Shift Registers

The concept of serial transmission may be further reinforced by implementing the hardware for a simplified TRANSMIT HOLD and TRANSMIT SHIFT REGISTER (Figure 5-12). D-type flip-flops are used here to implement both registers. BD0 through BD7 represent the buffered data byte from the CPU. The buffered data byte gets parallel loaded into the TRANSMIT SHIFT REGISTER when $\overline{\text{WR}}$ becomes active. If the TRANSMIT SHIFT REGISTER is empty, a Shift Register Load Pulse generated by the UART's TIMING GENERATOR and

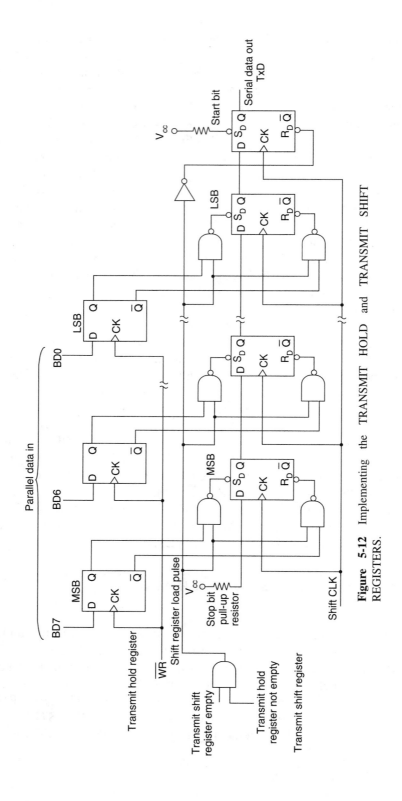

Figure 5-12 Implementing the TRANSMIT HOLD and TRANSMIT SHIFT REGISTERS.

120

CONTROL LOGIC will cause a *jam entry* via the direct set, $\overline{S_d}$, and direct reset, $\overline{R_d}$, inputs of the register. Without a Load Pulse, the AND gates depicted in the diagram remain inactive. A jam entry cannot occur. The flip-flops of the TRANSMIT SHIFT REGISTER can now serve as a shift register, providing a Shift CLK is generated.

The composite serial data stream begins as soon as the jam entry propagates to the output of the START BIT flip-flop. This is the Transmitted Serial Data Out, TxD. Note that the first bit to get shifted out of the register is the START BIT. This first flip-flop is always jam loaded with a logic 0, a SPACE, via the direct reset input, $\overline{R_d}$. The transmitted Serial Data Out changes at this time from a MARK, or IDLE condition, to a SPACE, that is, the beginning of the transmitted character's start bit. The Shift CLK, occurring at the programmed baud rate, is generated one bit time after the Load Pulse. This will allow the START BIT to remain LOW at the output of the first flip-flop for one bit time.

Since all flip-flops share the same Shift CLK, the data present at each of their D inputs will be shifted to their Q outputs. The START BIT has elapsed at this time, and the LSB of the data character is present at the serial output pin, TxD. The process will continue until all bits have been shifted out of the TRANSMIT SHIFT REGISTER. The last bit shifted out is the STOP BIT. A closer look at the shift register's most significant flip-flop will reveal a pull-up resistor at its D input. The logic HIGH produced by the pull-up resistor will ensure that on the *ninth* Shift CLK pulse (the jam entry held for one bit time accounts for the tenth bit) this HIGH will have been shifted through all flip-flops and will be present at the TxD output pin. This marks the beginning of the STOP BIT at this time. Continued clocking of the TRANSMIT SHIFT REGISTER will allow the length of the STOP BIT to be controlled. Once the Shift CLK is discontinued, the output of the TRANSMIT SHIFT REGISTER will remain a logic HIGH, which is the idle state.

5.3.5 Inserting Parity

The circuit in Figure 5-12 does not include provisions for inserting the parity bit. The preceding circuit would suffice; however, the generation of parity would depend on the software control of the MSB, BD7; that is, BD7 from the CPU *is* the parity bit. It can be set to even or odd parity by the program. It can also be simply set to zero at all times in cases where parity is deemed unnecessary. Parity generation by the UART is *disabled* through the command register.

If parity has been enabled through a command to the UART, the transmitter section must override the MSB of the data character and provide for the insertion of the parity bit. The circuit of Figure 5-13 depicts a method for doing this. If the parity enable, P.E., signal is active (LOW), BD7 would be inhibited and PARITY would be gated to the input of the next flip-flop. Conversely, if P.E. is inactive

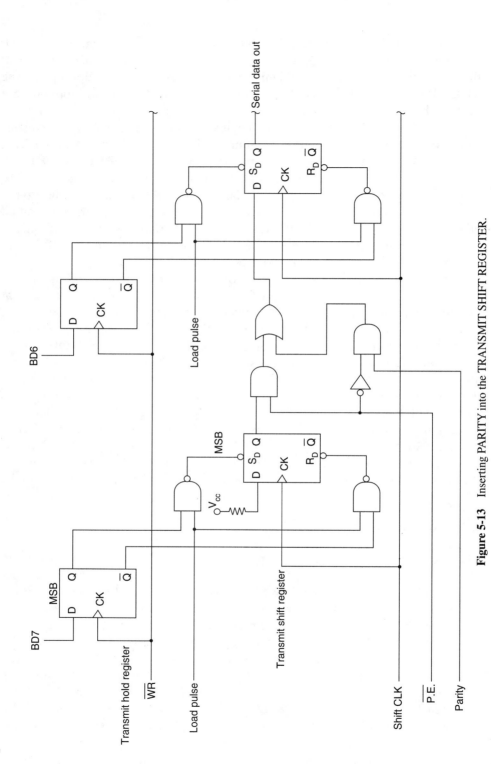

Figure 5-13 Inserting PARITY into the TRANSMIT SHIFT REGISTER.

(HIGH), then the parity bit would be disabled and BD7 from the CPU would be gated to the next flip-flop. The UART's Timing and Control Logic will ensure that the P.E. line is asserted at the appropriate time. Parity-generation and error-checking circuits are considered in Chapter 11.

PROBLEMS

1. Explain what is meant by *double buffering*.
2. What UART signal informs the CPU that a character is available and ready to be read from its RECEIVER HOLD REGISTER?
3. A character transmitted by a UART has a bit time of 104.16 μs.
   (a) Compute its baud rate.
   (b) If the character's length were 11 bits (seven-bit data, parity, start bit, and two stop bits), compute the maximum number of characters that can be transmitted per second.
4. What are the logic levels for the following conditions?
   (a) Mark
   (b) Idle
   (c) Space
   (d) Start bit
   (e) Stop bit
5. Refer to Figure 5-6. Compute the maximum sampling error (in seconds) for the three receiver clock frequencies shown in Figure 5-6b, c, and d. Assume a bit time of 104.16 μs.
6. Explain false start bit detection.
7. What UART signal informs the CPU that it is ready to accept a parallel word to be transmitted in serial format?
8. Refer to Figure 5-11b. How is it possible for two $\overline{WR}$ pulses to occur before the first character is even sent by the UART?
9. For the circuit shown in Figure 5-13, what are the logic levels necessary for the signals $\overline{P.E.}$ and PARITY to enable and set the parity bit to a logic LOW?
10. Explain how a logic 1 can be loaded into the MSB of the TRANSMIT SHIFT REGISTER shown in Figure 5-13.

6

Interfacing
the 8085A Microprocessor
to the 8251A USART

Interfacing a microprocessor to a peripheral device can be a difficult task without a reasonably in depth functional understanding of both the processor and the device itself. Often this level of understanding is beyond what the student can get out of manufacturers' specification data sheets. In Chapter 5 we presented the UART in a manner that was intended to give the student this functional insight. Now that we have developed this understanding, we are in a much better position to interface with it. In this chapter an actual asynchronous serial interface is implemented and discussed.

6.1 INTERFACING THE USART

Figure 6-1 depicts how the 8085A microprocessor and 8251A USART are interfaced together to transmit and receive serial data to and from a video display terminal (VDT). The circuit diagram is complete with the exception of the microprocessor, which depicts only that portion of the hardware required to interface to the USART. Both the 8251A and the 8085A have been widely accepted as industry standards in the field of data communications and processor applications. On this basis we have selected these two fundamental devices for our discussion.

6.1.1 Level Translation

Since the 8251A USART is a TTL device, its levels must be translated to standard RS-232-C levels in order for the exchange of serial data to occur between the terminal and USART. The device we have selected for this purpose is the

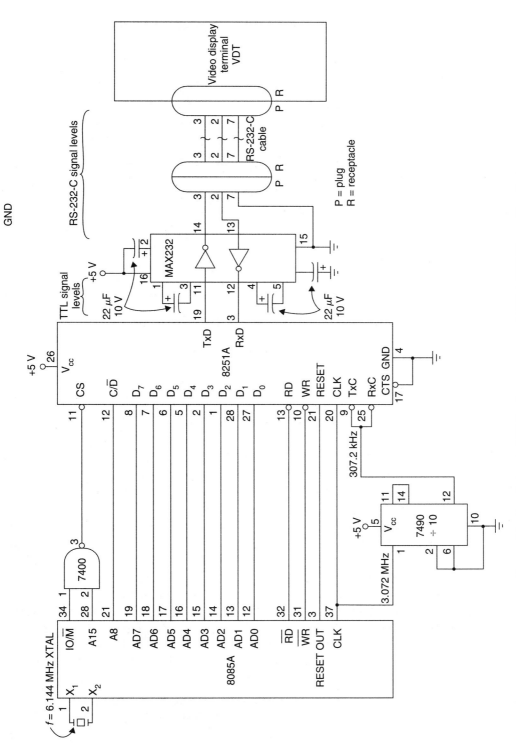

Figure 6-1 Interfacing the 8085A microprocessor and 8251A USART to a video display terminal.

125

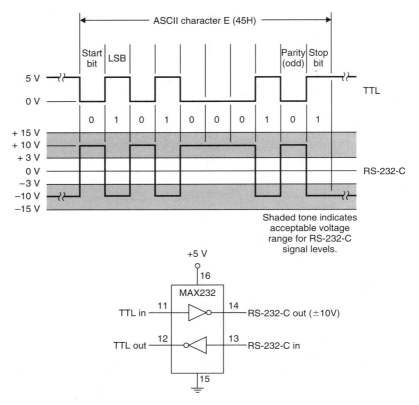

Figure 6-2 Level translation from TTL to RS-232-C for the ASCII character E.

single-chip MAX232 level translator.[†] Internal to the MAX232 are two TTL to RS-232-C level translators and two RS-232-C to TTL level translators. Notice in Figure 6-1 that it operates from a *single* +5 V power supply, unlike conventional level translators, which require multiple supplies. An internal charge pump generates ±10 V output swings for TTL level signals in. The acceptable RS-232-C voltage range for a logic 1 is −3 to −15 V. For a logic 0, it is +3 to +15 V. The ±10 V output swing from the MAX232 is well within the limits of the RS-232-C convention. The MAX232 is also capable of translating RS-232-C levels to TTL. Figure 6-2 illustrates the ASCII character E in asynchronous format. It is translated by the MAX232 from TTL to RS-232-C levels. The shaded zone indicates acceptable levels in RS-232-C format for a logic 1 and logic 0. Notice that the voltage levels are inverted by the MAX232. Logic levels, however, remain the same; that is, a TTL logic 1 into the MAX232 is also a logic 1 out. Only its *voltage level* has been translated.

[†] The MAX232 is manufactured by the MAXIM Corporation. Four external capacitors are required for operation. A MAX233 is available (at a higher price) that requires no external components for operation.

6.1.2 RS-232-C Interface to a Terminal

The interface cable depicted in Figure 6-1 uses the standard RS-232-C DB25 pin connector. A minimal interface (see Section 4.5) of three RS-232-C circuits will be used in this example:

Signal ground	AB	Pin 7
Transmitted data	BA	Pin 2
Received data	BB	Pin 3

The cable length selected must conform to the specifications outlined in Section 4.4.1. We have included the connector's sex in our illustration, denoted by P, male plug, and R, female receptacle. A direct connection from the level translator to the terminal can also be used. However, a cable assembly will allow us to interconnect to other RS-232-C terminals.

6.1.3 USART Transmit and Receive Clock

The selection of the USART's Transmit and Receive Clock frequency is not arbitrary. See inputs TxC and RxC of the 8251A shown in Figure 6-1. Careful attention must be given in selecting a frequency that is compatible to the features of the USART. The 8251A, for example, is capable of program dividing the Transmit and Receive Clock frequency by 1, 16, or 64. The resulting frequency is the data transmission rate that must be compatible in speed to the interfacing device.

An external oscillator can be used to generate the Transmit and Receive Clock for the 8251A. This is not necessary in our case. Instead, we will use the available CLK output of the 8085A microprocessor. Its frequency was intentionally selected for this purpose by using a 6.144-MHz crystal, shown in Figure 6-1, for the X1 and X2 inputs of the 8085A. Internally, the processor divides the crystal frequency by 2. The CLK frequency out of the 8085A is therefore 3.072 MHz (6.144 MHz/2). A 7490 Decade Counter has been configured to further divide this frequency by 10. The resulting frequency of 307.2 kHz will be used as our Transmit and Receive Clock. The 8251A USART can now be programmed to further divide this frequency by 16 or 64, resulting in a baud rate of 19.2 kbaud or 4800 baud, respectively. Lower baud rates can easily be attained by simply dividing the CLK output of the 8085A by a value greater than the 10 that we are using. Several single-chip ICs are available on the market to perform this task if desired.

6.1.4 Addressing the USART

For circuit simplicity, *partial address decoding,* the decoding of less than the total number of address bits, is used for port selecting the 8251A. The 7400 NAND gate depicted in Figure 6-1 enables the 8251A for communication. Since

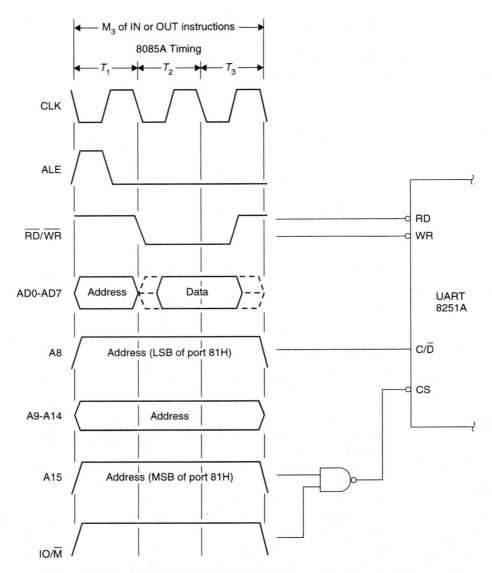

Figure 6-3 Timing for 8085A IOR/IOW cycle with partial address decoding to enable the 8251A.

port duplication occurs on the upper and lower address bits of the 8085A during an IOR or IOW machine cycle, the upper address bits can be used for decoding. By logically ANDing IO/$\overline{\text{M}}$ and A15 on an IN or OUT instruction to port 80H or above, we can enable the 8251A. This will occur on the *third* machine cycle (M_3) of instructions OUT 80H (or above) or IN 80H (or above), where IO/$\overline{\text{M}}$ and A15 are both a logic HIGH.

Figure 6-3 illustrates the timing produced by the 8085A for an IOR or IOW machine cycle. The timing input for the address decoder circuit is also illustrated. Notice that IO/$\overline{\text{M}}$ and A15 are both HIGH throughout the entire machine cycle. This will enable the 8251A by activating its Chip Select, $\overline{\text{CS}}$, input pin. Since $\overline{\text{RD}}$ or $\overline{\text{WR}}$, depending on whether the machine cycle is an IOR or IOW, both occur within this time frame, the 8251A can be commanded, or its status may be read by outputting or inputting from port 80H or above.

If one other address bit from the microprocessor is decoded, a distinction can be made by the 8251A as to whether it is being enabled for Control and Status or Data. Address bit A8, the LSB of our duplicated port address, is used for this purpose. It is connected to the Control/$\overline{\text{Data}}$ (C/$\overline{\text{D}}$) input pin of the 8251A. Addressing ports 80H and 81H will enable the 8251A for two separate functions:

1. Port 80H: Data

2. Port 81H: Control and Status

When C/$\overline{\text{D}}$ of the 8251A is HIGH (addressing port 81H) and $\overline{\text{WR}}$ is active, the Control Register of the 8251A is addressed. When C/$\overline{\text{D}}$ is HIGH and $\overline{\text{RD}}$ is active, the Status Register of the 8251A is addressed. C/$\overline{\text{D}}$ when LOW (addressing port 80H), allows access to the Data Register for sending and receiving data to and from the 8251A. Table 6-1 is a summary of the effect of control signals sent to the 8251A as a result of executing the 8085A instructions: IN 80H or OUT 80H.

Since *full address decoding,* the decoding of all eight port address bits, is not used in our address decoding scheme, a little thought will reveal that the ports we have selected, 80H and 81H, can be replaced in Table 6-1 with ports F0H and F1H, respectively, or E0H and E7H, and so on. In either case, the active state of the MSB and LSB of the addressed port must be considered when enabling the 8251A for communication.

TABLE 6-1 Summary of 8085A Control Signals to Enable the 8251A

$\overline{\text{CS}}$	$\overline{\text{WR}}$	$\overline{\text{RD}}$	Instruction	A15	A8, C/$\overline{\text{D}}$	IO/$\overline{\text{M}}$	Action
0	0	1	OUT 81H	1	1	1	Send control word to 8251A
0	1	0	IN 81H	1	1	1	Receive status from 8251A
0	0	1	OUT 80H	1	0	1	Send data to 8251A
0	1	0	IN 80H	1	0	1	Receive data from 8251A

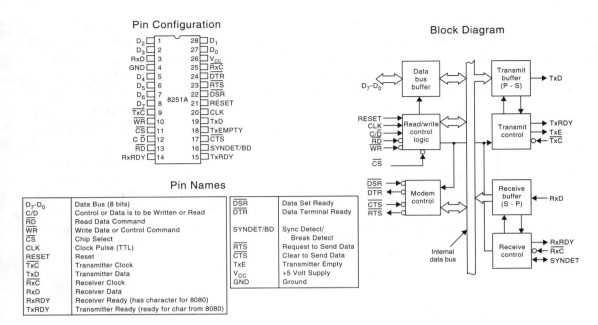

Figure 6-4 Block diagram and pin configuration for the 8251A. (Courtesy of Intel Corp.)

6.2 8251A BLOCK DIAGRAM AND PIN CONFIGURATION

A block diagram and pin assignment for the 8251A are shown in Figure 6-4. The internal functions of the 8251A can be grouped into the following categories:

1. Data bus
2. Read/write control logic
3. Transmitter section
4. Receiver section
5. Modem control

Table 6-2 lists the function of each of the 8251A's 28 pins by pin number and group.

6.3 INITIALIZING THE 8251A USART FOR COMMUNICATION

The 8251A must be programmed with a set of control words prior to transmitting or receiving serial data in either asynchronous or synchronous format. The control words define the complete functional definition of the 8251A and must be programmed immediately following an internal or external RESET.

TABLE 6-2 Description of 8251A Pins

Pin Name	Pin No.	Group	Function
D7–D0	8, 7, 6, 5, 2, 1, 28, 27	Data bus	Eight-bit bidirectional data bus used for control, status, and data.
RESET	21	Read/write control logic	A HIGH on this line resets the 8251A to its idle mode. Internal registers and flags are cleared.
CLK	20		**Clock input:** used to control the read/write timing between microprocessor and 8251A.
C/$\overline{\text{D}}$	12		**Control/data input:** When HIGH, data written to the 8251A are interpreted as **Control.** Data read are interpreted as **Status.** When LOW, data written and read from the 8251A are transmitted and received data, respectively. $\overline{\text{CS}}$ must be active with C/$\overline{\text{D}}$.
$\overline{\text{RD}}$	13		**READ:** A LOW on this line allows the CPU to read parallel data or status from the 8251A in accordance with the level of C/$\overline{\text{D}}$.
$\overline{\text{WR}}$	10		**Write:** A LOW on this line allows the CPU to write a control word or data word to the 8251A for serial transmission in accordance with the level of C/$\overline{\text{D}}$.
$\overline{\text{CS}}$	11		**Chip select:** A logic LOW on this pin enables the 8251A for communication.
TxD	19	Transmitter section	**Transmit data:** The transmitted serial data are output on this pin.
TxRdy	15		**Transmitter ready:** This line becomes active when the 8251A is ready to accept a data character from the CPU. It can be used for polling or interrupt I/O. TxRdy is reset when a byte is written to the Transmit Hold register.
TxE	18		**Transmit empty:** TxE will go HIGH when the 8251A has no further characters to transmit; that is, both the Transmit Hold and the Transmit Shift Registers are empty. TxE can be used to indicate the end of a transmission in half-duplex mode.
TxC	9		**Transmitter clock:** The frequency of this clock controls the baud rate of the serial data transmitted. The 8251A may be programmed to divide this clock frequency by 1, 16, or 64 to establish the desired baud rate.
RxD	3	Receiver section	**Receiver data:** Serial data are received on this pin.
RxRdy	14		**Receiver ready:** This line becomes active when the 8251A's Receiver Hold Register has a parallel data character ready for the CPU to read. It can be used by the CPU for polling or interrupt I/O. RxRdy is reset when the character is read by the CPU.

(continued)

TABLE 6-2 *(continued)*

Pin Name	Pin No.	Group	Function
RxC	25		**Receiver clock:** The frequency of this clock is a direct multiple of the baud rate of serial data in. The 8251A may be programmed to divide this clock by 1, 16, or 64 to equal the baud rate of serial data in.
SYNDET/BD	16		**Sync detect/break detect:** This pin is used in the synchronous mode only. It is used as an input or output pin as programmed through the Control Word. As an output pin, a HIGH indicates that the receiver has detected the SYNC character and character sync has been achieved. As an input pin, a rising edge on this pin causes the 8251A to start assembling data characters on the next rising edge of the Receiver Clock pulse, RxC.
$\overline{DSR}$	22	Modem control	**Data set ready:** This input is normally used to indicate the readiness of a modem or data set. Its condition can be tested by the CPU through a Status Read operation (bit D7 of the status word).
$\overline{DTR}$	24		**Data terminal ready:** This output is normally used to indicate to a data set or modem the readiness of the DTE. It can be set LOW by programming bit D1 of the command instruction word.
$\overline{CTS}$	17		**Clear to send:** This input signal is normally used to test if a modem is ready to receive and transmit a character from the 8251A to a communications channel. A LOW on this pin from a modem enables the 8251A to transmit serial data if the TxEN bit in the command word is set.
$\overline{RTS}$	23		**Request to send:** This output signal is normally used to request sending of serial data to a modem. $\overline{RTS}$ is the handshake line to $\overline{CTS}$. It can be set LOW by programming bit D5 in the command instruction word.
VCC	26	Power	+5-V supply
GND	4		Ground

Control Register	
Mode instruction (8)	Command instruction (8)

Figure 6-5 Control register format.

The control words are split into two separate bytes that make up the Control Register within the Read/Write Control Logic block (Figure 6-5). These two bytes are referred to as the 8251A's *Mode Instruction* and *Command Instruction*. Both the Mode and Command Instructions must follow a specific sequence in order for proper communication to occur. This sequence is listed in Figure 6-6. Notice that $C/\overline{D} = 1$ for all control words. Recall that our address decoding scheme discussed earlier uses the LSB of the port address for the 8251A's $C/\overline{D}$ input. Writing to PORT 81H will enable the 8251A for accepting a command from the CPU since $C/\overline{D} = 1$ when addressing PORT 81H.

After a RESET (internal or external), a Mode Instruction must then be written as the *first* command instruction to the 8251A. The 8251A automatically interprets this first word, after a RESET, as a Mode Instruction. It is the Mode Instruction that determines whether the 8251A is used for asynchronous or synchronous communications. If the Mode Instruction has been programmed for asynchronous communication, a Command Instruction must immediately follow it. If, on the other hand, the Mode Instruction is programmed for synchronous communication, one or two SYNC CHARACTERS must be written to the 8251A before a synchronous Command Instruction is issued. In each case, the 8251A's $\overline{WR}$ must be active and $C/\overline{D}=1$.

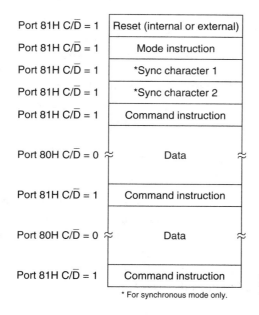

Port 81H $C/\overline{D} = 1$ — Reset (internal or external)
Port 81H $C/\overline{D} = 1$ — Mode instruction
Port 81H $C/\overline{D} = 1$ — *Sync character 1
Port 81H $C/\overline{D} = 1$ — *Sync character 2
Port 81H $C/\overline{D} = 1$ — Command instruction
Port 80H $C/\overline{D} = 0$ — Data
Port 81H $C/\overline{D} = 1$ — Command instruction
Port 80H $C/\overline{D} = 0$ — Data
Port 81H $C/\overline{D} = 1$ — Command instruction

* For synchronous mode only.

Figure 6-6 Command sequence for the 8251A. (Courtesy of Intel Corp.)

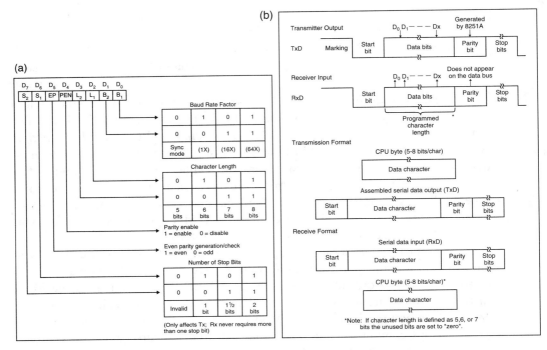

Figure 6-7 (a) Mode instruction format, asynchronous mode; (b) data format for an asynchronous serial character. (Courtesy of Intel Corp.)

6.3.1 Mode Instruction Definition for Asynchronous Communication

For asynchronous communication, the Mode Instruction word has the format illustrated in Figure 6-7a. Notice that the Mode Instruction determines the *baud rate factor, character length, parity,* and *number of stop bits.* Also, the least significant two bits of the Mode Instruction, D0 and D1, cannot be programmed LOW. If both bits are LOW, the synchronous mode will be programmed. Figure 6-7b illustrates the data format of an asynchronous serial character.

6.3.2 Mode Instruction Definition for Synchronous Communication

For synchronous communication, the Mode Instruction has the format illustrated in Figure 6-8a. The least significant bits, D0 and D1 here, are both LOW, indicating to the 8251A that this instruction is to be interpreted as a *synchronous* Mode Instruction. The remaining bits define the synchronous *character length* and *PARITY,* whether *syndet is internal or external,* and the *number of sync characters* (one or two). Figure 6-8b illustrates the data format of a synchronous serial character.

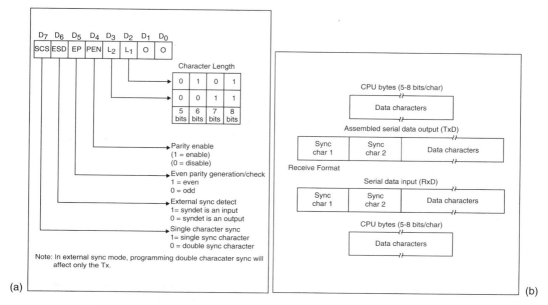

Figure 6-8 (a) Mode instruction format, synchronous mode; (b) data format for a synchronous serial character.

6.3.3 Command Instruction Definition

Once the 8251A's functional definition has been programmed by a Mode Instruction (and SYNC character/s have been loaded if the synchronous mode has been programmed), the Command Instruction is issued by the CPU. The Command Instruction format is the same for asynchronous and synchronous data communications (Figure 6-9). All control words written to the 8251A after the first Mode Instruction (and sync character if synchronous mode has been programmed) will be interpreted by the 8251A as Command Instructions. Functions of the Command Instruction can therefore be changed at any time simply by outputting a new byte to port 81H.

What happens if the baud rate, character length, or any other function of the Mode Instruction needs to be changed? Referring to Figure 6-8, bit D6 of the Command Instruction is the Internal Reset Bit, IR. Since the 8251A can accept a Command Instruction at any time after the Mode Instruction, an Internal Reset can be issued at any time. This will allow us to send a new Mode Instruction since the next command *after* a RESET is interpreted as such.

Once the Mode (and sync characters if synchronous mode has been programmed) and Command Instructions have been sent to the 8251A, serial transmission of data can occur by writing or reading to the 8251A from port 80H. $C/\overline{D} = 0$ in this case.

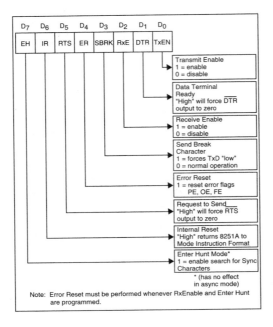

D₇	D₆	D₅	D₄	D₃	D₂	D₁	D₀
EH	IR	RTS	ER	SBRK	RxE	DTR	TxEN

Transmit Enable
1 = enable
0 = disable

Data Terminal Ready
"High" will force $\overline{DTR}$ output to zero

Receive Enable
1 = enable
0 = disable

Send Break Character
1 = forces TxD "low"
0 = normal operation

Error Reset
1 = reset error flags PE, OE, FE

Request to Send
"High" will force $\overline{RTS}$ output to zero

Internal Reset
"High" returns 8251A to Mode Instruction Format

Enter Hunt Mode*
1 = enable search for Sync Characters
* (has no effect in async mode)

Note: Error Reset must be performed whenever RxEnable and Enter Hunt are programmed.

Figure 6-9 Command instruction format. (Courtesy of Intel Corp.)

6.3.4 Status Read Definition

In data communication systems, it is often necessary to determine the status of the transmitting and receiving device. The 8251A USART can be polled at any time in order to ascertain if error conditions or other conditions that may require processor intervention exist. To accomplish this function, the CPU issues a Status Read command to PORT 81H; that is, $C/\overline{D} = 1$ and $\overline{RD} = 0$. A status word is put onto the bus for the CPU to read. The Status Read word definition is listed in Figure 6-10.

6.4 ASYNCHRONOUS 4800-BAUD COMMUNICATION PROGRAM FOR THE 8251A USART

A program will now be developed to transmit asynchronous serial data from the 8251A USART to a terminal at 4800 baud. The setup discussed earlier in Figure 6-1 will be used. Before we begin, we must first define the tasks of the program.

6.4.1 Program Task

1. Write an 8085A assembly language program to transmit your first and last name repeatedly to a terminal. The hardware setup is illustrated in Figure 6-1. Your name should occur on the display terminal once per line.
2. PORT 80H is to be used for data and PORT 81H is to be used for Control and Status.

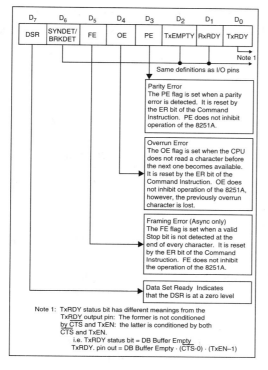

Note 1: TxRDY status bit has different meanings from the
TxRDY output pin: The former is not conditioned
by CTS and TxEN: the latter is conditioned by both
CTS and TxEN.
i.e. TxRDY status bit = DB Buffer Empty
TxRDY. pin out = DB Buffer Empty · (CTS-0) · (TxEN–1)

Figure 6-10 Status read format.
(Courtesy of Intel Corp.)

3. The program should be written under program Control I/O. Monitor the TxRdy bit of the Status Word for this task.

4. The following parameters should be specified in the initialization program:
   a. 4800 baud
   b. Seven-bit data character
   c. One stop bit
   d. Even parity generation

5. Assume that RAM exists for your program from 2000H to 20FFH (256 bytes). ORG your program at location 2000H. Store ASCII characters that represent your first and last name (including carriage return and line feed) starting at memory location 2040H. Use a NULL BYTE 00H to terminate the end of your name.

6. Accompany your program with a flowchart. Be sure to construct your flow-chart before you begin your program.

6.4.2 Mode Word Format

The *Mode Word* is 7BH. Bits D0 = 1 and D1 = 1 specify the asynchronous mode of operation with a baud rate factor of 64× the period of the CLK input. $1/[64 \times (1/307.2 \text{ kHz})] = 4800$. An easier approach is simply to divide the

8251A CLK input frequency by 64. (307.2 kHz)/64 = 4800. Bits D2 = 0 and D3 = 1 specify a seven-bit ASCII character. By setting D4 = 1 and D5 = 1, even parity will be generated for the parity bit. Setting D6 = 1 and D7 = 0, one stop bit will be sent with each character.

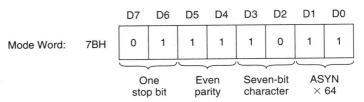

6.4.3 Command Word Format

The format of the Command Word has been configured to enable the transmitter section of the 8251A to send data. This is accomplished by setting bit D0 = 1 of the Command Word. Since our program task does not include the use of the receiver section of the 8251A, the receiver will be disabled, D2 = 0. This being the case, there is no need to monitor error-checking flags FE, OE, and PE of the Status Word. The Error Reset bit, D4 = 1, will be set as a precautionary measure to clear these flags. These flags are defined in Section 6.4.4.

D6, the Internal Reset bit, will initially be set in our program to internally reset the 8251A and to ensure that the first command sent to the 8251A is interpreted as a Mode Word. If this bit is always set, the Command Word will never be interpreted by the 8251A. Instead, a Mode Word will always be expected. Assuming that the 8251A has already been reset, we will set D6 = 0.

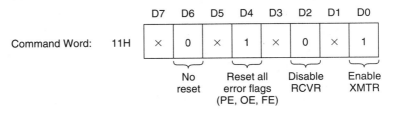

6.4.4 Status Word Format

The Status Word can be read into the 8085A accumulator by the instruction IN 81H (see Table 6-1). Once the 8251A has been initialized, our program must monitor the Transmit Ready bit, TxRDY, of the Status Word. When this bit is a logic HIGH, the transmitter section of the 8251A is ready and a parallel word can be sent from the 8085A to the 8251A. By outputting a data word to PORT 80H, our decoding circuit will enable the 8251A to accept the data word from the 8085A. The 8251A will then begin its internal procedure to transmit the character in serial form. The character will be transmitted serially in accordance with the Mode Word previously sent. Until the 8251A is ready to accept another character from the 8085A, TxRDY will remain a logic LOW. Our program can wait in an idle loop, continuously testing the state of the TxRDY bit, or it can be off performing other tasks, periodically testing the status of TxRDY. This concept is referred to as *polling*. In either case, the 8085A is under *program control I/O;* that is, the program is under the control of the peripheral device, in this case the USART. The flowchart and program to accomplish our overall task are illustrated in Figure 6-11. Directives (END, DB, and so on) should be added to the program if an assembler is used.

		D7	D6	D5	D4	D3	D2	D1	D0
Status Word:	01H	×	×	×	×	×	×	×	1

Transmitter ready

6.5 ERROR DETECTION: PE, OE, AND FE

The Status Word in our program has been masked of all bits, with the exception of D0, the TxRDY bit. The remaining bits are not checked. If our program task were to include the reception of characters from the terminal, we may want to check for the detection of error conditions. Problem 10, at the end of the chapter, includes a task to check error flags. A definition of these flags can be found in the Status Read Format of Figure 6-10. Additional clarification of these flags are necessary at this time in order to complete Problem 10.

6.5.1 PE: Parity Error

Perhaps the most simplified method of error checking is parity. Although it is not 100% accurate, it does allow some assurance that the transmitted data are equivalent to the received data. Parity can be set to either *even* or *odd* by a transmitting device. The receiving device must set itself accordingly. For even parity generation, the transmitter section of the 8251A sets the parity bit to a level that makes the total number of 1's in the character *even* (not including the start and stop bits). The opposite is true for odd parity. Consider the examples of both even and odd parity generation shown in Table 6-3.

In our example program in Figure 6-11, the Mode Word was set for odd parity. The 8251A will generate an odd parity bit for all characters transmitted to the terminal. Programming the 8251A for odd parity also enables the receiver section of the 8251A to perform an odd parity check on characters received from the terminal. Whether we are sending or receiving characters to or from the ter-

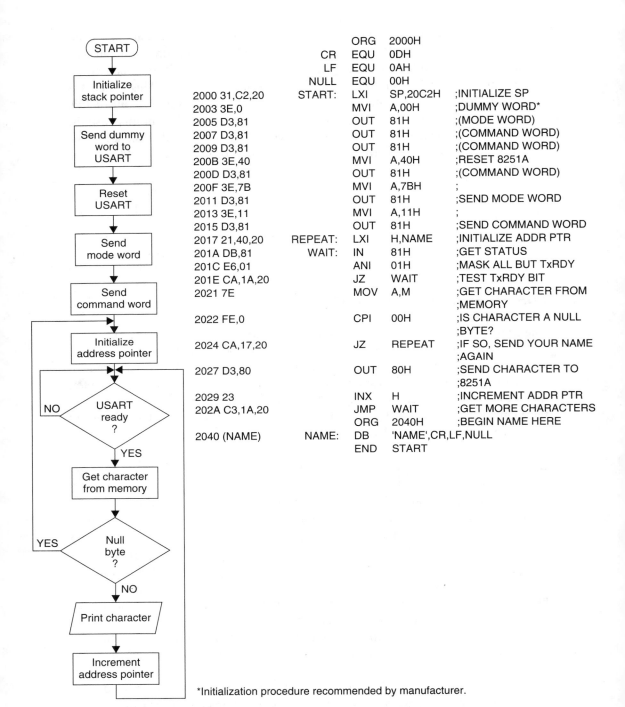

```
                              ORG   2000H
                    CR        EQU   0DH
                    LF        EQU   0AH
                    NULL      EQU   00H
2000 31,C2,20       START:    LXI   SP,20C2H      ;INITIALIZE SP
2003 3E,0                     MVI   A,00H         ;DUMMY WORD*
2005 D3,81                    OUT   81H           ;(MODE WORD)
2007 D3,81                    OUT   81H           ;(COMMAND WORD)
2009 D3,81                    OUT   81H           ;(COMMAND WORD)
200B 3E,40                    MVI   A,40H         ;RESET 8251A
200D D3,81                    OUT   81H           ;(COMMAND WORD)
200F 3E,7B                    MVI   A,7BH         ;
2011 D3,81                    OUT   81H           ;SEND MODE WORD
2013 3E,11                    MVI   A,11H         ;
2015 D3,81                    OUT   81H           ;SEND COMMAND WORD
2017 21,40,20       REPEAT:   LXI   H,NAME        ;INITIALIZE ADDR PTR
201A DB,81          WAIT:     IN    81H           ;GET STATUS
201C E6,01                    ANI   01H           ;MASK ALL BUT TxRDY
201E CA,1A,20                 JZ    WAIT          ;TEST TxRDY BIT
2021 7E                       MOV   A,M           ;GET CHARACTER FROM
                                                  ;MEMORY
2022 FE,0                     CPI   00H           ;IS CHARACTER A NULL
                                                  ;BYTE?
2024 CA,17,20                 JZ    REPEAT        ;IF SO, SEND YOUR NAME
                                                  ;AGAIN
2027 D3,80                    OUT   80H           ;SEND CHARACTER TO
                                                  ;8251A
2029 23                       INX   H             ;INCREMENT ADDR PTR
202A C3,1A,20                 JMP   WAIT          ;GET MORE CHARACTERS
                              ORG   2040H         ;BEGIN NAME HERE
2040 (NAME)         NAME:     DB    'NAME',CR,LF,NULL
                              END   START
```

*Initialization procedure recommended by manufacturer.

Figure 6-11   An 8085A communications program to transmit your name to a video display terminal.

TABLE 6-3 Examples of Even and Odd Parity

	Start bit	Seven-bit data	Parity bit	Stop bit
Odd parity	0	1 0 1 1 0 0 1	1	1
	0	0 1 0 0 1 0 1	0	1
Even parity	0	1 1 1 0 1 0 0	0	1
	0	0 0 1 1 0 1 0	1	1

minal, it is necessary to set the terminal features (baud rate, stop bit length, parity, and so on) to be compatible to what the 8251A has been programmed for.

The PE, *parity error,* flag will set if the 8251A receives a character whose parity bit does not correspond to the number of 1's set in the character. The PE flag can be monitored by reading the 8251A Status Word, bit D3. If an error has been detected, that is, PE = 1, the program must take the necessary steps to notify the operator of the error condition. The 8251A will otherwise continue with its operations. PE can be reset through a Command Instruction to the 8251A. SETTING the ER (*error reset*) bit, D4, of the Command Word and outputting it to port 81H will accomplish this. In Chapter 9, we take a closer look at the advantages and disadvantages of parity checking.

If a program were written to receive characters from a terminal (or DTE) and perform a parity check, an error condition should be forced in order to test if the error-checking capability of the 8251A is functioning. This can be accomplished by setting the parity of the terminal *opposite* to that for which the program is testing. If the 8251A has been programmed for odd parity and the terminal is set for even parity, a character received from the terminal will be tested for the number of 1 bits set. The parity bit of the received character will then be tested against this number of odd parity. Its state will be opposite the state for which the 8251A is testing (assuming that the received character has an even number of 1 bits). As a result, the 8251A SETS its PE flag.

6.5.2 OE: Overrun Error

When a serial character is received by the 8251A from the DTE, the processor is notified when the character has been assembled into parallel format and is ready to be read. This can occur by interrupting the processor with the 8251A's RxRDY pin or through the program's polling procedure whereby the 8251A's RxRDY, bit D1, or the Status Word is monitored. If the processor does not read the available character before the next one is available, an *overrun error* will occur. The previous character is destroyed. The 8251A will SET its OE flag, bit D4, of the Status Word.

Suppose that the 8251A were programmed to receive characters from DTE at a baud rate of 4800. Then the maximum number of characters per second (cps) that can be sent by DTE is 480, assuming a 10-bit character length:

$$4800 \text{ bps} \times \frac{1 \text{ character}}{10 \text{ bits}} = 480 \text{ cps}$$

The characters, in this case, would have to be sent in a contiguous manner; that is, the start bit of the next character must directly follow the stop bit of the preceding character. RxRDY would have to be polled at least every 2.083 ms (1/480 = 2.083 ms), or a character from DTE will be lost. Reading the character before the next one is available, however, will prevent OE from setting. A read will also reset the 8251A's RxRDY pin and status bit, D1, of the Status Word. This will allow the 8251A to inform the CPU when the next character is available.

The 8251A's OE bit of its Status Word can also be intentionally set for purposes of testing this error-checking feature. This can easily be accomplished by programming the 8251A to receive serial data from DTE and taking longer to read it than it takes DTE to send it. An example, using the setup described in Figure 6-1, would be to initialize the 8251A to receive data from the terminal. Once the initialization portion of the program is complete, a breakpoint or a delay loop can be executed. The 8251A, operating independently of the CPU at this time, will continue its operation, seeking characters typed in from the terminal. Since the CPU is not reading the available character before the next one has been assembled by the 8251A, OE will set. The Status Word can be checked at this time to verify if the overrun error condition has been detected. OE does not inhibit the operation of the 8251A. OE will remain set until it is cleared through a Command Instruction to the 8251A.

6.5.3 FE: Framing Error

In Chapter 5, we presented a description of how the UART's receiver section samples serial data sent from DTE to determine the level of each bit. The baud rate must be known beforehand in order for the receiver to sample the serial data at the center of each bit. The last bit sampled is the stop bit. Since the state of the stop bit (and start bit) is already known, the question may arise as to why this last sample is even necessary. The purpose of checking this last bit is to ensure that the character has been properly *framed*. Framing of an asynchronous character occurs when a UART's receiver section has synchronized with the start and stop bits of the character. If the state of the last bit is not a logic 1, opposite that of the start bit, a *framing error*, FE, has occurred. The 8251A will set its FE flag, bit D5, of its Status Word.

Framing errors can occur in several situations, many of which the 8251A is capable of flagging. Let us consider some of these conditions:

— The data sent by the DTE are faster than what the 8251A has been programmed to receive.

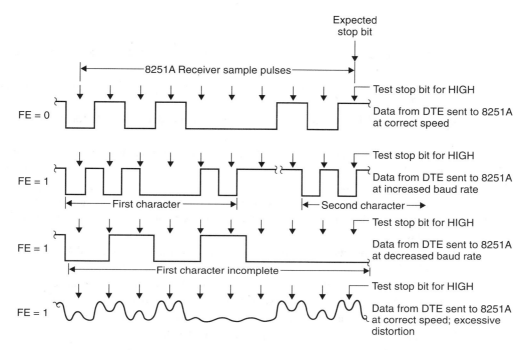

Figure 6-12 Conditions that will cause a FRAMING ERROR.

— The data sent by the DTE are slower than what the 8251A has been programmed to receive.

— The transmitted data from DTE becomes distorted en route to the 8251A.

— The 8251A's receiver clock, RxC, frequency is off.

Figure 6-12 illustrates conditions that would result in a framing error, FE. Bit D5 of the Status Word would be set. The occurrence of FE does not inhibit the operation of the 8251A. Here, again, it is up to the program to take immediate steps to notify the operator of the error condition. FE can be reset through a Command Instruction by setting the ER bit (error reset), D4, of the Command Word and outputting it to PORT 81H.

PROBLEMS

1. Draw the TTL waveform for the ASCII character @ as it would appear on an oscilloscope at the input and output of a TTL to RS-232-C level translator such as the MAX232 IC. Label all voltage levels. Use odd parity.

2. Define *partial address decoding*.

3. What function does the 8251A perform when its C/$\overline{D}$ line is:

(a) A logic HIGH?

(b) A logic LOW?

4. After resetting the 8251A, what is the first instruction that is necessary to condition the USART for communications?

5. What type of instruction is necessary to command the 8251A for asynchronous operation?

6. What type of instruction is necessary to command the 8251A for receiving data only?

7. Given the circuit shown in Figure 6-1, determine the mode word format for programming the 8251A for the following conditions: two stop bits, odd parity, seven-bit data, and 19.2 kbaud.

8. Given the circuit shown in Figure 6-1, what command word format would be necessary to enable both the transmitter and receiver sections of the 8251A and reset all error flags?

9. Explain why the internal reset bit is contained within the Command Word instead of the Mode Word.

10. For the program listed in Figure 6-11, what modifications would be necessary to halt the processor if a framing, parity, or overrun error occurred?

11. Define what is meant by a framing error, overrun error, and parity error.

12. Determine the parity bit ($\times$) for the following data words.

   (a) $110101\times$ (odd)

   (b) $100110\times$ (odd)

   (c) $111000\times$ (even)

   (d) $101110\times$ (even)

7

The Telephone Set
and
Subscriber Loop Interface

Alexander Graham Bell, inventor of the telephone, once theorized that if an electrical current could be made to vary in intensity precisely as the air waves vary in density during the production of speech, then speech could be transmitted over electrical wires. Bell, a Scotsman who emigrated to Canada, set out to invent such a device, called the *telephone.* In 1876, he transmitted a complete sentence to his assistant located in another room. Since then, the telephone set has emerged as one of the most widely used electrical devices in the world. It is estimated that there are over 1 billion telephone sets in operation worldwide today.

In this chapter we present the fundamental principles governing the theory and operation of the telephone set. The conventional rotary dial telephone will be looked at. Also, some of today's more modern telephones, which are gradually replacing the rotary dial with new technology, will be considered. These include the electronic pulse dial telephone; the Touch Tone or DTMF telephone, including its applications; the cordless telephone; and the cellular mobile phone.

7.1 BASIC FUNCTIONS OF THE TELEPHONE SET

Before discussing the operation of the telephone set, let us consider some of the basic functions that it serves.

1. The telephone set must notify the user of an incoming call through an audible tone such as a ring or bell.

2. The telephone set must transduce a caller's speech to electrical signals. Conversely, electrical signals must be transduced to audible speech signals.

3. A method of dialing subscriber numbers must be incorporated into the telephone set. This may be accomplished through dial pulses or tones.

4. The telephone set must regulate the speech amplitude of the calling party by compensating for the varying distances to the local telephone company, also known as the *central office*.

5. The telephone set must gain the attention of the central office when a user requests service by lifting the handset.

6. The telephone set must provide a nominal amount of feedback from its microphone to its speaker so that a user can hear himself or herself speaking. This feedback is called *sidetone*. Sidetone regulates how loudly one speaks.

7. When the telephone set is not in use, an open-circuit dc path must be provided to the central office.

8. In addition to receiving voice, the telephone set should also be capable of receiving call progress tones (busy, ringing, and so on) from the central office.

A block diagram of the conventional telephone set is shown in Figure 7-1. The dialing circuit shown is used to dial the person whom the caller wishes to speak to. A standard rotary dial switch or the more modern *dual-tone multifrequency* (DTMF) keypad is used. When placing or answering a call, the telephone is lifted off of its cradle, and the on-hook/off-hook circuit engages the telephone set to the telephone system. Power for the telephone set is derived from a −48 V dc supply located at the central office. The power is delivered to the telephone set via the *subscriber loop*. Since most subscriber loops are two-wire pairs, a *hybrid* circuit is necessary to transform the two-wire transmission line into four

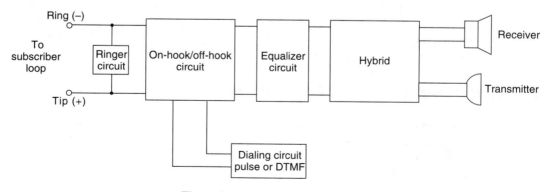

Figure 7-1 Block diagram of the telephone set.

wires, thus separating the telephone set's transmitted and received signals. Full-duplex operation is made possible. To compensate for the varying lengths of wire between the central office and its subscribers, *equalizer* circuits are incorporated into the telephone set to regulate voice amplitudes.

7.1.1 Telephone Transmitter

The transmitter for the telephone set is essentially a microphone. It is the part of the handset into which the person speaks. The function of the transmitter is to convert acoustical energy, generated from speech, into electrical energy, which is transmitted onto the subscriber loop. Figure 7-2 illustrates a cross-sectional view of the transmitter.

Dc current provided by the telephone system is passed through two electrodes separated by thousands of carbon granules. One electrode is attached to a diaphragm that vibrates in response to the acoustical pressures of sound. The opposite electrode is supported by the handset molding. Vibration of the diaphragm causes the contact resistance between the two electrodes to vary inversely with pressure. As the resistance varies, the current varies inversely, thereby translating the acoustical message into the electrical signal that is transmitted to the central office. The central office, in turn, routes the electrical signal to its destination.

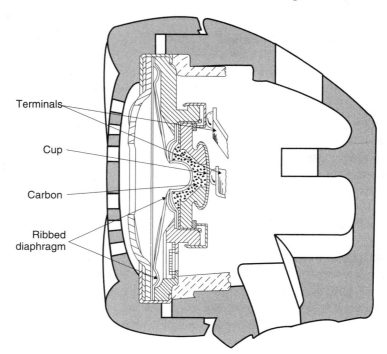

Terminals

Cup

Carbon

Ribbed
diaphragm

Figure 7-2 Cross-sectional view of the telephone transmitter. (Courtesy of Bell Laboratories.)

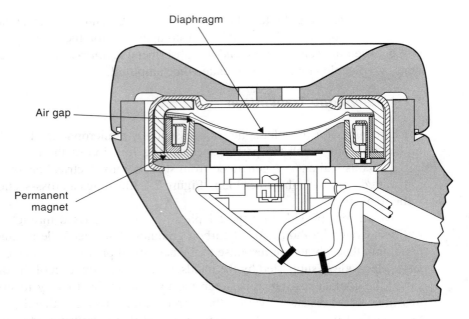

Diaphragm

Air gap

Permanent
magnet

Figure 7-3 Cross-sectional view of the telephone receiver. (Courtesy of Bell Laboratories.)

7.1.2 Telephone Receiver

Figure 7-3 is a cross-sectional view of the telephone set's receiver. The receiver is essentially a speaker used to transduce a voice-generated ac signal back to sound. A permanent magnet is used to produce a constant magnetic field. Insulated wire is wound around the permanent magnet to form a coil, which passes the ac signal. The magnetic field produced by the varying ac signal aids and opposes the existing permanent magnetic field. The resulting force causes the metallic diaphragm to vibrate. The vibrating diaphragm produces sound waves corresponding to the original sound waves delivered to the transmitter.

7.1.3 Telephone Ringer

The function of the *ringer* is to alert the party of an incoming call. The audible tone generated by the ringer must be loud enough for the party to hear from a distance. Several variations of ringers are used in today's telephone sets. The most popular type is the conventional bell type shown in Figure 7-4. Buzzers, horns, lights, and more recently, semiconductor sound generators with driven speakers are also used as ringers.

In the United States, telephone companies will *ring* the called party with an ac ringing signal, typically 90 V_{rms} at 20 Hz. The ring signal is superimposed

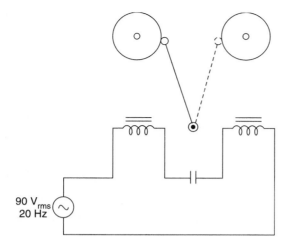

Figure 7-4 Telephone ringer.

upon the existing -48 V dc signal. Figure 7-5 illustrates the ring signal. The two coils shown in Figure 7-4 are wound in a manner that causes the pivoting hammer to strike each bell on alternate parts of the cycle. Two dissimilar metals are used for each bell to produce the familiar ringing sound. A capacitor is used to block dc current and pass the ac ringing current. Its value, combined with the coil inductance, is selected to provide a high impedance to voice frequencies.

7.1.4 Telephone Hybrid

The telephone set's *hybrid* is used to interface the transmitter and receiver's individually paired wires to a single pair for the subscriber loop. A multiple winding transformer is wound in a manner to electrically separate the transmitted and re-

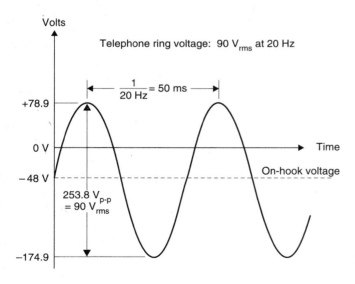

Figure 7-5 Ringing voltage for the telephone set.

ceived signals. This permits simultaneous transmission and reception of speech, or what is more commonly referred to as *full-duplex* operation.

7.1.4.1 Sidetone. A balancing network included in the hybrid circuit allows the manufacturer to adjust a small amount of feedback from the telephone set's transmitter to its receiver. This feedback is called *sidetone.* Sidetone allows the person speaking into the handset to hear himself or herself talking. Tests have shown that when the sidetone is adjusted properly a person can determine how loudly to speak based on the level of the sidetone presented to the receiver. If the sidetone is adjusted too small, the person talking tends to speak too loudly to compensate for the lack of hearing himself or herself at the receiver. Conversely, too much sidetone causes the speaker to lower his or her voice, consequently making it difficult for the receiving party to hear what is being said.

7.2 ROTARY DIALING WITH THE BELL 500 TYPE TELEPHONE

The Bell 500 telephone was first introduced in 1951. For over three decades it has been an industry standard. Although it is slowly being replaced by DTMF-type telephones, it still remains one of the most widely used telephones throughout the world. Figure 7-6 illustrates the Bell 500 model and its schematic diagram.

Switches S1 and S2 are the on-hook/off-hook switches that open and close when the handset is engaged and disengaged from its cradle. The two switches work in unison with each other. When the phone is resting in its cradle, the two switches are *open.* This is called the *on-hook* condition. Notice that the ringer circuit is connected to the telephone system, whereas the telephone set itself is disconnected. When a call is placed, the handset is lifted off its cradle and S1 and S2 *close.* This is called the *off-hook* condition. The normal −48 V dc on-hook voltage supplied by the telephone company drops to approximately −5 to −8 V dc due to the impedance that the telephone set presents to the line. Dc current begins to flow in the subscriber loop as a result of going off-hook. This current flow is detected by the central office, which in turn sends a dial tone to the caller indicating that service is available and a number may be dialed. The subscriber loop current can range anywhere from 20 to 120 mA, depending on loop length and the impedance of the telephone circuit.

The rotary dial has 10 finger positions and a finger stop. Each position on the dial is represented by a number or group of letters. With the exception of 0, the dialing process generates pulses equal to the number being dialed. Dialing a 1, for example, generates one pulse; dialing a two, generates two pulses; and so forth. Dialing a 0 produces 10 pulses. Pulses are generated by *making* and *breaking* contact with the line with switch D1.

(a)

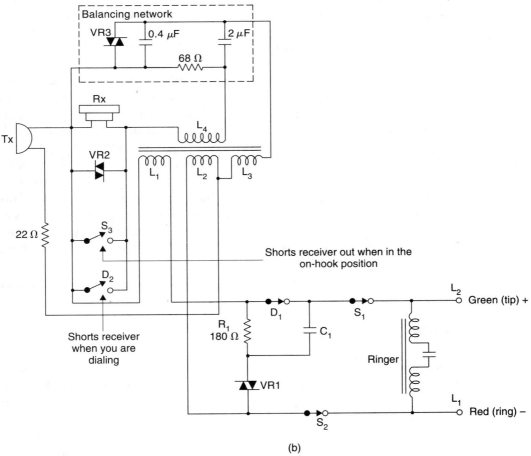

(b)

Figure 7-6 (a) Bell 500 telephone set; (b) schematic diagram of the Bell 500 telephone set.

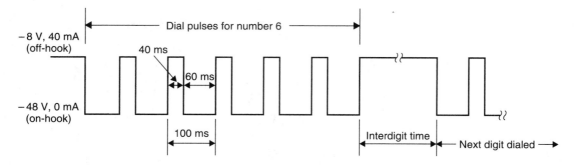

Figure 7-7 Dial pulses generated for the number 6.

Dialing is accomplished in the off-hook condition with the rotary dial switch, D1. Initially, D1 is closed. Going off-hook causes switches S1 and S2 to close and S3 to open. A current path is established to the receiver and transmitter and a dial tone can be heard. Rotating the dial wheel to the finger stop winds a spring within the dialing assembly. Switch D2 closes during the process. This shorts the receiver out and prevents clicking noise from being heard at the receiver. Dial pulses are not generated at this point. When the dial wheel is released from the finger stop, the internal spring unwinds, rotating the dial back to its original rest position. It is during this release time that the dial pulses are generated. A cam, driven by gears turned by the revolving dial shaft, opens switch D1 for 60 ms and closes it for 40 ms, thus creating a dial pulse. The number of times that the switch opens and closes is equal to the digit dialed or the number of dial pulses generated. A speed *governor* is included in the dial assembly to maintain a constant angular velocity as the dial returns to its rest position. This regulates the period of each pulse at 100 ms or a 10-Hz rate. Switch D2 opens at this time, activating the receiver until the next digit is dialed. Figure 7-7 illustrates the pulses generated when the number 6 is dialed. Nominal values for both current and voltage are shown. The idle time between digits is called the *interdigit time.* Phone companies limit the caller to approximately 10s before the next digit must be dialed.

7.2.1 Telephone Gain Control

Virtually every telephone set connected to the central office has a different subscriber loop length. As a result, a large variation in line resistance exists between each customer. Since each telephone is sourced from the same supply voltage at the central office, subscriber loop currents can range anywhere from 20 mA for longer loops to 120 mA for shorter loop lengths. The telephone set must compensate for this variation by employing some type of gain-controlling mechanism; otherwise, the signal amplitude at the telephone set's receiver would diminish with increasing distance from the phone company.

To maintain constant transmit and receive amplitudes, *varistors* VR1 through VR3 are used. The varistor is a semiconductor device whose resistance varies *inversely* with current. By placing the varistor across the receiver line, and balancing network as shown in Figure 7-6, large signals transmitted and received from shorter loop lengths are shunted through the varistor. For weak signals resulting from longer loop lengths, the varistor acts as a high impedance, thus permitting most of the signal to flow either through the receiver or out to the line.

The varistor also acts as a transient suppressor for excessive voltage spikes generated each time the dial switch makes and breaks contact with the line during dialing. These transients are produced by the interruption of line current through the ringer coil [$e_l = L(di/dt)$]. Capacitor C_1 and resistor R_1 further suppress the transients and prevent sparking across the dial contacts. This portion of the telephone circuit is often referred to as an *antitinkle* circuit because, without it, a tinkling noise produced by the high-voltage spikes across the ringer can be heard when dialing.

7.3 ELECTRONIC PULSE DIALING TELEPHONE

Many of the circuits discussed thus far are gradually being replaced with semiconductor integrated circuits (ICs). Single-chip ICs are now available that perform dialing and ringing functions and more. Memory and control circuits are included in these ICs for storing and automatically dialing telephone numbers. The *electronic pulse dialing telephone* incorporates much of this technology. This type of telephone is compatible with electromechanical switching facilities at the central office. Instead of a rotary dial switch, a keypad is used to enter the telephone number to the controlling IC. Although numbers can be rapidly entered with a keypad, the dial pulses generated are identical to those produced by the rotary dial switch.

Electronic ringing circuits have replaced the old electromechanical bell with piezoelectric transducers driven by oscillators. These circuits are less costly and take up considerably less space. A major disadvantage, however, is that the ringing intensity produced by some of the transducers are significantly less than that of the conventional bell type. In an effort to bolster the electronic ringer, some manufacturers have equipped telephones with multitone oscillators and power amplifiers that drive speakers.

Memory within the IC permits the *redial* function. This function is useful when a line is busy and the caller wants to redial the last number entered at some later time, even if the phone has been placed back on-hook. By depressing the redial key, the telephone automatically redials the last phone number entered.

7.4 DUAL-TONE MULTIFREQUENCY (DTMF)

A much more efficient means of providing the dialing function of the telephone set is through the use of *dual-tone multifrequency* (DTMF). DTMF is also known as *Touch Tone*. Most central offices are equipped to handle both Touch Tone and dial pulses. In a Touch Tone telephone, a push-button keypad is provided for entering digits, instead of a rotary dial. The arrangement of the keypad is shown in Figure 7-8. There are 12 keys corresponding to the numbers 0 through 9 and the characters * and #. Some keypads include four additional keys, A through D, for special control functions. Four rows and four columns of keys form a frequency matrix consisting of a *low band* and a *high band* of frequencies. The frequencies for each row and column are separated by a difference of approximately 10%, whereas the two bands of frequencies are separated by a difference of approximately 25%.

When a key is depressed, two tones are generated and sent to the telephone company for processing: one from the low band of frequencies and one from the high band. The digit 6, for example, generates a 770-Hz tone and a 1477-Hz tone. Figure 7-9 illustrates the waveform generated by the electrical sum of these two tones. Frequency tolerances are specified at ±1.5% for the telephone set's DTMF generator and ±2% at the receiving central office. These stringent tolerances prevent the telephone company from detecting erroneous digits or characters. For the same reason, DTMF frequencies have been carefully selected so that they are not *harmonically* related to each other. The second and third harmonics for the digit 1, for example, include the following low- and high-band frequencies:

	Fundamental	Second harmonic	Third harmonic
High-band frequency (Hz)	1209	2418	3627
Low-band frequency (Hz)	697	1394	2091

Notice that the harmonics are not equal to the tones generated by other keys. Nor are they equal to any call progress tones generated by the phone company that may get cross-coupled into the line and misinterpreted as a digit.

The major advantage of Touch Tone dialing over rotary dialing is speed. Keys need only be depressed for a minimum of 50 ms in order for telephone companies to detect and decode the digit. The minimum interdigit time is also 50 ms. Each digit can therefore be sent in 100 ms. A seven-digit telephone number can be sent in less than 1 s, whereas on a rotary dial phone, to dial the number 0 alone, would take 1 s (100 ms/pulse × 10 pulses). This does not include the time that it would take to rotate the dial to its finger stop position. Figure 7-10 is a circuit diagram of a Touch Tone telephone using National Semiconductor's

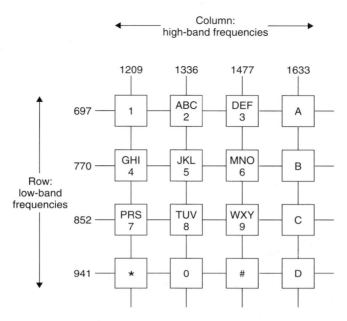

Figure 7-8 DTMF frequency and keypad layout.

TP5650/TP5700 Ten-Number Repertory DTMF Generator. Another popular IC is the TP5088 DTMF Generator shown in Figure 7-11. A four bit word and a *tone enable* pulse generates 1 of 16 DTMF tones.

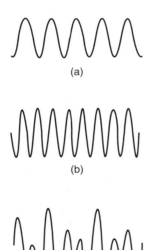

(a)

(b)

(c)

Figure 7-9 DTMF waveforms: (a) 770-Hz low-band frequency; (b) 1336-Hz high-band frequency; (c) electrical sum of the low- and high-band frequency produces the DTMF tone for the digit 5.

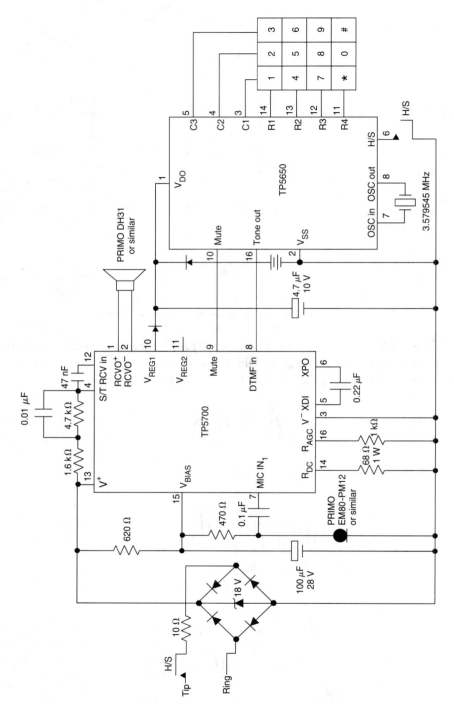

Figure 7-10 Circuit diagram of a Touch Tone telephone using National Semiconductor's TP5650 and TP5700 ICs. (Courtesy of National Semiconductor, *Telecommunications Data Book*, 1990, p. 3-16.)

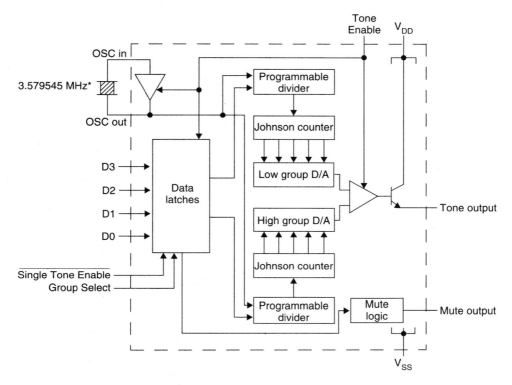

*Crystal Specification: Parallel Resonant 3.579545 MHz, $R_S \leq 150 \, \Omega$, L = 100 mH, C_0 = 5 pF, C_1 = 0.02 pF.

(a)

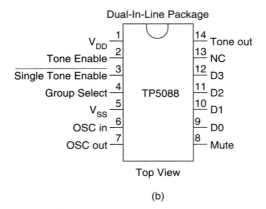

(b)

Figure 7-11 National Semiconductor's TP5088 DTMF generator: (a) block diagram; (b) pin-out; (c) functional truth table; (d) microprocessor interface. (Courtesy of National Semiconductor, *Telecommunications Data Book,* 1990, pp. 3-3 to 3-6.)

Sec. 7.4   Dual-Tone Multifrequency (DTMF)   157

Keyboard	Data inputs				Tone	Tones out		Mute
equivalent	D3	D2	D1	D0	enable	f_L (Hz)	f_H (Hz)	
X	X	X	X	X	0	0V	0V	0V
1	0	0	0	1	⤵	697	1209	O/C
2	0	0	1	0	⤵	697	1336	O/C
3	0	0	1	1	⤵	697	1477	O/C
4	0	1	0	0	⤵	770	1209	O/C
5	0	1	0	1	⤵	770	1336	O/C
6	0	1	1	0	⤵	770	1477	O/C
7	0	1	1	1	⤵	852	1209	O/C
8	1	0	0	0	⤵	852	1336	O/C
9	1	0	0	1	⤵	852	1477	O/C
0	1	0	1	0	⤵	941	1336	O/C
*	1	0	1	1	⤵	941	1209	O/C
#	1	1	0	0	⤵	941	1477	O/C
A	1	1	0	1	⤵	697	1633	O/C
B	1	1	1	0	⤵	770	1633	O/C
C	1	1	1	1	⤵	852	1633	O/C
D	0	0	0	0	⤵	941	1633	O/C

(c)

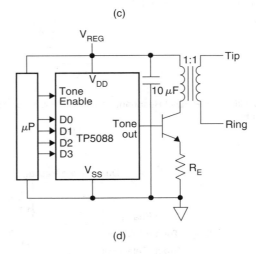

(d)

Figure 7-11   *(continued)*

7.5 CORDLESS TELEPHONE

The *cordless* telephone has gained considerable popularity in the last decade. Its declining cost coupled with the convenience of operating without the restrictions of an attached cord have made it a common household item.

As its name implies, the cordless telephone is operated without an attached cord. There are two units that makes this possible: a *base unit* and a *portable unit.* The base unit is powered by the 117-V ac household line, whereas power for the portable unit is derived from an internal rechargeable battery. Each unit contains an FM transmitter and receiver. The base unit is directly connected to the subscriber loop through a standard telephone jack. It transmits and receives all signals between the portable unit and the central office. This includes both voice and control such as ringing and dial tone. The portable unit can be used from a remote location by placing it in the standby mode.

Figure 7-12 illustrates one method used to achieve cordless operation of the telephone. The 117-V ac, 60-Hz household electrical wiring is used as a transmitting antenna for the base unit. This permits the portable unit to receive signals uniformly throughout the house and around the perimeter of the house. A carrier frequency in the range 1.6 to 1.8 MHz is frequency modulated by the base unit and transmitted onto the ac line. Capacitive coupling is used to couple the transmitted signal onto the line, while at the same time preventing 60-Hz line currents from damaging the base unit. A loop stick antenna built internal to the portable unit is used to receive the base unit's transmitted signals. This antenna is similar to the type used in a hand-held transistor radio.

Signals transmitted from the portable unit to the base unit have a carrier frequency in the range 49.8 to 49.9 MHz. FM is the modulation technique employed. A telescoping whip antenna is used by both the portable and base units for transmitting and receiving these signals, respectively. To operate from a remote location, the portable unit is normally placed in the standby mode. In this mode, the portable unit's receiver is ON and its transmitter is OFF. The base unit maintains the on-hook condition to the subscriber loop. For incoming calls, the low-power receiver section of the portable unit actively awaits ring signals from the base unit. Since the portable unit receives these signals by its loop stick antenna, there is no reason for its whip antenna to be pulled out at this time. When ringing is detected from the central office by the base unit, it transmits its own ring signals to the portable unit. These signals are detected and used to drive a speaker built into the portable unit. Ringing prompts the called party to answer the telephone call. The whip antenna is pulled out at this time and the transmitter is activated by depressing the TALK/HANG-UP switch located on the portable unit. A signal is sent back to the base unit, which answers the phone call by going off-hook. Simultaneous conversations can now take place on the high- and low-frequency channels.

The portable unit can also be used to place calls from a remote location. With the whip antenna pulled out, the TALK/HANG-UP switch is depressed, which sends a signal to the base unit, causing it to go into the off-hook condition. The dial tone received from the central office is transmitted from the base unit back to the portable unit. Dialing is accomplished through the use of the keypad on the portable handset. Its carrier frequency is modulated with DTMF tones or dial pulses representing the key depressed on the handset. The base unit demodulates these tones and places them onto the subscriber loop where they are sent to the central office for processing. Return signals generated from the central office, including call progress tones and voice, are modulated onto the 117-V ac, 60-Hz line frequency and sent back to the portable unit for the caller to hear. When the

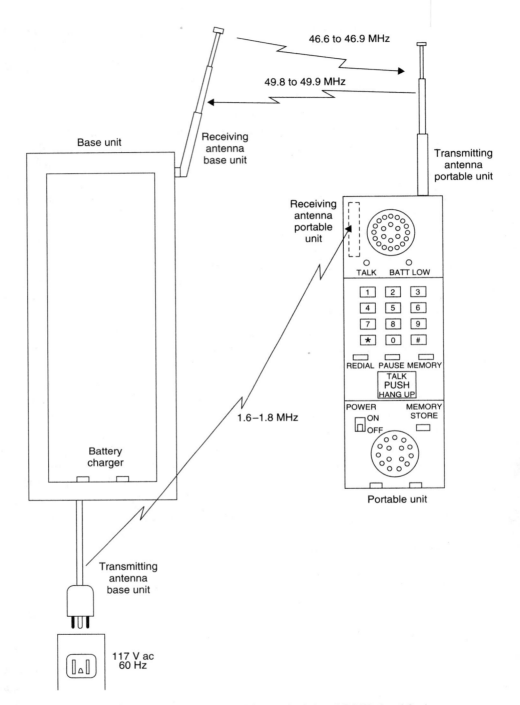

Figure 7-12 Cordless telephone. Earlier models use the 1.6-to-1.8 MHz band for base-to-portable unit transmission over the 117-V ac line. Current models use the 46.6-to-46.9 MHz band for base-to-portable unit transmission.

portable unit is not in use, it should be placed back into the base unit for recharging its batteries.

As of January 1984, the Federal Communications Commission (FCC) allocated a new band of frequencies for cordless telephones in the 46- and 49-MHz region. Ten pairs of carrier frequencies are used between the base unit and portable handset. They are listed in Table 7-1 along with the earlier 1.6- to 1.8-MHz band. Unlike the earlier base units that transmitted in the 1.6- to 1.8-MHz region over the 117-V ac line, the whip antenna of the base unit is used to transmit to the portable unit in the new 46-MHz band and receive from the portable unit in the 49-MHz band. Most cordless telephones today operate in this mode. The overall performance has proved to be superior to earlier technology. Transmitting over the 117-V ac line is inconsistent, particularly in cases where electrical conduit is used.

TABLE 7-1 Cordless Telephone Frequencies (MHz)

1.6- to 1.8-MHz band	49-MHz band	46-MHz band
1.665	49.670	46.610
1.690	49.770	46.710
1.695	49.830	46.770
1.710	49.845	46.630
1.725	49.860	46.670
1.730	49.875	46.730
1.750	49.890	46.830
1.755	49.930	46.870
1.770	49.970	46.970
	49.990	46.930

7.6 THE LOCAL LOOP

For individual telephones to be useful, they must be interconnected to other telephones to establish a communications link. Figure 7-13 illustrates a simplistic method of interconnecting six parties together. The noticeable problem here is the overwhelming number of interconnecting lines necessary for each party to have the ability to call any one of the other parties. To provide service for n parties, the number of lines required for this method of interconnection is governed by the following equation:

$$\text{Number of interconnecting lines} = \frac{n(n - 1)}{2}$$

where n is the number of parties. For the six parties illustrated, the number of interconnecting lines is 15:

$$\text{Number of interconnecting lines} = \frac{6(6 - 1)}{2} = 15$$

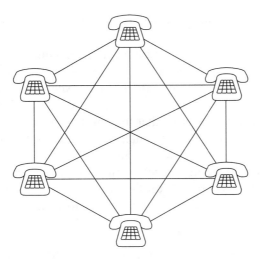

Figure 7-13 Method for interconnecting six telephone parties. Each party has access to any of the other five.

Each telephone in this setup must have the ability to switch to any of the other five. Imagine a telephone network of this type having to provide service to 50,000 subscribers! This would be quite impractical. Furthermore, it is not necessary for telephone systems to assume that every telephone connected to the network is in use 100% of the time. Normally, fewer than 10% are. Clearly, an alternative method for providing telephone service is needed.

7.6.1 A Need for Centralized Switching

Given the situation just presented, it makes sense to devise a centralized form of switching that is capable of establishing a temporary connection between two parties wishing to communicate with each other. This, indeed, has been the established method since the days of the first telephone networks. Only the manner in which the connection is made has become more sophisticated. Each telephone subscriber is connected to a *central office* through a twisted pair of wires used as the transmission medium. This pair of wires is referred to as the *subscriber loop* or *local loop* (Figure 7-14). It is here at the central office where a temporary connection is made between parties.

The first telephone networks used a *switchboard* to terminate subscriber loops. Switching was actually performed at the switchboard by a telephone operator who manually connected two subscriber loops together. Telephones were individually powered with batteries and were part of a system referred to as the *local battery system.* The calling party signaled the operator for service by cranking a magneto (a hand generator) located within the telephone set. The resulting ac signal activated a lamp at the switchboard, notifying the operator that a connection was desired. The operator, shown in Figure 7-15a, then determined from the caller which party to contact. A *patch cord* was used to interconnect the two parties' subscriber loops. The patch cord is made up of two telephone jacks

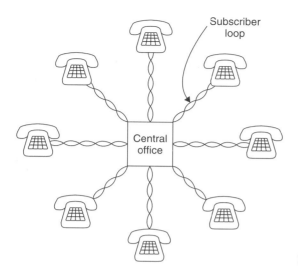

Figure 7-14   Central office.

separated by a cord (Figure 7-15b). The *tip* and *ring* contacts of the jack, on each end of the cord, connect the two subscriber loop lines.

A major drawback with the local battery system was having to maintain the condition of the battery. Phone calls could not be made on weak or dead batteries. Other inherent problems associated with local battery systems were the lack of privacy and also having to staff the switchboard 24 hours a day.

(a)

Figure 7-15   (a) Telephone switchboard operator; (b) patch cord. (Courtesy of AT&T Archives.)

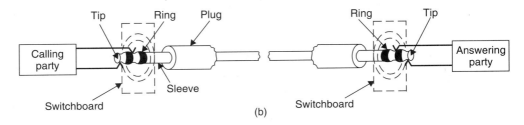

Today's telephone sets receive power from the central office for signaling. This is referred to as *common battery*. Switching is performed automatically. For traditional reasons, the names *tip* and *ring* are still used today to identify the two-wire subscriber loop lines.

7.6.2 The Local Feeder Network

Figure 7-16 illustrates how telephone lines are distributed to a community from the central office through a *feeder network*. Thousands of twisted-pair wires are brought out to the community in bundled cables that are fanned out to a number of servicing areas. The number of subscriber loop pairs is planned ahead of time to exceed the number of subscribers in a service area. This is to allow additional customers to be serviced at a later time.

7.6.3 Operating Specifications and Call Procedures

The central office supplies -48 V dc (typical) to the ring and ground to the tip side of each loop, as shown in Figure 7-17. A negative voltage is used to minimize *electrolytic corrosion* of the subscriber loop wires. Signaling the central office is performed in one of two ways: DTMF signaling or pulse dialing. Connection to the calling party is done automatically through computers and relays.

When a caller goes off-hook by lifting the handset off its cradle, a dc current path is provided by the telephone circuit. Current, I_s, flows from the central office, through the telephone, and returns via the subscriber loop. The amount of current that flows depends on such factors as wire size and type, length of the subscriber loop, and telephone impedance. Typical subscriber loop currents, I_s, range from 20 to 80 mA and subscriber loop resistances range from 0 to 1300 Ω, as shown in Table 7-2. Telephone impedance ranges from 500 Ω to 1 kΩ. The loop current is therefore directly affected by several variables.

Current initially flowing in the subscriber loop as a result of a caller going off-hook is an indicator to the central office that service is required. A dial tone is sent to the caller to acknowledge the connection. A phone number may then be dialed. The central office translates and processes the dial pulses or DTMF tones sent by the caller. If current is flowing in the subscriber loop of the party being called, a *busy tone* is sent to the party placing the call. The caller termi-

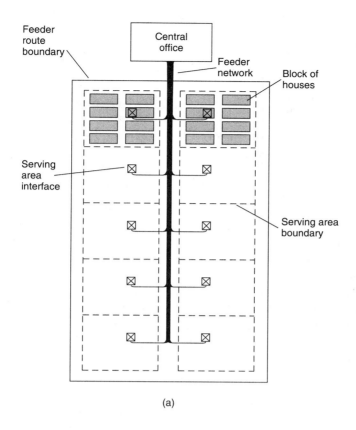

(a)

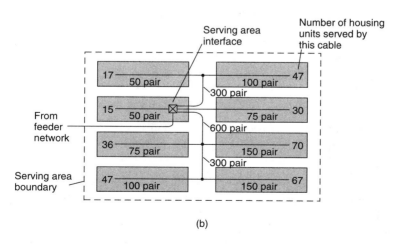

(b)

Figure 7-16 Feeder network distributing service to a community: (a) local distribution area; (b) detail of a serving area. (Courtesy of Bell Laboratories.)

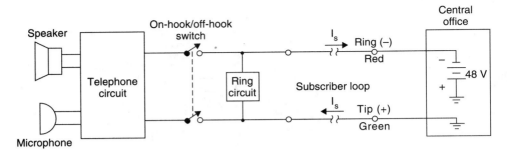

Figure 7-17 Dc connection to the central office.

nates the procedure and attempts to place the call at a later time. If, however, no current is flowing in the subscriber loop of the party being called, its phone is on-hook. An ac *ring signal* of 90 V_{rms} at 20 Hz is sent by the central office to alert the called party that a call is waiting. At the same time that the ring signal is being sent, a *ring tone* is sent back to the original caller as an indicator that the central office is ringing the requested party. Both ring signals are sent until either current flow has been detected in the called party's loop, that is, the phone has gone off-hook and has been answered, or the caller simply hangs up. The central office then interconnects the two parties for service. The connection is made un-

TABLE 7-2 Subscriber Loop Operating Parameters

Parameter	Typical U.S. values	Operating limits	Typical European values
Common battery voltage	-48 V dc	-47 to -105 V dc	Same
Operating current	20 to 80 mA	20 to 120 mA	Same
Subscriber loop resistance	0 to 1300 Ω	0 to 3600 Ω	
Loop loss	8 dB	17 dB	Same
Distortion	-50 dB total	NA	
Ringing signal	20 Hz, 90 V_{rms}	16 to 60 Hz, 40 to 130 V_{rms}	16 to 50 Hz, 40 to 130 V_{rms}
Receive sound pressure level	70 to 90 dBspl[a]	130 dBspl	Varies
Telephone set noise		Less than 15 dBrnC[b]	

[a]dBspl = dB sound pressure level.

[b]dBrnC = dB value of electrical noise referenced to -90 dBm measured with C message weighting frequency response.

Source: From the book *Understanding Telephone Electronics*, 3rd edition by Stephen Bigelow, copyright © 1991. Published by SAMS Publishing, a division of Prentice Hall Computer Publishing. Used by permission of the publisher.

til one of the parties goes on-hook (hangs up). Some central offices will release the line only when the calling party hangs up. These are merely a few of the many tasks that the central office performs.

7.6.4 Call Progress Tones

Many of the signals received by the telephone set are acknowledgments and status indicators for the calling process. It is necessary for the caller to be able to recognize the sound of these tones to place a call properly. Most of these tones are frequency pairs that are turned on and off at varying rates. They are listed in Table 7-3. *Dial tone*, for example, is a composite of 350- and 440-Hz sine waves that are electrically summed together and sent by the central office to the caller. It is an acknowledgment to the caller going off-hook and informs the caller that service is available and a phone number may be dialed. To minimize costs, the central office presents dial tone to a caller for a limited amount of time, typically 20 sec. A telephone number must be entered within this time frame or a recorded help message is played.

The *ring-back* signal informs the caller that the dialed party is being rung. This signal is often mistaken for the actual ringing signal (90 V_{rms} at 20 Hz) used to ring a caller's bell. They are two separate signals entirely. The ring-back signal may not necessarily be in sync with the ringing signal. Because of this, the situation may arise where the party being called may answer the phone before the caller even hears the ring-back tone.

The *busy tone* indicates that the party being called is off-hook and therefore the party's line is in use. The phone company is able to determine this by the dc current flowing in the called party's line.

During peak periods of the day, it is possible for the central office to become overburdened with calls. This generally occurs around 8:00 A.M. to 10:00 A.M. and 3:00 P.M. to 5:00 P.M. When this happens, a *congestion tone* is sent to the caller. This tone sounds the same as the busy tone except that its on-off rate is twice as fast. The call is said to be *blocked*. Blocking is discussed further in Chapter 8.

TABLE 7-3 Telephone Call Progress Tones

Tone	Frequency (Hz)	On time (sec)	Off time (sec)
Dial	350 + 440	Continuous[a]	
Busy	480 + 620	0.5	0.5
Ring back	440 + 480	2	4
Congestion	480 + 620	0.2	0.3
Receiver off-hook	1400 + 2060 + 2450 + 2600	0.1	0.1

[a]Dial tone is presented to the customer for a limited amount of time, typically 20 sec.

The *receiver off-hook* signal creates a very loud tone at the telephone receiver. Its purpose is to alert the customer that the telephone's handset has accidentally gone off-hook and must be placed back onto its cradle. The tone is loud enough to hear from a distance (across the room) and lasts approximately 40 sec. Usually, a voice-recorded help message is sent to the caller before this tone is generated.

7.7 LINE CHARACTERISTICS

The quality of communications over the PSTN is determined by several factors, such as bandwidth, impedance, transmission media, and line length. In the past, these line characteristics were considered primarily in terms of their effect on speech between two parties. The PSTN, as a result, has been tailored toward voice-grade communications. Because there are literally millions of miles of existing voice-grade lines, it is necessary for us to study these line characteristics to better understand their effects on voice, as well as the limitations they impose on the transmission of data.

7.7.1 Bandwidth Constraints

Voice includes frequency components ranging from approximately 100 Hz to as high as 8 kHz. Most of one's voice energy is distributed in frequencies ranging from 400 to 600 Hz, with a nominal voice frequency of 500 Hz. Figure 7-18 illustrates these characteristics. Extensive studies on human speech have indicated that the relative importance of voice-frequency components in terms of intelligibility is distributed differently from their energy distribution. As shown in Figure 7-18, voice frequencies below about 200 Hz and above 2 kHz play only a

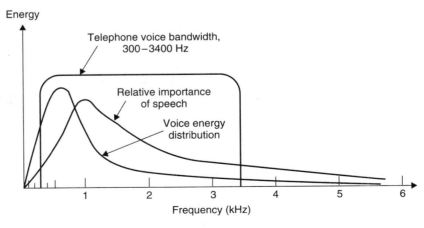

Figure 7-18   Telephone system bandwidth versus voice characteristics.

minor role in determining intelligibility. The telephone lines used for subscriber loops have been tailored to capture these characteristics with a bandwidth ranging from 300 to 3400 Hz. At the same time, these bandwidth constraints limit the noise originating from 60-Hz lines, dial pulses, thermal and shot noise, and the like, which would otherwise degrade the quality of reception.

7.7.2 Loop Resistance

The subscriber loop resistance is governed primarily by the type of wire used. Typically, copper wire is used in sizes ranging from 19 to 26 gauge. For 26-gauge copper wire, the attenuation is approximately 3 dB/mile. However, for 19-gauge copper wire, the attenuation is approximately $1\frac{1}{4}$ dB/mile. Thus, roughly twice the transmission distance can be attained with the larger-diameter wire. Attenuation is generally kept at a value of less than 8 dB in the subscriber loop, with a maximum permissible loop resistance of 1300 Ω. The loop resistance also includes the resistance of the telephone, typically in the order of 120 Ω. The smallest wire used for the subscriber loop is 26 gauge, which has a dc resistance of 40.8 Ω at 68°F per 1000 ft. For 1300 Ω, less the resistance at the telephone, this is just over 6 miles (3-mile pair). The average customer loop is about 2 miles; thus 26-gauge wire satisfies the needs of most subscriber loops.

Figure 7-19 illustrates the model of a transmission line. The distributed values of L, R, and C make it more evident why there are signal losses and distortion in wire transmission media. The model can be used to represent the subscriber loop. Since the values of L, R, and C can be measured at any point along the subscriber loop, experiments have been conducted to determine methods to improve the transmission characteristics of the phone lines. The most practical method has been found to be adding *series inductance* to the line at various intervals between the subscriber and the central office. This is referred to as *loading*. The loading coil is depicted in Figure 7-20. The effect of loading increases the line impedance, consequently decreasing the overall attenuation. Longer distances can therefore be achieved through loading, at the same time preserving the dc characteristics of the line. Loading also results in the sharp cutoff frequency at approximately 3.4 kHz. This is undesirable for high-speed digital transmission.

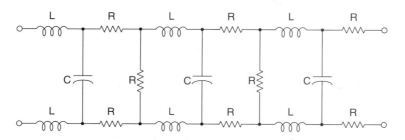

Figure 7-19   Model of a transmission line.

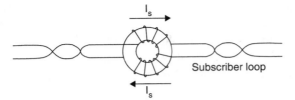

Figure 7-20 Loading coil in the sub-scriber loop.

Table 7-4 depicts how telephone cable pairs are labeled. The letters H and B correspond to series inductance added to the line every 6000 or 3000 ft, respectively. The designator, D, is also used for inductances added every 4500 ft. The value of the loading coil in millihenrys (mH) is also specified. Typical values are 135, 88, 44, 22, and 7.5 mH. For example, a 26-gauge cable pair with loading coils of 88 mH every 6000 ft is labeled 26H88. Most cable pairs use 88-mH coils with a spacing of 6000 ft.

TABLE 7-4 Loading Coil Designation

Designation[a]	Nominal cutoff frequency, fc	Use
H88	3,500–4,000	Message trunks and long subscriber lines
H44	5,000–5,600	Certain data circuits
B22	10,000–11,200	Program networks

[a]The letter designates the spacing: $H = 6000$ ft, $B = 3000$ ft. The number gives the inductance of each loading coil in millihenrys.
Source: Standard Handbook for Electrical Engineers, 10th ed., McGraw-Hill, New York, 1969, p. 25-49.

Table 7-5 lists the electrical characteristics for common wire sizes used by the PSTN. A comparison of nonloaded versus H88-loaded cables is made. Notice, under the attenuation column, that the decibel per mile specification is reduced considerably through the use of loading coils.

7.7.3 The Private Line

Of the millions of telephone lines connected to the central office, the overwhelming majority are two-wire, twisted-pair loops. These subscriber loop lines, which connect our homes to the central office, are referred to by telephone companies as *switched* or *dial-up* lines since they require the switching services of the central office.

The two-wire loop can also be a dedicated *leased* or *private* line. These types of lines are direct hard-wired connections between two locations through the central office, offering 24-hour service. The switching matrix at the central

TABLE 7-5 Electrical Characteristics of Exchange-type Cables at 68°F[a]

Gauge (AWG)	Primary constants (at 1000 Hz)					Secondary constants (at 1000 Hz)									
	C (μF/mi)	R (Ω/mi)	L (mH/mi)	G (S/mi)		Attenuation		Phase shift		Characteristic impedance		Midsection image impedance		Cutoff frequency (Hz)	Speed of propagation (mi/s)
						dB/mi	Np/mi	rad/mi	deg/mi	$R(\Omega)$	$X(\Omega)$	$R(\Omega)$	$X(\Omega)$		
19	0.084	86	0.886	1.219	Nonloaded	1.27	0.146	0.156	8.9	296	−276				40,000
					H88–loaded	0.42	0.049	0.519	29.8			1013	−93	3440	12,000
22	0.082	173	0.870	1.190	Nonloaded	1.81	0.208	0.214	12.3	417	−403				29,000
					H88–loaded	0.79	0.091	0.519	29.7			1035	−180	3480	12,000
24	0.084	274	0.950	1.219	Nonloaded	2.31	0.266	0.272	15.6	516	−503				23,000
					H88–loaded	1.21	0.140	0.536	30.7			1045	−272	3440	12,000
26	0.079	440	0.995	1.146	Nonloaded	2.85	0.329	0.332	19.0	671	−660				19,000
					H88–loaded	1.79	0.206	0.542	31.0			1121	−425	3540	12,000

[a]Np = neper = $0.5 \ln A_V$ (Np)
dB = decibel = $10 \log A_V$ (dB)
dB = 8.69 Np or Np = 0.115 dB
Source: *Standard Handbook For Electrical Engineers*, 10th ed., McGraw-Hill, New York, 1969, p. 25–50.

office is bypassed; hence no digits need to be dialed. The private line can also be a four-wire circuit offering full-duplex data transmission on dedicated pairs. There are several advantages that the private line has over the two-wire switched lines:

— Line characteristics are consistent since the same signal path is used at all times.
— The line is less prone to impulse noise generated from central office switching circuitry.
— Line conditioning is available to improve on signal attenuation and delay distortion.
— There are higher data transfer rates with lower error rates.
— A private line is less expensive than a switched line if utilization is high.
— Unlike the unbalanced switched line, the private line is a balanced circuit, making it more suitable for line conditioning.

7.8 LINE CONDITIONING

Although the standard public switched lines are used primarily for voice, they may also be used for the transmission of data, provided that special modulation techniques are used. These switched lines are classified by telephone companies as the basic *3002 voice-grade lines*. The Federal Communications Commission (FCC) has set tariffs governing the amount of distortion allowed on voice-grade lines. To compensate for this distortion, the PSTN offers private line services for an additional cost. These lines may be specially treated or *conditioned* to improve on the quality of transmitted data. Higher transmission rates with a reduced number of errors are achieved through line conditioning.

7.8.1 3002 Unconditioned Voice-Grade Lines

The standard public switched telephone lines are classified as 3002 unconditioned voice-grade lines. As described earlier, the bandwidth is limited to approximately 3 kHz (300 to 3400 Hz). For data transmission, these lines are normally used for speeds ranging from 300 to 9600 bps. Speeds beyond 9600 bps are now possible over unconditioned voice-grade lines. However, they are not guaranteed by the phone company.

Voice and data transmission can be impaired by the switched lines because the signal path is not fixed through central and toll offices. Signal paths estab-

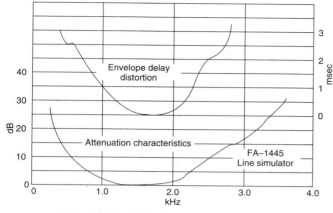

Because different frequencies encounter different amplitude-attenuation and propagation-delay times through the telephone network, not all of the bandwidth can be utilized for transmission of digital data. These differences are largely immaterial in voice communication but can be detrimental to data transmission, particularly at speeds faster than 2400 bps.

ENVELOPE DELAY DISTORTION AND ATTENUATION

Figure 7-21   Envelope delay and attenuation distortion characteristics for the 3002 unconditioned voice-grade line. (Reprinted with permission from Racal-Vadic.)

lished during one call are likely to be different at another time between the same two parties; hence line characteristics differ for each connection. For this reason, switched lines cannot be conditioned by telephone companies beforehand.

Two important electrical parameters that must be considered, particularly in the transmission of data, are *envelope delay distortion* and *attenuation distortion.* The propagation time for a signal to travel across a transmission medium varies with frequency. Some frequencies of the transmitted signal arrive ahead of others. This phase distortion is not readily noticeable for voice. However, for data it may well render the voice channel completely unusable. The same is true for amplitude variations across the passband of the channel. Attenuation distortion and envelope delay distortion are particularly severe at the breakpoints of the channel's passband. Figure 7-21 illustrates a typical response curve for a 3002 unconditioned voice-grade line. Envelope delay and attenuation distortion are defined below.

Envelope Delay Distortion.   Envelope delay distortion is the phase variation that occurs as a function of frequency over the passband of a given transmission medium. This specification is determined by measuring the *propagation delay time* or *phase delay* of the *envelope* of an AM wave measured throughout the medium's passband. Ideally, a linear or flat response is desired. In this case the time that it would take for the signal to propagate from source to destination would be the same for all frequencies. Envelope delay distortion is measured in microseconds.

Attenuation Distortion.   Attenuation distortion is the frequency response of the transmission medium. Amplitude variations throughout the medium's passband must not exceed specified limits. These limits are relative to the gain measured at the approximate center of the medium's passband. Ideally, a linear or flat response is desirable so that all frequencies throughout the passband encounter the same gain. Attenuation distortion is measured in decibels (dB).

7.8.2 Conditioning the Private Line

The permanent connection of the private line allows telephone companies to compensate for some of the distortion characteristics mentioned above. Phase and amplitude *equalizer* circuits are placed in individual leased lines. These circuits contain inductors and capacitors that are adjusted to flatten or *equalize* the line characteristic for envelope delay and attenuation across the band. The process of equalizing the line is referred to as *line conditioning* (Figure 7-22). Signal attenuation and propagation delay are made relatively constant for all frequencies within the passband.

7.8.2.1 C- and D-type line conditioning. Two types of line conditioning are offered by telephone companies: *C-type* and *D-type* line conditioning. A customer can request one or both types at an extra monthly fee. C-type line conditioning sets the maximum limits on the amount of attenuation distortion measured on the line. FCC Tariff No. 260 describes these limits for private line services in North America. These limits are referenced to the channel gain at a standardized frequency of 1004 Hz. C-type conditioning is available in five types: C1 through C5. In general, C-type conditioning is used for frequency-shift keying (FSK) and phase-shift keying (PSK) modulation. Specifications are shown in Table 7-6.

D-type conditioning is a more recent type of conditioning introduced by AT&T for 9600 bps over voice-grade lines. Limits are set on the amount of *signal-to-C notched noise ratio* and harmonic distortion measured on the line. C-notched noise is a standard measurement used to determine the channel background noise power level. D-type conditioned lines meet the following specifications:

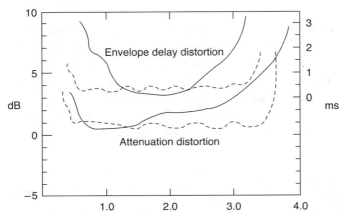

Figure 7-22 Effect of line conditioning through the use of equalizer circuits. The solid lines are before equalization and the dashed lines are after equalization. (Reprinted with permission from Racal-Vadic.)

Signal-to-C notched noise ratio = 28 dB
Signal-to-second harmonic ratio = 35 dB
Signal-to-third harmonic ratio = 40 dB

7.8.3 Equalization

Compensation for amplitude and phase delay distortion encountered on telephone lines can now be performed within the modem through the use of phase and amplitude equalizer circuits. A *compromise equalizer* provides *pre-equalization* of the signal. It is implemented in the transmitter section of the modem. An *adaptive equalizer* provides *post-equalization* of the signal and is implemented in the receiver section of the modem. These circuits are standard features in modem designs and have become internationally recognized by CCITT modem recommendations.

Pre-equalization is best suited when the characteristics of the line are fixed and known beforehand. Since the private line offers this advantage, the compromise equalizer can be manually adjusted to compensate for the given line condition. In switched voice-grade lines, however, line characteristics are unpredictable since the signal path is subject to change with each call. Post-equalization with adaptive equalizers is more effective in this situation.

TABLE 7-6 C1– Through C5–Type Line Conditioning for Private Leased Lines

Conditioning type	Envelope delay distortion		Attenuation distortion	
	Frequency range (Hz)	Specification limits (μs)	Frequency range (Hz)	Specification limits (dB)
C1	1000–2400	1000	300–2700	−2 to +6
			1000–2400	−1 to +3
C2	500–2800	3000	300–3000	−2 to +6
	600–2600	1500	500–2800	−1 to +3
	1000–2600	500		
C3 (access lines)	500–2800	650	300–3000	−0.8 to +3
	600–2600	300	500–2800	−0.5 to +1.5
	1000–2600	110		
C3 (trunks)	500–2800	500	300–3000	−0.8 to +2
	600–1600	260	500–2800	−0.5 to +1
	1000–2600	260		
C4	500–3000	3000	300–3200	−2 to +6
	600–3000	1500	500–3000	−2 to +3
	800–2800	500		
	1000–2600	300		
C5	500–2800	600	300–3000	−1 to +3
	600–2600	300	500–2800	−0.5 to +1.5
	1000–2600	100		

Adaptive equalizers automatically adjust their gain and phase characteristics in response to the received signal. Feedback techniques are used in the modem's receiver to track the phase and amplitude variations in the line, thus permitting equalization to be adjusted continuously during the course of reception.

When modems are installed in a private line multidrop configuration, each station's link exhibits a different line characteristic, due to the separate location of each modem. Compromise equalizers are manually adjusted to best suit the multidrop configuration. Adaptive equalizers compensate for the different characteristics of each line. Here, adaptive equalization is most effective.

7.9 MOBILE TELEPHONY

Mobile telephony has been around for many decades, providing the luxury and convenience of placing calls through the telephone company directly from one's automobile. The conventional mobile setup illustrated in Figure 7-23 includes a single base station connected to the central office and the mobile unit in which the mobile telephone set is installed. The base station is capable of transmitting and receiving on several UHF channels in succession. A high-power transmitter delivers 200 to 250 W to the base station's antenna, typically elevated on a tower or building. The mobile unit can travel within a 30-mile radius of the base station and reliably communicate with a transmission power output of up to 25 W. Unfortunately, such extreme amounts of power have caused interference between adjacent channels when mobile units are in proximity to each other or the base.

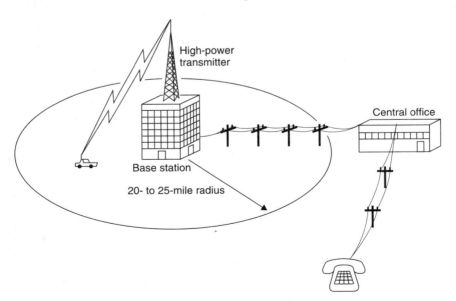

Figure 7-23 Conventional mobile telephone setup.

In addition, only one conversation can be held at a time on the limited number of frequency channels available within a given service area. Under these circumstances, the use of the mobile telephone has been limited to a chosen few.

7.9.1 Cellular Technology

Congested frequency bands coupled with increased demands for mobile communications have given rise to a new technology called AMPS (advanced mobile phone service). AMPS was developed by AMPS, Inc., a subsidiary of AT&T. AMPS has also become known as *cellular telephone.* As of 1983, the Federal Communications Commission (FCC) has been granting approval to implement cellular telephone service throughout the United States. The basic concept behind this new technology is to divide heavily populated areas into many small regions called *cells.* As depicted in Figure 7-24 each cell is linked to a central location called the *mobile telephone switching office* (MTSO). The MTSO

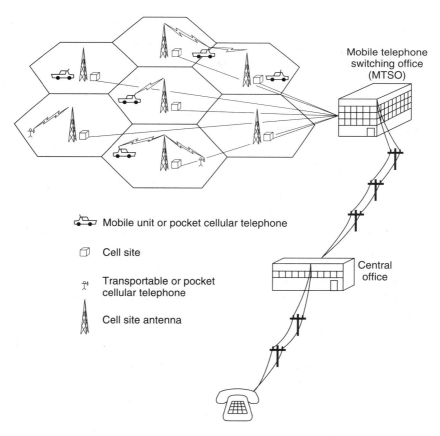

Figure 7-24 Cellular network.

coordinates all mobile calls between an area comprised of several cell sites and the central office. Time and billing information for each mobile unit is accounted for by the MTSO.

At the cell site, a base station is equipped to transmit, receive, and switch calls to and from any mobile unit within the cell to the MTSO. The cell itself encompasses only a few square miles, thus reducing the power requirements necessary to communicate with cellular telephones. This permits the same frequencies to be used by other cells, since the power levels emitted diminish to a level that does not interfere with other cells. In this manner, heavily populated areas can be serviced by several transmission stations, rather than one, as used by conventional mobile techniques.

7.9.2 Theory of Operation

For cellular communications, the FCC has apportioned 40 MHz of the frequency spectrum ranging from 825 to 845 MHz and 870 to 890 MHz. Full-duplex operation is possible by separating transmit and receive signals into separate frequency bands. Cellular phone units transmit in the lower band of frequencies; 825 to 845 MHz, and receive in the higher band; 870 to 890 MHz. The opposite frequency bands are used by the base units at the cell sites. Within these two bands, 666 separate channels (333 channels per band) have been assigned for voice and control. Each channel occupies a bandwidth of 30 kHz.

When a cellular phone is turned on, its microprocessor samples dedicated *setup* channels and tunes to the channel with the strongest signal. A closed loop is effectively established between the mobile unit and cell site at all times.

To place a call from the cellular phone to the MTSO, a local seven-digit number or a 10-digit long-distance telephone number is entered via the keypad. A quick glance at the numerical display confirms that the numbers have been entered correctly. The send button is now depressed, causing a burst of data to be transmitted onto a setup channel. These data include the cellular phone's identification number and the subscriber number being called. From the cell site, the data are forwarded to the MTSO along with the cell site's identification number. An unused voice channel is established by the MTSO's controller. This information, along with the cellular phone's identification number, is sent back through the cell site to the mobile unit for processing. The microprocessor within the cellular phone adjusts its frequency synthesizer for transmitting and receiving on the designated voice channel. Once the MTSO detects that the cellular phone's carrier frequency is on the designated channel, the call is placed to the central office for processing. Ringing can now be heard at the cellular phone's receiver.

If the cellular phone's signal strength significantly diminishes as a result of traveling outside one cell and entering another, the MTSO polls through all of its cell sites to determine the new cell that it is located in. This will be the cell in which the maximum signal strength is received. A new voice channel is automatically assigned to the cellular phone by the MTSO. This is called a *handoff*. Handoffs are transparent to the user. Conversation can go uninterrupted as the mobile unit travels from cell to cell.

In addition to controlling handoffs, the MTSO also regulates the amount of power transmitted by the cellular phone. Cellular phones transmitting in close proximity to the cell site are commanded by the MTSO to reduce power to prevent interference with other channels.[†] Power is reduced or increased (as need be) in steps of 4 dB with a maximum permitted power of 7 watts, or 8.45 dBW.

When a call is placed from a land line to a cellular phone, a connection is made between the central office and the MTSO. After determining if the subscriber number is valid, the MTSO begins a regional search through each cell to establish the location of the cellular phone. This process is referred to as *paging*. The MTSO pages a cellular phone by sending its identification number to every cell site in a given service area over every setup channel, one of which the cellular phone is constantly listening to. If there is no response from the system-wide page, a recorded message is sent back to the caller indicating that the cellular phone is not available. If the mobile unit responds to the page, an idle trunk between the serving cell site and MTSO is seized. An unoccupied voice channel is set up by the cell site, as instructed by the MTSO. The cellular phone tunes its transmitter and receiver accordingly and communication begins. Signal strength is monitored by the MTSO and handoffs are made as required.

Normally, the cellular phone is used only within the metropolitan area in which the cellular phone is registered. This may include several cities or counties. Frequently, there is a need to operate a cellular phone outside the home area. This is called *roaming*. Roaming is possible anywhere throughout the country, provided that cellular services are available and a prearranged agreement has been made between telephone companies and their users. A roam LED indicator on the cellular phone will light when the cellular phone travels outside the home area. With new cell coverage being implemented everyday, a cellular phone can be used in virtually every major city throughout the United States and Canada. Calls can be placed to anywhere in the world.

Most of today's cellular phones offer dozens of microprocessor-controlled features, such as alphanumeric directories with scrolling displays, LCD and LED displays, programmable security codes, horn alert, scratch pad dialing, recall dialing, hands-free operation in automobiles, 32-digit dialing, paging, and even vibrators for pocket phones. Some models even include voice-recognition circuits capable of automatically dialing a number through voice-activated commands. Many of the mobile telephone units can be used as a *transportable unit*. Transportable cellular phones offer the convenience of operation outside the au-

[†]Stan Prentiss; *Introducing Cellular Communications: The New Mobile Telephone System* (Blue Ridge Summit, Pa.: TAB Books, 1984).

Figure 7-25 Motorola's MicroTAC Ultra Lite™ cellular pocket flip phone and features. (Courtesy of Motorola.)

Figure 7-26 OKI Telecom's model 492 transportable unit. (Courtesy of OKI Telecom.)

MICRO T•A•C *Lite*

Call Placement Features

- **Memory Linking/Pause in Memory**—Enables easy placement of calls to standard phone numbers that require additional number sequences such as a credit card or pager number.
- **Auto Answer**—After two rings, the telephone will automatically answer. Just say "Hello".
- **Super Speed Dialing**—Provides faster memory dialing with a minimum of keystrokes.
- **Incoming Call Screening**—Gives user a visual and audible indication of incoming calls.
- **101 Memory Locations**—Allows you to store 99 32-digit phone numbers plus last number redial and scratchpad.
- **Memory Autoload**—Automatically stores a new number into the next available memory location.
- **Automatic Redial**—Automatically attempts to complete a system busy call for 4 minutes after first attempt.
- **Memory Overwrite Protection**—Warns user that memory location is already filled.
- **832/2412 Channel Capability**—Full spectrum cellular service.*

Usage Control Features

- **Automatic Lock**—Helps control unwanted usage by automatically locking phone each time it is turned off.
- **5 Call Timers**—Includes Resettable, Individual, Audible, Auto Display and Cumulative for monitoring phone expenses by tracking airtime usage.
- **6 Levels of Call Restrictions**—Enhances control of phone expenses and security of the numbers in memory.
- **6 Levels of Selectable System Registration**—Allows the user the choice of operating systems.

* 2412 channel capability will be available in some cities in late 1991. Ask your service provider to determine if your city will offer this service.

Accessories

In-Vehicle

Digital Hands Free Adapter— Provides full duplex hands free operation, battery rapid charging in the vehicle and a glass mount antenna for expanded in vehicle reception.

Travel Battery Saver—Conserves the MicroTAC Lite™'s battery by powering the phone through the vehicle's cigarette lighter.

Vehicular Charger—Charges any Digital Personal Communicator battery in just one hour while in the vehicle. Don't just drive, when you can also charge your battery.

Battery Saver/Charger—Combines the features of both the Travel Battery Saver and the Vehicular Charger into one accessory.

Extended System—Boosts power to 3 watts, rapidly charges battery, provides full duplex hands-free operation, glass mount antenna and a mobile handset.

The Portable Cellular Connection™ (See your cellular sales representative for more details)

Extra Batteries

Talk PAK XT Battery—High capacity battery provides up to 150 minutes of continuous talk time. Continuous standby time is 24 hours.

Introducing two new batteries!

XT Battery—The XT Battery offers you up to 70 minutes of continuous talk time. Continuous standby time is 12 hours.

Standard Battery—The smallest, slim-line battery provides up to 45 minutes of continuous talk time. Continuous stand-by time is 8 hours. Weighing less than 4 ounces, it is convenient to carry as a spare.

tomobile with up to three watts output power. A battery pack permits several hours of operation in the field. Figure 7-25 illustrates Motorola's *MicroTAC Ultra Lite*™ cellular pocket phone and its features. Introduced in 1992, the MicroTAC Ultra Lite™ was the first pocket phone to break the six-ounce weight barrier. Several new pocket phones, even smaller and lighter in weight, are now available. Figure 7-26 depicts OKI Telecom's Model 492 Transportable Unit.

PROBLEMS

1. What part of the telephone circuit interfaces the two-wire subsciber loop to the four-wire transmit and receive circuit?
2. Explain the mechanism used in the telephone set to convert acoustical signals to electrical signals.
3. Explain how electrical signals are converted by the telephone set to acoustical signals.
4. What is the ring potential and the frequency sent by the central office to ring the telephone set?
5. Why is the telephone ring voltage superimposed on -48 V dc?
6. Explain *sidetone*.
7. What is the difference between a ringing signal and a ring-back signal?
8. What frequencies are associated with a busy tone and what is its on–off rate?
9. Refer to Figure 7-6b.
   (a) What is the purpose of D1?
   (b) What is the purpose of D2?
   (c) When the telephone is said to be on-hook, what are the conditions of switches S1 and S2?
   (d) Which components are used to regulate voice amplitude?
10. For a rotary or electronic-pulse dial-type telephone, how many pulses are generated when the digit 0 is dialed?
11. What is the pulse width, period, and repetition rate for a dial pulse?
12. Assume that the off-hook dc potential is measured to be -5 V dc. Draw the waveform that would appear on an oscilloscope if the digit 8 were dialed.
13. What is *interdigit time*?
14. What are the DTMF frequencies for the digit 7?
15. For a cordless telephone:
   (a) What is the frequency range used for transmitting from the base unit to the portable unit?
   (b) What is the frequency range used for transmitting from the portable unit to the base unit?
16. What types of antennas are used for earlier models of base units and portable units of a cordless telephone? Explain whether they are used for transmit only, receive only, or both.

17. How many pairs of carrier frequencies are available for use in today's cordless telephones?

18. Refer to Figure 7-13. How many lines would be necessary to interconnect 50 telephone parties using this method?

19. What is the tip-to-ring potential for a telephone in its on-hook position?

20. What is the nominal tip-to-ring dc subscriber loop current?

21. Describe the *busy* signal in terms of frequency and on–off times.

22. Describe the *receiver off-hook* tone.

23. What is the bandwidth of the switched telephone lines?

24. What is the nominal voice frequency?

25. Refer to Table 7-5. Given a cable pair labeled 19H88:
    (a) What is the attenuation in dB/mi?
    (b) Compute the ac resistance for a 5-mile loop length.
    (c) What is the cutoff frequency?

26. Assume that a subscriber loop has a total dc resistance of 950 Ω. This includes a telephone resistance of 130 Ω. Compute the following (assume 68°F).
    (a) Subscriber loop current
    (b) Off-hook dc potential measured at the telephone
    (c) Distance to the central office for 19-AWG copper wire having a dc resistance of 8.33 Ω per 1000 ft at 68°F

27. Repeat Problem 26 for a subscriber loop having a total dc resistance of 800 Ω. The telephone resistance is given as 142 Ω.

28. Explain the difference between envelope delay distortion and attenuation distortion.

29. Refer to Table 7-6. A private line uses C4-type conditioning. For a 600-Hz to 3-kHz test tone:
    (a) What is the maximum envelope delay distortion that could occur?
    (b) What is the maximum attenuation distortion that could occur?

30. Explain the difference between a compromise equalizer and an adaptive equalizer.

31. What is the primary advantage of cellular technology over conventional mobile telephone?

32. What does MTSO stand for and what function does it serve?

33. What is a *handoff*?

34. What is *roaming*?

8

The Telephone Network

The invention of the telephone in 1876 led to an explosive outgrowth of engineering developments. These developments continue to thrive to this date. This proliferation, through what has become known today as "the largest industry in the world," has led to the existing informational era in which we live. Despite the Bell System divestiture (breakup) on January 1, 1983, the telecommunications industry still exists, employing millions of people. Divestiture has clearly established a distinction between the principal entities or segments of the telecommunications industry. Over 1400 independent telephone companies exist today, making up what has become known as the *public switched telephone network* (PSTN).

The PSTN has been dubbed the "world's most complex machine." It is truly one of the modern wonders of the world. In the United States alone, the PSTN interconnects hundreds of millions of telephones through the largest network of computers in the world. Despite its electrical and mechanical sophistication, even the most unskilled person can utilize its services. At the touch of a dial, any two people, virtually anywhere in the world, can communicate with each other in a matter of seconds. In this chapter we examine the basic structure of the PSTN and how it has evolved over the years. The most recent technological advances are introduced.

8.1 THE PUBLIC SWITCHED TELEPHONE NETWORK (PSTN)

Since the invention of the telephone, the PSTN has grown proportionately with the increased demands to communicate. Switching services beyond metropolitan areas were soon developed, increasing the size and complexity of the central of-

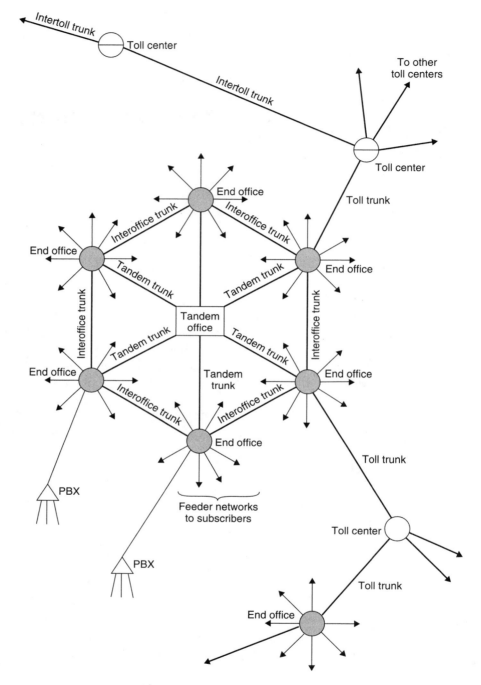

Figure 8-1 Interconnection of switching exchanges in North America.

fice. New methods of switching were required to interconnect central offices through the use of *interoffice trunks* and *tandem trunks.* Figure 8-1 depicts how today's central offices are connected through the use of trunks. In largely populated areas, a *tandem office* is used to minimize the number of trunks that a call must be routed through to reach its destination. Outside the local area, *toll trunks* are used to connect the central office to *toll centers.* Toll centers may be located in adjacent cities outside the local area. To achieve even longer distances, toll centers are interconnected by *intertoll trunks* as shown.

8.1.1 Switching Hierarchy of North America

A hierarchy of switching exchanges evolved in North America to accommodate the demand for longer-distance connections. The PSTN has classified these exchanges into five *levels* of switching, as depicted in Table 8-1. At the lowest level, class 5, is the central office or *end office.* A large metropolis may require several end offices for service, whereas in a rural area a single end office is usually sufficient. When calls are made outside the local area, they are routed through class 4 centers, *toll centers,* and possibly higher levels of switching, depending on the destination of the call and the current traffic volume within the PSTN. To aid in the volume of traffic between toll centers, *primary centers, sectional centers,* and *regional centers*, classes 3, 2, and 1, respectively, are used. Figure 8-2 illustrates the possible routes that a toll call can take throughout the switching hierarchy. The best route is the shortest route or the route utilizing the smallest number of switching centers. A call may not always take this route, however. It depends on the availability of trunk circuits when the call is placed. Several alternative routes can be taken up and down the switching hierarchy. The selection of a route is under program control at the switching center.

Typically, trunks used to interconnect higher levels of switching are designed for high-speed transmission using the multiplexing techniques that will be

TABLE 8-1 Hierarchy of Switching Exchanges in North America

Class 1	Regional center	□
Class 2	Sectional center	△
Class 3	Primary center	○
Class 4	Toll center	⊖
Class 5	End office	◉

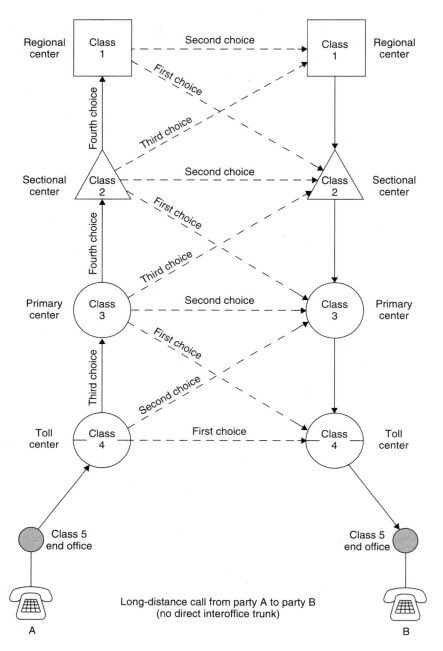

Figure 8-2 Long-distance switching hierarchy within the PSTN depicting possible routes to complete a call between parties A and B.

The Telephone Network   Chap. 8

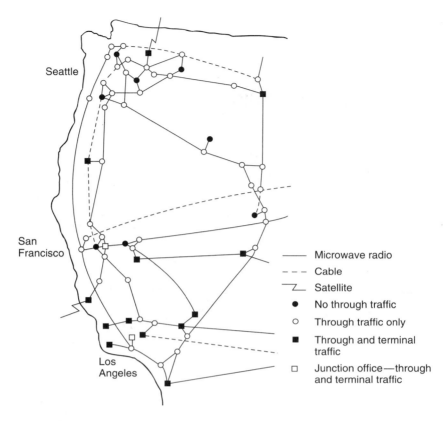

Seattle

San
Francisco

Los
Angeles

——— Microwave radio

- - - Cable

⌐ Satellite

● No through traffic

○ Through traffic only

■ Through and terminal
traffic

□ Junction office—through
and terminal traffic

Figure 8-3 Long-haul network depicting various trunk media. (Courtesy of
Bell Laboratories.)

described in Section 8.2. These higher levels of switching make up the *long-haul
network*. Calls placed onto the long-haul network are subject to toll fees. Figure
8-3 illustrates the long-haul network. To sustain higher transmission rates within
the long-haul network, wideband trunk media are used. This includes coax, microwave ground and satellite links, and fiber-optic cables. Figure 8-4 illustrates
the distribution of switching centers throughout the United States.

8.1.2 Two-Wire Versus Four-Wire Circuit

In two-wire circuits, whether the line is switched or private, the transmitted and
received signals share the same two lines. When two parties talk at the same
time, their signals are superimposed upon each other. Although each party hears
one another, they probably do not understand what is being said. Intelligent
voice transmissions on two-wire circuits from this standpoint can be regarded as

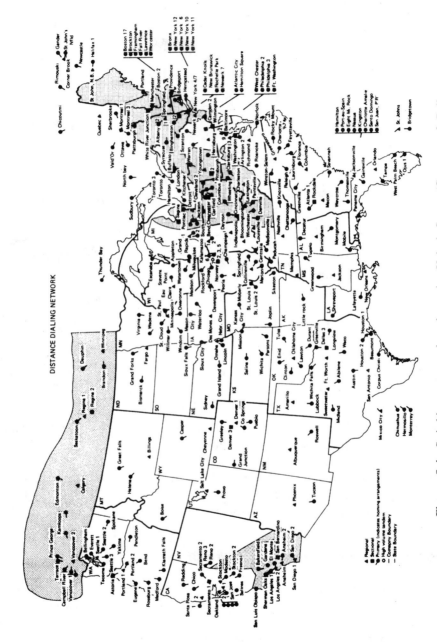

Figure 8-4 Switching centers throughout the United States. (Courtesy of Bell Laboratories.)

half-duplex. For data communications, this presents a serious problem, unless a modulation technique is used to separate full-duplex signals.

Another inherent problem with the two-wire circuit is providing consistent transfer characteristics for bidirectional signal flow, since energy from a transmitting source falls off (as in any transmission medium) with distance. Telephone signals transmitted beyond more than a few miles must be amplified and restored to their original condition. A bidirectional amplifier is not a practical solution. It is much more desirable to physically separate transmitted and received signals by employing a four-wire line so that directional amplifiers may be used in each signal path.

The PSTN uses a variety of transmission mediums to interconnect switching centers. These interconnecting links are referred to as *trunks*. Trunks normally carry multiple telephone signals, often several thousand simultaneously. Class 5 switching centers are typically linked with *four-wire circuits* for trunks. As shown in Figure 8-5, four-wire circuits allow the transmitted and received signals to propagate on physically separate pairs of wires. This allows the use of *repeaters* in each direction. Repeaters amplify and condition the signals for longer transmission distances. The multiplexing techniques discussed in Section 8.4 are best performed on transmitted and received signals that are physically separate from each other.

Four-wire circuits may not always mean four physical wires. Two physical wires can also serve as a four-wire circuit by partitioning the transmitted and received signals into separate frequency bands. Full-duplex operation is made possible by transmitting on one band of frequencies and simultaneously receiving on the other. This is referred to as a *derived four-wire circuit*. In most cases, however, four physical wire circuits are used for interoffice trunks.

8.1.3 Hybrids

The two-wire subscriber loop connection at the central office must be converted to a four-wire circuit for interfacing to a trunk circuit. This conversion is performed by the *hybrid circuit*. The hybrid circuit, as illustrated in Figure 8-6, con-

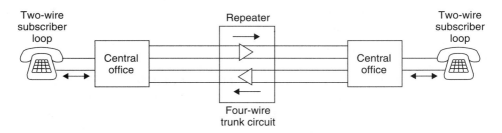

Figure 8-5 Four-wire trunk circuit with separate amplifiers for each direction.

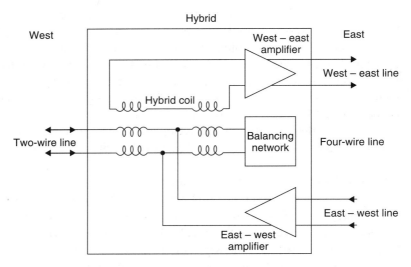

Figure 8-6 Hybrid two- and four-wire terminating circuit.

sists of two separate amplifier circuits. The top amplifier circuit amplifies signals traveling in the west–east direction. The bottom amplifier amplifies signals traveling in the opposite direction from east to west. The hybrid coil is a balanced cross-coupled transformer assembly with equal windings in each coil. Signals in the east–west direction are canceled in the west–east direction, and vice versa. A *balancing network* is used to properly match the subscriber loop impedance to that of the hybrid. When precisely matched, maximum power is transferred in the appropriate direction, and no portion of the transmitted signal is returned to its source. The splitting of signal power causes a 3-dB (half-power) signal loss. Amplifiers within the hybrid make up for this loss.

8.1.4 Echo Suppressors and Echo Cancellers

Long-distance transmission of signals over the PSTN often suffers from impedance mismatches. This occurs primarily at the hybrid interface. Balancing networks within the hybrid can never perfectly match the hybrid to the subscriber loop due to temperature variations, degradation of transmission lines, and other variables. As a result, a signal transmitted in the east–west direction does not completely cancel itself in the secondary of the hybrid coil. A small portion of the signal is returned in the west–east circuit. This returned signal is known as an *echo*. In general, for distances of less than 1000 miles between two parties, echoes are unnoticeable. In fact, if strong enough, they can serve the same purpose as a telephone's sidetone, reinforcing the caller's own voice. When the round-trip delay time between two parties exceeds about 45 ms, an echo of one's voice can be heard. For transcontinental calls and international calls, the round-trip delay time can be several hundreds of milliseconds. Telephone calls via satellite links can take up to as much as a half-second. The resulting echoes are extremely annoying, making it difficult to converse.

To circumvent the problem of echoing in long-distance communications, *echo suppressors* are used. Figure 8-7 illustrates the echo suppressor. In most cases, two people conversing do not talk at the same time. The operation of the echo suppressor relies on this principle. When a caller from the west begins to speak, the speech detector for the west–east direction is activated, which in turn causes the logic control circuit to disable the amplifier for the east–west direction. The returned echo is effectively *suppressed* by approximately 60 dB. Likewise, when a caller from the east begins to speak, the speech detector for the east–west direction becomes activated, thus causing the logic control circuit to disable the amplifier in the return west–east channel. Echo suppressors are used in cases where the round-trip delay time of the signal exceeds 45 ms. Geosynchronous satellites are positioned 22,300 miles above earth. The round-trip delay time for satellite communications in this case is nearly a quarter of a second (22,300 miles $\times$ 2/186,000 miles/s). Echo suppressors are required in this case.

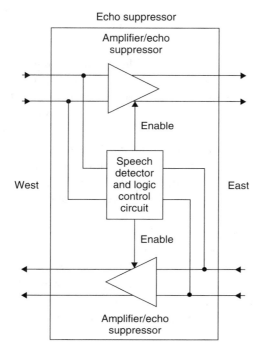

Figure 8-7 Echo suppressor.

What happens when two long-distance parties talk at the same time when echo suppressors are employed? You may want to attempt this experiment the next time you are talking long distance beyond a few thousand miles. Echo suppressors in *both* directions are enabled and you will not be able to hear each other. Only one person can talk at a time; thus long-distance communication is half-duplex.

To achieve full-duplex operation with voice or data, more modern devices called *echo cancellers* are used. Echo cancellers eliminate the return echo by electrically subtracting it from the original signal rather than disabling the return amplifier circuit. This allows both speakers to talk and listen simultaneously.

Echo suppressors present a problem when data are transmitted over the PSTN in both directions simultaneously (full-duplex mode). The data are lost as the echo suppressors are enabled. Telephone companies have equipped echo suppressors with a means of disabling themselves. Upon receiving a tone within the frequency range of 2010 to 2240 Hz for a duration greater than 400 ms, the echo suppressor disables itself. The disabling tone can be sent from either direction. No other signal should be present during this interval. Once disabled, both amplifiers within the echo suppressor circuit remain active as long as there is signal energy within the frequency range of 300 to 3000 Hz. An interruption of more than 50 ms will automatically enable the echo suppressor to its normal mode of operation.

8.1.5 Analog Companding

One of the characteristics of voice is the wide dynamic range over which its power varies. This range can be as high as 60 dB. A signal with such a large dynamic range is difficult to transmit over long distances because of the required use of several repeaters and amplifiers. Large signals tend to saturate amplifiers en route, and small signals eventually get lost in the amplified noise. The problem can be overcome through the use of a *compandor*.

A compandor is a circuit that performs two functions: *com*pressing a signal's amplitude range before it is transmitted and ex*panding* a signal's amplitude range back to its original condition when it is received. The overall process is referred to as *companding*. The compandor circuit includes logarithmic amplifiers (log amps) with nonlinear transfer characteristics.

A comparison between transmission systems with and without companding is shown in Figure 8-8. The diagram depicts power loss versus transmission distance. A 20-dB repeater is used to amplify the signal between the transmitting and receiving stations. No companding is used in the system shown in Figure 8-8a. The SNR is degraded from 80 to 55 dB for peak amplitudes. Low-level

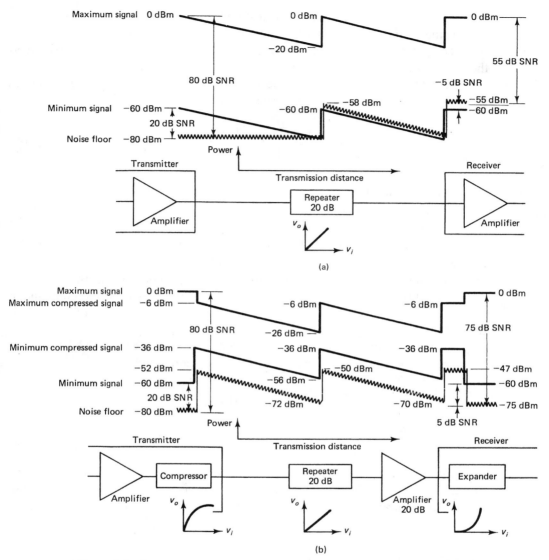

Figure 8-8 Comparison between a transmission system with and without companding. The diagram depicts signal strength versus transmission distance: (a) transmission system without companding; (b) transmission system with companding.

amplitudes, originally at 20 dB SNR, eventually fall below the amplified noise power to −5 dB SNR. Improved SNR and a reduction in the likelihood of saturating amplifiers are attained through the process of companding. This is shown in Figure 8-8b. Strong and weak signals are *compressed* from a 60- to a 30-dB dynamic range at the transmitter and restored to 60 dB at the receiving end. The SNR is maintained above unity throughout the transmission system.

The preferred method of transmission onto trunk circuits today is to convert the analog voice signal to digital form. Before the conversion process, voice signals are first compressed by compandors. In the United States, the μ-*law* is used for companding. The μ-law is governed by the equation

$$v_o = \frac{\ln(1 + \mu |v_i|)}{\ln(1 + \mu)} \qquad 0 \le |v_i| \le 1$$

where ln = natural log
v_o = output voltage
v_i = input voltage
μ = mu, compression factor, typically 100 or 255 used in U.S. PSTN

Figure 8-9a depicts a graph of the μ-law equation for various values of μ. Notice that the gain, v_o/v_i, is much larger for lower-amplitude signals than it is for higher-amplitude signals. This is the nonlinear characteristic of the compandor. As μ is increased, the degree of compression is increased. A value of $\mu = 0$ corresponds to linear amplification.

In Europe, voice signals are companded in accordance with CCITT's *A-law* for companding and is governed by the following equation:

$$v_o = \frac{A |v_i|}{1 + \ln A}, \qquad 0 \le v_i \le \frac{1}{A}$$

$$= \frac{1 + \ln(A |v_i|)}{1 + \ln A}, \qquad \frac{1}{A} \le v_i \le 1$$

where A is the compression factor (CCITT recommendation: $A = 87.6$). As shown in Figure 8-9b, there is very little difference between μ-law and A-law companding, except that at low amplitudes the gain is greater for μ-law companding. A compression factor, A, of 87.6 is recommended by CCITT.

8.2 TRANSMISSION MEDIA FOR TRUNKS

Several types of transmission media are used for trunk circuits, ranging from twisted-pair wire to coaxial and fiber-optic cable. Satellite communication is also used as a transmission medium for trunk circuits. In each case the purpose of the trunk is to interconnect switching centers to achieve longer-distance communications. Trunk circuits carry multiple voice channels, and in many cases, particularly with the latest fiber-optic technology, thousands of voice channels are multiplexed together and sent over a single cable. Various trunk media and their characteristics will now be considered.

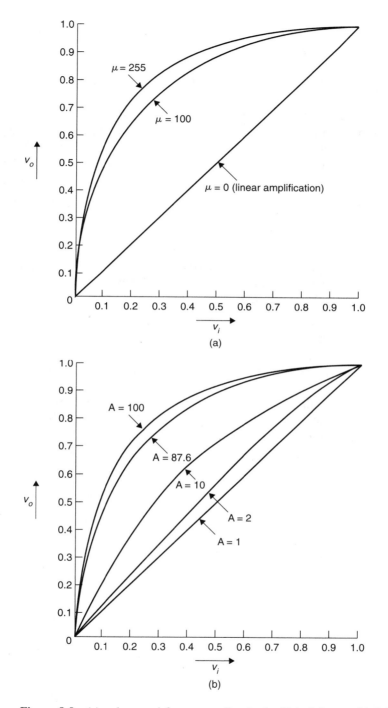

Figure 8-9 (a) μ-law used for companding in the United States; (b) A-law used for companding in Europe.

8.2.1 Open-Wire Pairs

Open-wire pairs were used in the early years of the telephone system. Although they are still used in some rural communities and older sections of towns, for the most part they have become part of our historical past.

A pair consists of two *open wires* suspended on telephone poles. The wires are separated approximately 1 ft to minimize capacitance. Glass insulators mounted on wooden crossbeams of the telephone pole suspend the wires between poles. Open wire is usually made of steel, coated with copper. Steel is used for the strength necessary to withstand varying weather conditions and to withstand the suspension weight of the wire between poles. Since most ac flows on the outer portion of a conductor due to *skin effect*, coated copper, the better of the two conductive metals, carries the electrical signals.

Distances of up to 50 km can be achieved with open-wire pairs before repeaters are necessary. Several disadvantages of open-wire pairs have caused it to become obsolete. It is unsightly and bulky, subject to crosstalk and severely affected by weather conditions.

8.2.2 Twisted-Pair Wires

Twisted-pair wires are insulated pairs of copper wire that have been bundled together and insulated to form a single cable. Individual pairs of wires are twisted together to minimize crosstalk. Cables can contain several hundreds of twisted pairs, ranging from 18- to 26-gauge copper wire. The cables are laid throughout cities in underground tunnels. Since the twisted pairs are bound so closely together, they suffer from crosstalk even more so than open-wire pairs. Due to the small diameter of the wires, resistance contributes significantly to signal loss. The use of repeaters every 3 to 4 miles is required. Twisted pairs are used for short-haul trunks with typically 12 or 24 multiplexed voice channels per twisted pair.

8.2.3 Coaxial Cable

For frequencies above 1 MHz, it becomes desirable to use coaxial cable. Radiation losses and adjacent channel interference are virtually eliminated by coaxial shielding. The greater available bandwidth offered by coax over wire pairs makes it much more suited for high-capacity trunk circuits used between cities and states and for national trunk routes. Coaxial cable also has a high propagation velocity rating with excellent phase delay response.

Coax is manufactured with a hollow cylindrical copper tube used for the shield. Insulating material (dielectric) separates the conductive tube from the copper center conductor. Several individual coaxial tubes are often bound together with insulating material and steel reinforcement to produce a high-capacity trunk cable. Each coaxial tube can carry several thousands of voice channels.

8.2.3.1 The Bell System's L5 coaxial carrier.

The Bell System's *L5 carrier* is a long-haul trunk that includes 22 coaxial tubes bound together to form a single cable. Figure 8-10a illustrates a cross-sectional view of the L5 carrier. A total of 108,000 simultaneous two-way voice conversations can be carried by the cable. Ten tubes carry 108,000 voice channels in one direction and 10 tubes are for the opposite direction. Two coaxial tubes are for backup spares. Repeaters for the L5 system are spaced at 1-mile intervals maintaining an overall system bandwidth of 58 MHz. The dc power required for the repeaters is supplied through the cable. Power supply stations spaced at 120 km (75 repeaters apart) feed the adjoining cable with a dc potential of 1350 V. To minimize the possibility of arcing across the insulation, +675 V is supplied by a station to one end of the cable and −675 V is supplied by the adjacent station. Figures 8-10b through 8-10d illustrate the cable laying of the L5 carrier system.

8.2.4 Microwave Links

An alternative to coaxial cables for high-capacity long-haul trunks are microwave radio links. Several thousand voice channels are modulated onto microwave carrier frequencies and transmitted to repeater stations spaced 20 to 30 miles apart. Repeater antennas are typically perched on towers, hilltops, and huge skyscrapers so that there are no obstructions in the *line-of-sight* transmission path between two repeater stations. Figure 8-11a illustrates the Bell System's distribution of broad-band microwave routes throughout the United States. A single microwave link interconnecting Cape Charles to Norfolk is shown in Figure 8-11b.

Microwave radio links are expensive. However, the costs over coaxial media are offset by several advantages:

— Fewer repeaters are necessary for amplifying signals.
— Distances between switching centers are shorter.
— Underground facilities are not necessary.
— Multiple channels can be transmitted over a single link.
— Delay times are minimal.
— There is minimal crosstalk.
— Fewer repeaters mean increased reliability and less maintenance.

(a)

(b)

(c)

(d)

Figure 8-10 (a) Cross-sectional view of the Bell System's L5 coaxial carrier system; 108,000 simultaneous two-way voice conversations can be carried by the L5 system; (b) to (d) laying of the L5 carrier system. (Courtesy of AT&T Archives.)

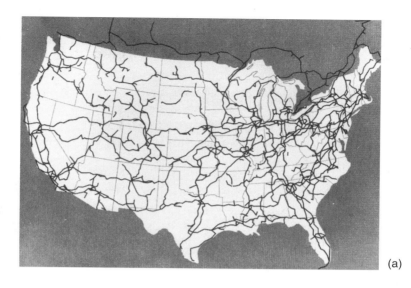

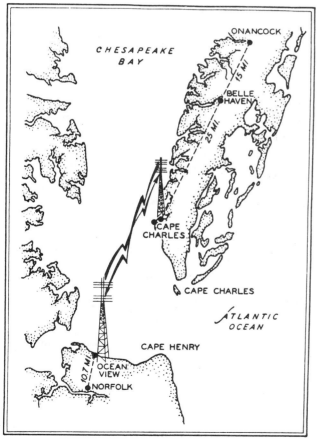

Figure 8-11 (a) Bell System's broadband microwave routes across the United States; (b) microwave route linking the Cape Charles to Norfolk radio telephone system. (Courtesy of AT&T Archives.)

Since both coaxial media and microwave links are well suited for high-capacity trunks, additional factors must be taken into account in determining which is best suited for a given application. Whereas coaxial cable is less prone to RF interference, microwave radio links rely on the absence of physical as well as electrical obstructions between transmitting and receiving stations. Microwave antenna beam widths can be as narrow as 1 degree. Trees, buildings, mountains, and even airplanes can interfere with signal transmission. Varying weather conditions such as rain and intense heat can alter the direction of the beam, thus causing the signal to fade.

Over the years, the telecommunications highways through the air have become heavily congested with electrical signals. Microwave beams in large metropolitan areas will often intersect with each other, causing interference. Frequency channels allocated by the Federal Communications Commission (FCC) are limited in the amount of skyway traffic that can be handled.

8.2.5 Submarine Cables

The transmission of voice signals overseas has become possible through the use of submarine cables. Submarine cables are coaxial cables specially designed to withstand the rugged oceanic floor conditions throughout the world. It took nearly a century of progress before the first voice-grade cable was laid across the Atlantic Ocean floor in 1956. The TAT-1 (Transatlantic) Cable System, developed by AT&T, spanned a distance of 2200 nautical miles. The construction of the cable includes several layers of insulation and armored steel reinforcement surrounding the conductor to protect it from corrosion, temperature changes, and leakage. Repeaters are constructed in a similar manner to prevent the damage of internal circuitry. Figure 8-12 shows an undersea repeater unit. Dc power for the repeater, as in coaxial trunk media used on land, must also be housed within the cable.

A considerable amount of engineering, testing, and oceanic research is performed before and during the laying of intercontinental submarine cables. Consider some of the major factors involved in this engineering feat:

— The cable must be protected from saltwater corrosion and leakage.

— Suboceanic terrain conditions and ocean depth must be considered.

— Temperature and pressure changes from sea level to ocean floor must be determined.

— The weight of cable material and rate of descent to the oceanic floor are critical parameters.

— Off-coast trenches must be dug in shallow waters to bury and protect the cable from fishing trawlers and anchors.

Figure 8-12   Construction of the submarine repeater unit.

— Electrical circuits must be environmentally tested at temperature extremes exceeding those of the ocean floor.

— Repeater units must be x-ray tested for faulty welds and leaks.

— Performance tests must be exercised constantly while laying cable to determine immediately the location of a fault.

The first-generation submarine cables laid in the mid-1950s carried 48 voice channels on two separate conductors that measured 0.62 in. in diameter. Vacuum tube amplifiers were used within the repeater units. Repeaters were separated by a distance of 39 nautical miles. The overall bandwidth for the 2200-nautical mile TAT-1 cable was 164 kHz. The latest generation of coaxial submarine cables span distances of up to 4000 miles, interconnecting every continent in the world. Solid-state amplifiers are used for repeater units. By decreasing the separation between repeater units and increasing conductor size, overall bandwidths of 28 MHz have been achieved. As many as 4000 voice channels are carried on center conductors measuring $1\frac{1}{2}$ inches in diameter.

In recent years, AT&T has installed the TAT-8 SL (Submarine Lightwave) cable shown in Figure 8-13a. The single fiber-optic cable doubles the number of existing transatlantic cable circuits that are available. Figure 8-13b depicts a comparison between the TAT-8 and the coaxial copper cable used for the TAT-7. The fiber cable shown is half the size and one-third the weight of the TAT-7 coaxial copper cable.

8.2.6 Satellite Communications[†]

In 1965, the first commercial satellite used for telecommunications was launched into space from Cape Kennedy. The Intelsat I (called Early Bird) was designed to handle an average of 240 voice channels. Since then, several generations of

[†]James Martin, *Telecommunications and the Computer,* 2nd ed. (Englewood Cliffs, N.J.: Prentice Hall, 1976), p. 280.

(a)

(b)

Figure 8-13 AT&T's TAT-8 SL (submarine lightwave) fiber-optic cable more than doubles the previously laid transatlantic cable circuits. (Courtesy of AT&T Archives.)

satellites have been deployed throughout space. A dramatic increase in the number of TV and voice channels has transcended with each generation, consequently lowering the cost of this transmission medium. Satellite communications has become a major facet of the telecommunications industry. Thousands of long-distance transoceanic telephone calls are now placed throughout the world via the satellite link.

The satellite is essentially a microwave relay station placed in orbital space. Telephone and television broadcast signals are beamed up to the satellite from an *earth station* through the use of a large, highly directive microwave dish antenna that is synchronized to the position of the satellite. A device called a *transponder* is used on board the satellite to receive the weak microwave signal, amplify and condition it, and retransmit the signal back to another earth station in a different location on earth.

To prevent the transponder's strong transmitted signal from interfering with the earth station's weak received signal, most commercial satellite links separate, transmit, and receive carrier frequencies by about 2 GHz. Earth stations typically transmit their signals to satellites on carrier frequencies in the 6-GHz band, ranging from 5.92 to 6.43 GHz. This is called the *up-link* frequency. The satellite's transponder down-converts these signals to a 4-GHz band, ranging from 3.7 to 4.2 GHz. These frequencies are referred to as the *down-link* frequencies. Earth station receivers are tuned to these frequencies.

Modern telecommunication satellites are positioned in orbit at an elevation of approximately 22,300 miles above the equator. This is referred to as *geosynchronous orbit.* In this orbit the satellite is made to travel at a velocity necessary to maintain a fixed position relative to a point above the equator (i.e., it rotates radially with the surface of the earth and therefore remains fixed in the sky 24 hours a day). At an altitude of 22,300 miles, 40% of the earth is exposed. The satellite's antenna is designed to emit a radiation pattern that covers this entire *exposed* portion. Satellites positioned in geosynchronous orbit, 120 degrees apart from each other, can cover the entire surface of the earth, with the exception of the polar caps. Hence the advantage to geosynchronous orbit is that it permits line-of-sight tracking by earth stations 24 hours a day. Earlier satellites had elliptical orbits and were useful for only short periods of the day. Figure 8-14 depicts the Telstar Models I and II communications satellites and a tracking microwave dish antenna.

Example 1[†]

A satellite is to be launched into a geosynchronous orbit over the equator (i.e., it should rotate at the same rate as the earth so that it remains over a fixed point on the earth). What should be its altitude?

Constants needed:

$g = 32.2$ ft/sec^2 = gravitational acceleration at sea level

$R = 3960$ miles = radius of the earth

M = mass of earth (data not needed)

G = universal gravitation constant (data not needed)

F = gravitational force of attraction between two bodies

[†]The geosynchronous altitude problem was derived and presented by Ron Fischer, Mathematics Department instructor and Center Coordinator for Evergreen Valley College, San Jose, Calif.

(a)

(b)

(c)

Figure 8-14 (a) One-half of Bell Laboratories' "telephone terminal to outer space" at Crawford Hill, Holmdel, N.J.; (b) complete Telstar Model I satellite on worktable; (c) the Telstar III is capable of relaying tremendous amounts of information. It is able to carry up to 94,000 simultaneous two-way telephone conversations, or 360 video teleconferences, or 24 color television programs, or billions of bits per second of high-speed data and facsimile signals. The Telstar III satellite will operate longer than any of its predecessors—ten years, instead of seven. (Courtesy of AT&T Archives.)

Solution: Newton's law of universal gravitation states that the gravitational force between the earth, of mass M, and a body, of mass m, is $F = GMm/r^2$, where r is the distance between the centers of mass of the masses. But when $r = R$, at sea level, $F = mg$. Therefore, $GMm/R^2 = mg$ and hence

$$GM = gR^2$$

(*Note:* This formula makes it unnecessary to look up G and M.) We therefore have

$$F = \frac{GMm}{r^2} = \frac{gR^2 m}{r^2} = gm\frac{R^2}{r^2}$$

Now for rotational motion, the acceleration is $\omega^2 r$, where ω equals angular velocity, so

$$F = ma = m\omega^2 r = gm\frac{R^2}{r^2}$$

Solving for r^3 gives

$$r^3 = g\frac{R^2}{\omega^2}$$

Now all quantities must be placed in consistent units. To get r in miles we need g in miles/hr^2, R in miles, and ω in rad/hour.

$$g = 32.2 \ \frac{\text{ft}}{\text{sec}^2} \cdot \frac{1 \text{ mile}}{5280 \text{ ft}} \cdot \frac{(3600)^2 \text{ sec}^2}{1 \text{ hr}^2} = 79{,}036.36 \text{ mi/hr}^2$$

$$\omega = \frac{2\pi \text{ rad}}{24 \text{ hr}} = (\pi/12) \text{ rad/hr}$$

Therefore,

$$r^3 = g\frac{R^2}{\omega^2} = (79{,}036.36)\frac{(3960)^2}{(\pi/12)^2} = 1.8083 \times 10^{13}$$

or $r = 26{,}247.83$ miles. The altitude is then

$$h = r - R = 26{,}247.83 - 3960 = 22{,}287.83 \text{ miles}$$

Note: In solving for r above using dimensional analysis, the rad^2 term in the denominator is dimensionless and therefore disregarded.

8.3 CENTRAL OFFICE SWITCHING SYSTEMS

Switching at the central office is necessary to establish a connection between two parties. For common battery systems, this is performed *automatically*. Common battery switching systems are classified into three basic categories:

1. Step by step (SXS)
2. Crossbar
3. Electronic switching system (ESS)

8.3.1 Step-by-Step Switching (SXS)

By the 1920s, reliance on switchboard operators within the PSTN eventually reached the point where the services provided did not meet the needs of the general public. Alleged eavesdropping and suspicion of business malpractice on the part of telephone companies led to the invention of the first automatic switch. Alman B. Strowger, an undertaker concerned that the telephone company may have been diverting his business calls to competitors, invented an automatic dial switch which became the basis for telephone switching for the next 70 years. The *Strowger switch*, depicted in Figure 8-15, is a step-by-step (SXS) system that performs switching in two dimensions: vertically and horizontally. The switching action is a direct result of the dial pulses generated by the rotary dial telephone. The switch shown here has ten rows and ten columns, making it possible for a caller to connect to 100 other subscribers with two dialed numbers.

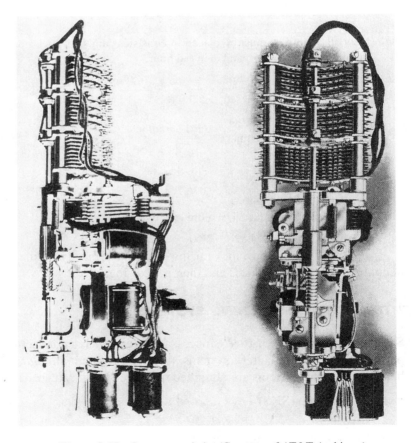

Figure 8-15　Strowger switch. (Courtesy of AT&T Archives.)

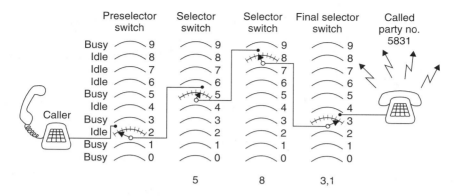

Figure 8-16 Simplified SXS call procedure to subscriber: 5831.

Strowger switch exchanges are still being used throughout the world today. Many are currently being replaced with more modern computerized switching exchanges.

Figure 8-16 illustrates a simplified version of a 10,000-line SXS central office switch. There are 10,000 subscribers that can be represented here by the numbers 0000 through 9999. When a caller goes off-hook, current is detected in the subscriber loop and the *preselector* switch becomes active. The preselector switch advances to a level that seizes an idle line and sends a dial tone to the caller. Let us assume that the party being dialed is **5831**. When the number **5** is dialed, the resulting electrical impulses cause the electromechanical relay of a *selector switch* to step in the vertical direction to a level equal to 5. The wiper then advances, step by step, in the horizontal direction until it seizes an idle line available on the next selector switch. The movement of the wiper contact is similar to the way in which a television's channel selector switch operates. An **8** is then dialed and the procedure is repeated with the next selector switch. The connection is further advanced to its final destination. The *final selector switch* is capable of handling the last two digits, **3** and **1.** When the **3** is dialed, the final selector switch advances vertically to the third level and horizontally to the first position. The called party at **5831** is tested for a busy condition and the ring potential is applied if the party line is idle. Once the party answers, the lines are further supervised until the conversation is terminated. Additional switches can be used to extend the procedure, thus increasing the sophistication of the network.

8.3.2 Crossbar Switch

It was not long after the Strowger switch came into use before a faster and more sophisticated system was developed. This system is called the *crossbar* switch. Crossbar switches are still serving several metropolitan and rural areas throughout the world. The crossbar switch, as its name implies, is a lattice of crossed bars that make and break contact. Figure 8-17 shows the crossbar switch.

Crossbar switches, like Strowger switches, are electromechanically activated and rely on moving parts. A detailed view of the crossbar switch is shown in Figure 8-18. The switch contains sets of contact points or *crosspoints* with three to six individual contacts per set. Magnets cause vertical and horizontal bars to cross each other and make contact at coordinates determined by the number being called. Each switch typically has either 100 or 200 crosspoints. The lattice structure of the crossbar switch has 10 *horizontal select bars* and either 10 or 20 *vertical hold bars.* The horizontal and vertical hold bars are activated by magnets. Any individual crosspoint within the matrix can make contact by activating one horizontal select bar and one vertical hold bar, similar to a rectangular coordinate system.

8.3.2.1 Blocking.

By combining crossbar switches, the number of possible signal paths can be increased, consequently lessening the likelihood of a signal path being *blocked.* As mentioned earlier, in most cases less than 10% of all telephones are in use at the same time. It is therefore not economical to provide a signal path between every subscriber. *Blocking* occurs during heavy traffic vol-

Figure 8-17 Crossbar switch.
(Courtesy of AT&T Archives.)

The Telephone Network   Chap. 8

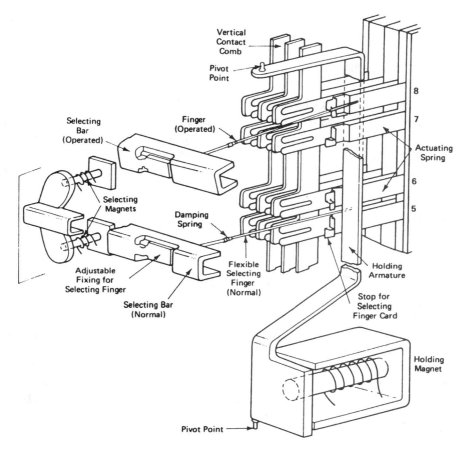

(a) Principle of Operation of Crossbar Switch

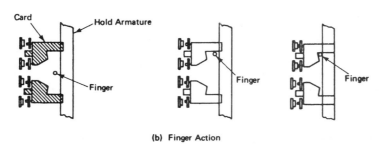

(b) Finger Action

Figure 8-18 Detailed views of the crossbar switch: (a) principle of operation; (b) finger action. (Courtesy of Northern Telecom.)

ume when the central office switches are fully utilized. A distinctive busy tone is sent to the caller from the central office if the connection cannot be made. The blocked call has nothing to do with the party being called and whether or not that line is busy or idle. The call will have to be postponed until a later time when traffic subsides. The busiest times of the day are from 8 A.M. to 10 A.M. and 3 P.M. to 5 P.M. Holidays such as Christmas and New Year's Day are also peak operating times for phone companies. These are times when one is likely to experience blocking. Blocking has also been known to occur during emergency situations involving entire communities, such as earthquakes and fires.

Figure 8-19 illustrates a simplified matrix with 25% blocking. Callers 4 and 7, 2 and 8, and 5 and 6 are on line with each other. When caller 3 attempts to make a call to 1, it is blocked. All available lines are used in the switching arrangement. This is referred to as 25% blocking; that is, one out of four calls is blocked.

8.3.3 Electronic Switching System (ESS)

The computer revolution has brought forth major developments within the central office. These developments have carried the industry forward toward much faster and more reliable switching systems capable of handling more calls and offering new customer services. Although this progress has been slow due to the enormous size and complexity of the PSTN, eventually all the existing electro-mechanical switching systems will be replaced with computerized switching sys-

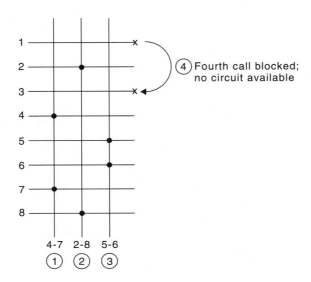

Figure 8-19 Crossbar switch with 25% blocking.

tems. These switching systems are referred to as *ESS,* an acronym for *electronic switching system.*

8.3.3.1 No. 1 ESS (electronic switching system). In 1965, the Bell System introduced the first computer-controlled switching system to be used in the PSTN. The *No. 1 ESS* digital computer uses *stored program control* (SPC) to perform switching, signaling, and administrative tasks. Customer service with the No. 1 ESS has proved to be far superior to the conventional techniques of the crossbar and SXS switching facilities. The No. 1 ESS can handle from 10,000 to as many as 70,000 subscribers. Several metropolitan areas currently use the No. 1 ESS. It has served as a steppingstone to more advanced ESS models developed over the years. The control panel for the No. 1 ESS system is shown in Figure 8-20. Switching and memory are described as follows.

No. 1 ESS Switching. The switching arrangement for the No. 1 ESS is made up of *sealed dry* reed switches that are activated or deactivated by a 300-μs current pulse. The magnetic material within the reed relay is called a *ferreed.* No power is consumed by the switch while it is in its quiescent state of either opened or closed. The current pulses determine the remanent state of the switch.

Figure 8-20 Front panel control for the No. 1 ESS. (Courtesy of AT&T Archives.)

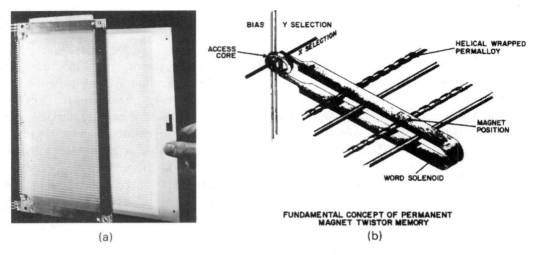

(a)　　　　　　　　　　　　　　　　　　　　**(b)**

Figure 8-21 Twistor memory card. Background information: No. 1 ESS semiperma-
nent twistor memory operates as follows: Information is stored in the vicalloy magnet
next to the intersection of the twistor wire and the copper solenoid. A current pulse sent
through the *x* and *y* wires will cause the ferrite core to send a pulse through the copper
"word" solenoid. If the magnet at the intersection of the solenoid and the twistor is not
magnetized, the pulse will reverse the direction of the magnetic field that links the intersection and generate a "read-out" pulse in the twistor wire. If the magnet at the intersection is magnetized, its field will prevent the reversal at the intersection and there will be
no significant voltage produced in the twistor wire. (Courtesy of AT&T Archives.)

No. 1 ESS Memory. Two types of memory are used for the No. 1 ESS:
twistor memory and *ferrite sheet memory*. Twistor memory is superpermanent
memory that stores the control program and any data that are not likely to
change. Memory is thus *read-only* (ROM) and therefore nonvolatile. Power
failures and read errors cannot alter the twistor memory contents. Only an operator can change its contents by manually extracting the card and reprogramming
its contents with a special device. The twistor memory card is shown in Figure
8-21.

The twistor memory card is made of aluminum sheets measuring approximately 11 by $6\frac{1}{2}$ in. Each sheet contains rows of magnetic spots that are either
magnetized or demagnetized prior to installation. The magnetic state of each
spot represents one bit of stored data. Each card stores 64 words that are 24 bits
in length. The state of each bit is determined by sending a pulse down an *interrogating loop* of wire. Permalloy magnetic tape is spirally wound around copper
sensing wires placed over the magnetic spots. The interrogating loop intersects
the magnetic spot and sensing wire at right angles. If the spot has not been magnetized, the interrogating pulse will cause the permalloy magnetic tape to become
magnetized for the duration of the interrogating pulse. This results in a small

(a)

(b)

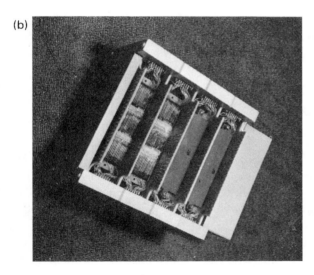

Figure 8-22 (a) Ferrite sheet memory card; (b) ferrite sheet memory module.

current that is sensed by amplifiers. When the spot is already magnetized, no current is produced.

Ferrite sheets are used for temporary storage of data (read–write memory) related to the processing and administration of a call. The ferrite sheet, shown in Figure 8-22a, consists of 256 perforated holes on a 1-in.-square board. Ferrite material surrounds each hole. Three lines are threaded through each hole to allow reading and writing to each hole in a manner similar to *core memory*. Each hole acts as a core and stores one bit of information. The ferrite sheets are stacked in a module, and four modules make up a *Call Store,* each holding 196,608 bits of read–write information. Figure 8-22b depicts one module of ferrite sheet memory. An office of about 10,000 lines generally requires two or more Call Stores (eight modules). A 65,000-line office with a high calling rate may contain 40 Call Stores. The No. 1 ESS can address over 1 million bits of memory.

8.3.3.2 No. 2 ESS.
The No. 2 ESS is capable of handling 1000 to 10,000 lines. Although its capacity to handle several lines is not as great as the No. 1 ESS, its attractiveness lies in its ability to provide extremely reliable service to smaller communities at an economical cost. Also, the No. 2 ESS is fully operational from a remote location. Many rural areas employ the No. 2 ESS, whereby traffic volume, maintenance requirements, and performance of administrative tasks are monitored and performed in a more populated area. An upgrading of logic, from the diode–transistor logic (DTL) used in the No. 1 ESS to resistor–transistor logic (RTL), has increased the speed and reduced the size of the control circuitry.

8.3.3.3 No. 3 ESS.
LSI technology in the 1970s brought forth further enhancements to the ESS. The No. 3 ESS employs bipolar LSI ROMs for microprogram control. Under microprogram control, the SPC is executed from a unique set of *microinstructions* stored in ROM. The No. 3 ESS was designed for small offices serving rural communities and small cities of 100 to 1000 subscribers.

8.3.3.4 No. 4 ESS.
The No. 4 ESS was first installed within the PSTN in 1976. It was the first all-electronic exchange with digital circuit technology employed in its computerized control and switching matrix. Over 10,000 trunk circuits can be handled by the No. 4 ESS, which uses a combination of *time-division multiplexing* (TDM) and *space-division multiplexing* (SDM) (see Section 8.7).

In time-division multiplexing, incoming analog voice signals are digitized and converted to PCM (pulse code modulation) signals. PCM signals from several voice channels are multiplexed together and loaded into a memory buffer. From here, the stored program control selects an available path through the SDM switch. The multiplexed signal is then routed through the outgoing buffer to the appropriate line.

8.3.3.5 No. 5 ESS.
The No. 5 ESS is the most advanced and versatile central/toll switching unit developed in the Bell System's ESS product line. It uses the latest technology in integrated circuits, fiber optics, and software design. Metropolitan areas with as many as 100,000 subscribers can be serviced as well as rural areas with as few as 1000. A fully remote controlled unit, the ESS No.

Figure 8-23 No. 5 ESS. (Courtesy of AT&T Archives.)

5A, has been developed to serve as an economical central office that can be operated by a host No. 5 ESS system from as far as 100 miles away. Up to 4000 subscribers per unit can be serviced remotely over digital T1 carrier facilities.

The hardware design of the No. 5 ESS relies heavily on large-scale integration technology (LSI) specifically designed to handle high-voltage potentials up to 500 V. Testing of ringing functions can therefore be performed without the large risks of damaging components. The addition of new technology to the No. 5 ESS was kept in mind by Bell Laboratory's design engineers. A modular design concept was followed that permits rapid growth of the switching system as well as ensuring that the system fits smoothly into the Bell System's existing facilities.

Software technology incorporated into the No. 5 ESS was also designed to permit the rapid addition of new features as the hardware technology advances. Most of the controlling software for the No. 5 ESS are written in a language called "C," developed by Bell Labs. The operating system, *UNIX*, also developed by Bell Labs, provides an abundance of tools specifically tailored to support software design and testing. Figure 8-23 illustrates the No. 5 ESS. Listed below are some of the custom calling features of the No. 5 ESS.[†]

Call forwarding: Transfers a call automatically to where you can be reached.

[†]Compiled from the *Sprint Central Telephone–Nevada First Source Phone Book,* Area Code 702 Las Vegas Telephone Directory, Reuben H. Donnelley and Centel Directory Company, July 1993.

Call waiting: Acts as a home or business second line. While you are using the phone, a short tone warns you of an incoming call. The dial switch can be pressed and released to answer the second caller. The same action allows you to switch back and forth between callers.

Cancel call waiting: Allows you to cancel Call Waiting so that you can enjoy uninterrupted or important calls.

Call within: Allows you to use internal telephones within your household as an inexpensive intercom. By dialing your own number and hanging up, all the phones connected to the same line will ring and you may talk to other members that answer.

Speed call: Allows you to program several telephone numbers that can be dialed by pressing a one- or two-digit code. A two-digit access code is typically dialed first to enable Speed Call.

Three-way calling: Allows you to call two other parties and have a three-way conversation.

Return call: If you cannot get to your telephone when it is ringing, Return Call will allow you to dial a special code and the telephone network will announce the last number that called you, unless it is a private number.

Redial call: Allows you to redial the last number you called, whether the call was answered, unanswered, or busy.

Call trace: Call Trace allows you to trace the last call that you received. By activating the trace with a special code number entered from your Touch Tone keypad, the telephone company will store the calling number, your own number, and the date and time the last call took place. This feature should only be used for threatening or obscene phone calls. Only an authorized agency is provided with trace results.

Caller ID: Allows you to screen your calls by viewing a special telephone number display that shows you the telephone number of the person calling you. You can choose to answer or not, depending on the number being displayed. Caller ID is available in some states and not in others for privacy and other reasons.

Caller ID block: Allows you to prevent your telephone number from appearing on a Caller ID display device when you are making a call. Instead, the party you are calling will see "Private Number" on their display when their telephone is ringing.

8.4 MULTIPLEXING

In communications, *multiplexing* is the process of combining two or more signals and transmitting them over a single transmission link. *Demultiplexing* is the reverse process of separating the multiplexed signals at the receiving end of the transmission link. The link can be any of the transmission media discussed earlier. Multiplexing results in the efficient use of the communications link. Trunk circuits used within the PSTN use multiplexing techniques to combine several signals. The number of signals that can be multiplexed together is directly related to available *space, time,* and *bandwidth.* Space is required for the medium to exist in. Time is required for multiple signals to be transmitted and received. Perhaps the single most important factor in multiplexing is bandwidth. Bandwidth is necessary to accommodate the related frequency components of the multiplexed signals. The greater the bandwidth, the greater the number of signals that can be multiplexed. For digital signals, transmission speed is directly related to the available bandwidth. Wideband transmission media such as coaxial cable, microwave, and fiber-optic links, for example, have the capacity to multiplex several thousand voice signals together. There are three fundamental classifications of multiplexing:

1. Space-division multiplexing (SDM)
2. Frequency-division multiplexing (FDM)
3. Time-division multiplexing (TDM)

Multiplexing has become an essential part of the communication system today. It is necessary for us to understand the concept of multiplexing and how it results in efficient use of the communications channel.

8.4.1 Space-Division Multiplexing (SDM)

Space-division multiplexing (SDM) is simply the combining of physically separate signals into a bundled cable. It would not make sense to have a unique set of telephone poles or underground trenches for each telephone subscriber. Subscriber loop and trunk circuits are therefore combined. The shared use of space is attained in space-division multiplexing. Since large volumes of space are not always available to contain cables, time-division multiplexing or frequency-division multiplexing are often combined with space-division multiplexing.

8.4.2 Frequency-Division Multiplexing (FDM)

Frequency-division multiplexing (FDM) utilizes a channel's available bandwidth to send multiple signals simultaneously. By modulating several subcarriers with independent telephone signals, a composite signal can be generated and modu-

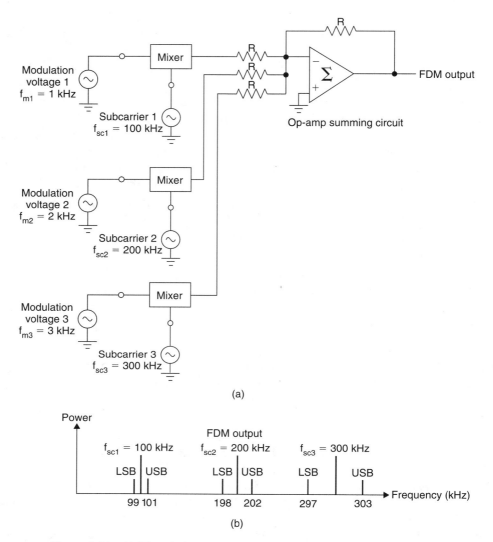

(a)

(b)

Figure 8-24 (a) Three independent sine waves are mixed with subcarrier frequencies to produce the frequency-division multiplexed output; (b) frequency spectrum of the FDM output signal.

lated onto a main carrier. The carrier is then transmitted onto a channel. Space-division multiplexing is consequently minimized. Figure 8-24a depicts frequency-division multiplexing of three sine waves: 1, 2, and 3 kHz. Three subcarriers (100, 200, and 300 kHz) are mixed with the three independent modulating voltages. The output of each (nonlinear) mixer is electrically summed together to produce the given FDM output. The resulting frequency spectrum is shown in Figure 8-24b.

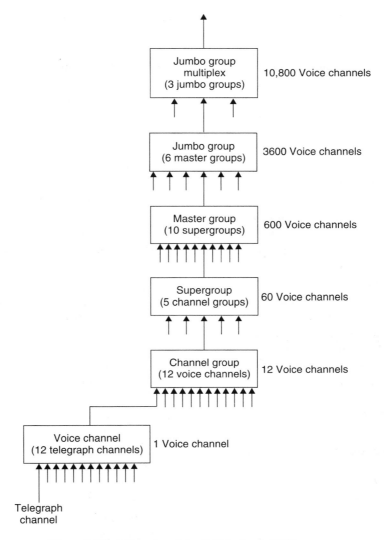

Figure 8-25 Hierarchy of the Bell System's FDM groups.

8.4.2.1 Hierarchy of the Bell System's FDM groups. To standardize the telecommunications highways of the world, the Bell System has formed a hierarchy of *groups* that classify the number of voice channels that are multiplexed together before they are sent onto a trunk circuit. See Figure 8-25. The trunk circuit may be microwave, fiber, or coax. Groups may also be multiplexed together to form higher groups in the hierarchy prior to transmission. The 300- to 3400-Hz telephone voice channel is regarded as the fundamental building block of the hierarchy. A single voice channel can be further divided into 12 telegraph channels.

At the bottom of the hierarchy is the *telegraph channel.* A telegraph channel contains messages sent by telegraph equipment such as a Teletype terminal. The electrical impulses from the Teletype are used to frequency modulate a tone. Up to 12 tones can be modulated by 12 independent Teletypes to produce a composite signal that is sent over a single voice channel. The composite signal has no frequency components outside the voice-grade channel of 300 to 3400 Hz.

Twelve voice channels make up a *channel group.* By multiplexing five channel groups, a total of 60 voice channels is combined to form a *supergroup.* Ten supergroups form a *master group* containing 600 voice channels. Six master groups form a *jumbo group* of 3600 voice channels. At the top of the hierarchy is the *jumbo-group multiplex.* The jumbo-group multiplex is formed by multiplexing three jumbo groups together for a total of 10,800 voice channels. The Bell System's L5 Carrier, discussed earlier, contains 20 coaxial tubes, each tube carrying a jumbo-group multiplex signal. Ten are used for transmitted signals and 10 are for received signals. Thus 108,000 (10 transmit and 10 receive jumbo-group multiplex signals) simultaneous telephone calls are possible through a combination of both space-division multiplexing and frequency-division multiplexing on the L5 carrier.

Figure 8-26a illustrates how 12 voice channels are frequency-division multiplexed together to form the basic channel group of composite signals. Each voice channel is mixed by a balanced modulator with subcarriers separated by 4 kHz. This technique is referred to as *single-sideband suppressed carrier* (SS-BSC); 4-kHz bandpass filters are used to separate adjacent voice channels and filter the lower sideband (LSB) of each signal. This prevents sidebands from spilling into adjacent channels. The 12 lower sidebands are electrically summed together and further filtered to produce the FDM channel group output. A *pilot tone* of 104.08 kHz is sent with the channel group output signal for monitoring and demodulation purposes at the receiving end.

The distribution of frequencies for the channel group is shown in Figure 8-26b. A triangle is typically used to denote the distributed frequency spectrum as a result of mixing. The triangle, when facing as shown, denotes the selection of the LSB signals. LSB signals are said to be *inverted sidebands;* that is, the higher frequencies of the voice channel become the lower frequencies of the translated frequency spectrum.

Figure 8-27 illustrates the frequency distribution of signals for the supergroup. Five channel groups that make up a supergroup are translated to a frequency band occupying the range from 312 to 512 kHz, a bandwidth of 240 kHz (5 channel groups × 48 kHz each). The FDM process is repeated here. The upper-sideband (USB) frequencies are selected in the formation of a supergroup. Notice that the triangle for the supergroup is opposite that of the channel group

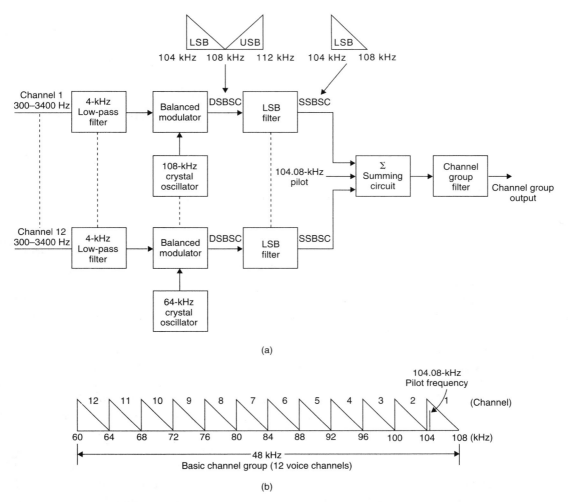

Figure 8-26 (a) Formation of the Bell System's channel group; (b) frequency distribution of the channel group.

due to the selection of the USB. Higher frequencies appear at the higher end of the distributed spectrum.

8.4.3 Time-Division Multiplexing (TDM)

A third form of multiplexing is called *time-division multiplexing* (TDM). Prior to 1960, telecommunications was predominantly analog transmission with FDM serving as the major form of multiplexing. Since then, time-division multiplexed PCM (pulse-code modulation) has dominated the scene and has to this date become the preferred method of transmission onto PSTN trunk circuits.

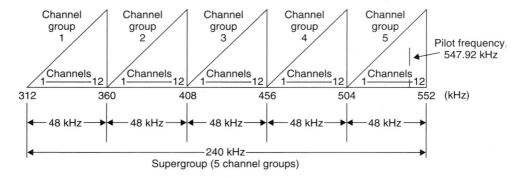

Figure 8-27 Frequency distribution of the Bell System's supergroup.

In contrast to FDM, TDM involves the distribution of multiple signals in the *time domain*, whereas in FDM these signals are distributed in the *frequency domain*. Another major distinction between these two forms of multiplexing is that FDM is an analog process, whereas TDM is a digital process. In TDM, several analog signals are sampled and converted to digital bit streams through the use of *analog-to-digital (A/D) converters*. The process of converting the analog signal into an encoded digital value is referred to as *pulse-code modulation* (PCM). In time-division multiplexing, signals from several sources are digitized and *interleaved* to form a PCM signal. The time-division multiplexed PCM signal is then transmitted onto a single channel. When the digital bit stream is received, the reverse process is performed. The bit stream is *demultiplexed* and converted back to the original analog signals. The analog signals are routed to their final destination.

Figure 8-28 illustrates the concept of TDM. Four analog signals are contiguously sampled and digitized by an eight-bit A/D converter. The interleaved binary serial bit stream is transmitted onto the communications channel. One of CH2's sampled and digitized values has been extracted to depict the binary signal actually transmitted.

8.5 PULSE-CODE MODULATION (PCM)

Pulse-code modulation (PCM) was developed in 1937 at the Paris Laboratories of AT&T. Alex H. Reeves has been credited with its invention. Reeves conducted several successful transmission experiments across the English Channel using various modulation techniques, including pulse-width modulation (PWM), pulse-amplitude modulation (PAM), and pulse-position modulation (PPM). At the time, the circuitry involved was enormously complex and expensive. Although the significance of Reeve's experiments was acknowledged by Bell Laboratories, it was not until the semiconductor industry evolved in the 1960s

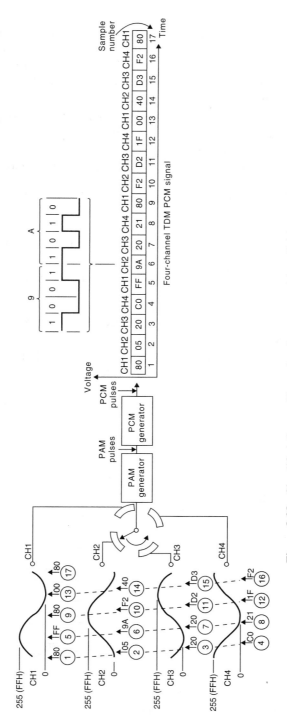

Figure 8-28 Simplified diagram illustrating the concept of time-division multiplexing.

that PCM became more prevalent. Currently, in the United States and the United Kingdom, PCM is the preferred method of transmission within the PSTN.

PCM is a method of serially transmitting an approximate representation of an analog signal. The PCM signal itself is a succession of discrete numerically encoded binary values derived from digitizing the analog signal. The maximum expected amplitude of the analog signal is first *quantized*; that is, divided into discrete numerical levels. The number of discrete levels depends on the resolution (number of bits) of the A/D converter used to digitize the signal. If an eight-bit A/D converter is used, the analog signal range is quantized into 256 (2^8) discrete levels. The quantizing range is governed by the equation

$$\text{quantizing range} = 2^{(\text{no. of A/D converter bits})}$$

Example 2

Given a 12-bit A/D converter, compute the quantizing range (number of discrete levels that a signal can be divided into).

Solution: Quantizing range = 2^{12} = 4096 discrete levels.

Figure 8-29 illustrates the quantization of a sine wave. For simplicity, a four-bit A/D converter is used. At any instant in time, the sine wave can be approximated to the closest quantized level from 0 to 15 (2^4 = 16). This assumes that the signal does not exceed the dynamic range of the A/D converter. If an A/D converter with a dynamic range of 0 to +10 V is used, the input signal shall not exceed this range.

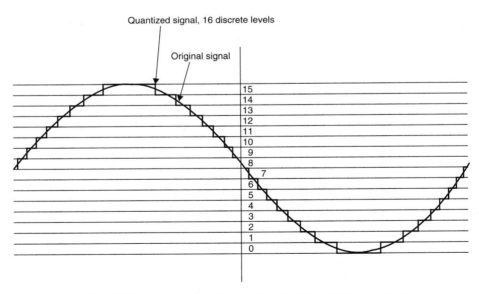

Figure 8-29 Quantization of an analog signal into 16 discrete levels.

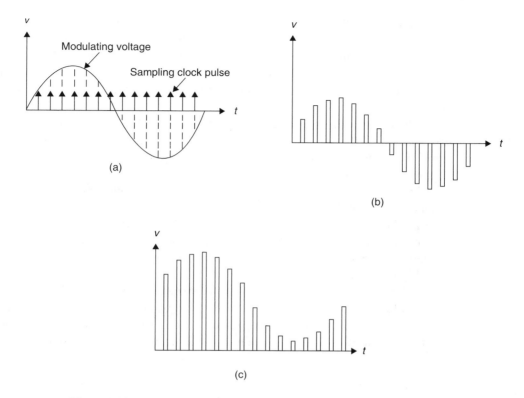

Figure 8-30 PAM: (a) original signal; (b) double-polarity PAM; (c) single-polarity PAM.

If the resolution of the A/D converter is increased, a closer approximation of the signal can be attained, since there would be a greater number of quantized levels. However, the quantized value would require a larger encoded binary value. A larger bandwidth would be required to transmit the signal, as we shall see.

Once the range of quantization has been established, the analog signal is *sampled* at equally spaced intervals in time. Sampling is the process of determining the instantaneous voltage at these given intervals in time. A technique called *pulse-amplitude modulation* (PAM) is used to produce a pulse when the signal is sampled. As shown in Figure 8-30, the pulse's amplitude is equal to the level at the time in which the analog signal was sampled. The amplitude of the pulses in a PAM signal contains the intelligence or modulating voltage. Each pulse of the resulting PAM signal is digitized by an A/D converter to the closest quantized value at the time of the sample. PAM serves as a preliminary step toward generating the PCM signal.

Since the A/D converter takes a finite amount of time to complete its conversion process, it is necessary to *hold* the sampled amplitude while the conversion process occurs (hence the reason for the pulse width in PAM). A *sample-and-hold* amplifier is used to perform this task. If the analog voltage at the input to the A/D converter were allowed to change during this time, an erroneous digital word would be produced.

The encoded binary value produced by the A/D converter represents an *approximation* of the amplitude at the time of the sample. It is an approximation since the instantaneous amplitude at the sampling interval can lie anywhere within the quantized level. The binary value representing the amplitude is then serialized to form the PCM signal and transmitted onto the communications channel. A block diagram of a PCM system is shown in Figure 8-31.

At the receiving end of the communications channel, the reverse process is performed. The PCM signal is received, decoded, and reconstructed by a D/A converter. The resulting signal is an approximation of the original signal.

Figure 8-32 depicts the recovered signal for various sampling rates. The higher the sampling rate, the closer the recovered signal approaches the original signal. Ideally, an infinite sampling rate would be desirable in terms of reproducing the original signal. This is not practical, however, due to the bandwidth limitation on the large amounts of data that would need to be transmitted.

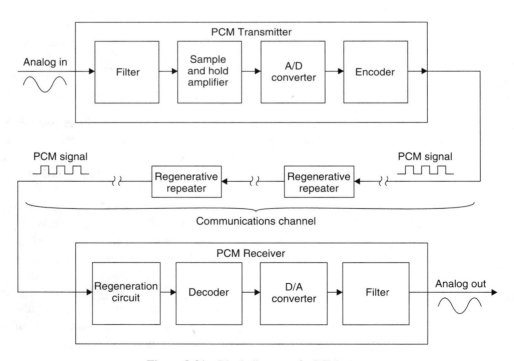

Figure 8-31 Block diagram of a PCM system.

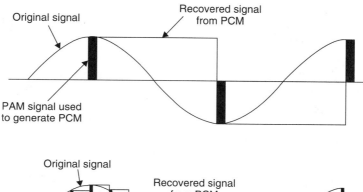

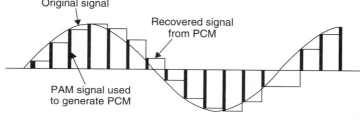

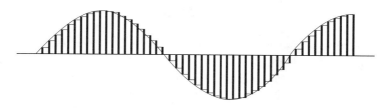

Figure 8-32 Recovered signals for various sampling rates.

In telephony, voice signals are sampled at 8 kHz. The 8-kHz sampling rate is based on the *Nyquist* sampling rate theorem.

Nyquist Sampling Theorem. *If a signal is sampled at a rate that is at least twice the highest frequency that it contains, the original signal can be completely reconstructed.*

A signal containing frequency components up to 4 kHz can therefore be recovered with minimal distortion. Since the bandwidth of the telephone lines is 300 to 3400 Hz, 8 kHz is easily above twice the highest frequency component within this range.

8.5.1 Regenerative Repeater

The advantage of PCM lies chiefly in the fact that it is a digital process. Digital signals have a high noise immunity. That is, it is much easier for a receiver to distinguish between a 1 and a 0 than to reproduce faithfully a continuous wave signal (or a composite of several continuous waves) when both are subjected to the same noise environment. The effect of noise on a PCM signal can be removed entirely through the use of *regenerative repeaters.* Transmission media carrying PCM signals employ regenerative repeaters that are spaced sufficiently close to each other (approximately 1 mile) to prevent any ambiguity in the recognition of the binary PCM pulses. Figure 8-31 depicts how the regenerative repeater fits into the PCM system.

When pulses arrive at the repeater, they are attenuated and distorted. The regenerative repeater conditions the received pulses through preamplifiers and equalizer circuits. The signal is then compared against a voltage *threshold* as shown in Figure 8-33. Above the threshold is a logic 1 and below the threshold is a logic 0. The resulting signal is said to be *threshold detected.*

Timing circuits within the regenerative repeater are synchronized to the bit rate of the incoming signal. The threshold detected signal is sampled (see arrows) at the optimum time to determine the logic level of the signal. The resulting code is used to regenerate and retransmit the new equivalent signal. Threshold detection is an art in itself; entire books have been written on the subject.

8.5.2 Distortion in a PCM Signal

Distortion in a PCM signal lies principally in the encoded signal itself. When an analog signal is quantized, the encoded value is approximated within the limitations of the A/D converter. This quantizing error results in the signal distortion that was shown in Figure 8-32. The quantizing error can be as much as one-half of a quantizing level. An eight-bit A/D converter, for example, can have a maximum quantizing error of $\frac{1}{512}$th of the signal ($\frac{1}{2}$ of $\frac{1}{256}$); that is, the instantaneous voltage at the time of the sample lies precisely at midpoint between quantized levels. The least significant bit of the A/D converter in this case can be either a 1

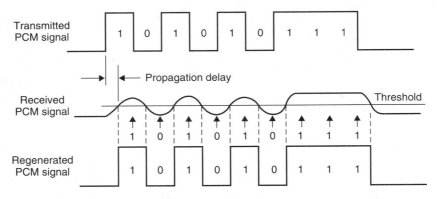

Figure 8-33 Regeneration of a PCM signal with a regenerative repeater.

The Telephone Network   **Chap. 8**

or 0. The noise resulting from this distortion is referred to as *quantization noise*. Figure 8-34 illustrates the transfer function for a linear quantizer. The quantized signal is a staircase. Subtracting the quantizing error from the quantized staircase signal results in the original signal. The error can be reduced by increasing the number of quantized levels, consequently decreasing the quantizing noise.

The magnitude of quantization noise power is directly related to the number of quantization levels and therefore related to the number of binary bits used to represent a quantized level. To give an indication of this magnitude, for a sinusoid whose peak-to-peak amplitude utilizes the full dynamic range of the A/D converter, the rms signal-to-quantization noise ratio is given by

$$\text{SNR} = 1.8 + 6n \qquad \text{dB}$$

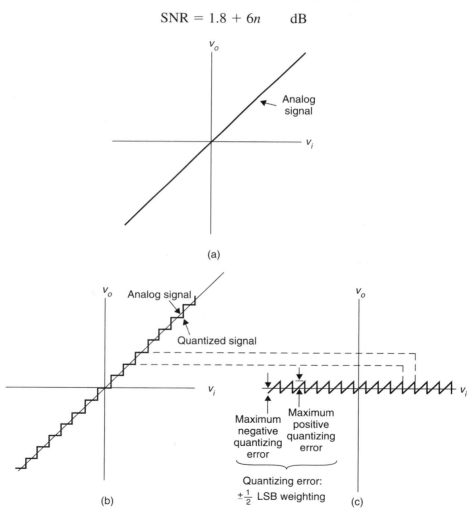

Figure 8-34 (a) Linear transfer function; (b) quantization of analog signal; (c) quantizing error.

TABLE 8-2 Signal-to-Quantizing Noise Ratio Versus Number of Binary Bits Used to Represent a Quantized Level

Number of binary bits, n	Number of quantization levels, 2^n	Signal-to-quantizing noise ratio (dB)
4	16	25.8
5	32	31.8
6	64	37.8
7	128	43.8
8	256	49.8

where n equals the number of binary bits used to represent the quantized level. Table 8-2 shows how the signal-to-quantizing noise ratio increases with increasing quantization bits. Eight bit quantizing (256 levels) is most commonly used in today's PCM systems to represent voice. This satisfies the trade-offs between signal resolution and transmission time.

8.5.3 Voice Compression in a PCM System

The discussion of quantization has been simplified through the use of a sine wave. In the examples given earlier, the sine wave utilized the full dynamic range of the A/D converter. The signal-to-quantizing noise ratio is optimum in this case. Voice, however, does not utilize the entire range of the A/D converter in a uniform manner, as the sine wave presented earlier does. Instead, voice variations are sporadic. They contain low- to high-amplitude variations throughout the quantizing range. The ratio of this variation, or, more properly stated, the *dynamic range* of voice can be as high as 60 dB. An inherent problem exists here when attempting to use the common eight-bit A/D converter for quantization. Only 256 quantized levels are possible. This is suitable for a signal with a dynamic range of 48 dB ($20 \times \log 256$). For voice, however, much of its characteristics can be lost within quantized levels. Figure 8-35 depicts this anomaly and how it is resolved in part by *compression* of the signal.

The voice pattern shown in Figure 8-35 has been quantized and reconstructed through the PCM process. For simplicity, a four-bit A/D converter is used again. Figure 8-35a depicts the reconstructed signal without compression. Figure 8-35b shows the effect of compression on the same voice pattern. Notice in Figure 8-35a that the dynamic range of the voice pattern exceeds that of the A/D converter used for quantization. Low and high amplitudes are not resolved due to the lack of quantization levels. As a result, the characterization of the

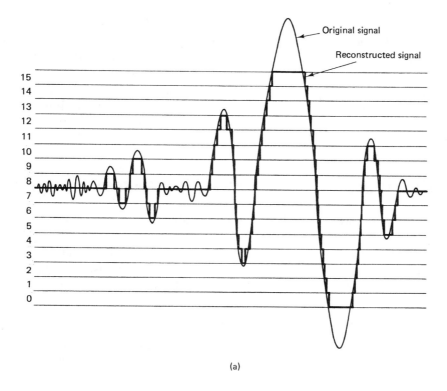

(a)

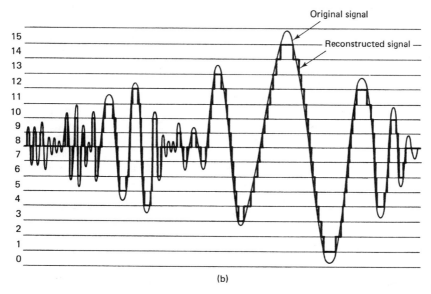

(b)

Figure 8-35 (a) Reconstructed voice pattern without compression; (b) reconstructed voice pattern with compression is much closer to the original signal.

voice pattern through the PCM process is poor. In Figure 8-35b, the dynamic range of the signal has been compressed to within the quantizing range of the A/D converter. Low-amplitude variations of the signal are amplified to a greater extent than higher-amplitude variations. The compressed voice pattern is a much closer representation of the original signal. The recovered signal can now be *expanded* to its original level. The PCM system shown earlier in Figure 8-31 can be improved by the addition of compandors placed before and after the A/D and D/A conversion process.

8.6 NORTH AMERICAN DIGITAL MULTIPLEX HIERARCHY

Prior to divestiture, the Bell System had established one of the most widely used time-division multiplexed PCM systems in Northern America. This transmission system is known as the *T-carrier* system. A hierarchy of this TDM PCM structure is shown in Figure 8-36. Table 8-3 lists the characteristics of each T-carrier signal. T-carriers are used in most major trunk circuits today.

TABLE 8-3 North American Digital Multiplexing Hierarchy

Line type	Digital signal no.	Number of TDM voice channels	Transmission rate (Mbps)	Transmission media
T1	DS-1	24	1.544	Wire pair
T1C	DS-1C	48 (2 T1 lines)	3.152	Wire pair
T2	DS-2	96 (4 T1 lines; 2 T1-C lines)	6.312	Wire pair, fiber
T3	DS-3	672 (28 T1 lines; 7 T2 lines)	44.736	Coax, microwave fiber
T4	DS-4	4032 (168 T1 lines; 6 T3 lines)	274.176	Coax, microwave fiber
T5	DS-5	8064 (336 T1 lines; 2 T4 lines)	560.160	Coax, fiber

8.6.1 The Bell System's T1 Carrier

Since Bell System's *T1 carrier* is the fundamental building block of the TDM multiplex hierarchy, we will limit our discussion to the T1 carrier. In 1984, more than 120 million voice miles were transmitted on T1 carriers in the United States alone. Your telephone conversations today, beyond the local loop or class 5

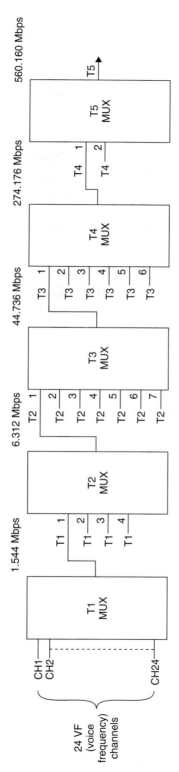

Figure 8-36 Block diagram of the North American digital multiplexing hierarchy.

switching centers, have most likely been digitized, encoded into a PCM signal, and multiplexed onto a T-carrier together with several other callers.

The T1 carrier consists of 24 voice channels that are sampled, digitized, and encoded into a TDM PCM signal. Each sample is encoded into an eight-bit digital word (sign plus seven-bit data) that represents the voice amplitude at the time of the sample. Coding of the PCM signal for a single sample is shown in Figure 8-37.

The transmission rate onto the T1 carrier is 1.544 Mbps (million bits per second). DS-1 (digital signal 1) is the designation for the signal number and its transmission rate. The 1.544-Mbps transmission rate is established by sampling each of the 24 voice frequency channels contiguously at an 8-kHz rate. Twenty-four channels at eight bits per sample yield a total of 192 bits.

Since the sampling sequence of a single sweep of all 24 channels occurs in a predetermined order, the receiving equipment must employ a method of synchronizing with the serial bit stream. For this purpose, a *framing bit* (F-bit) is added to the beginning of the 192 bits to make up the *T1 carrier frame*. The T1 carrier frame therefore consists of 193 bits. The time duration of a frame is 125 μs ($\frac{1}{8}$ kHz); 193 bits $\times$ 8 kHz results in the bit rate of 1.544 Mbps.

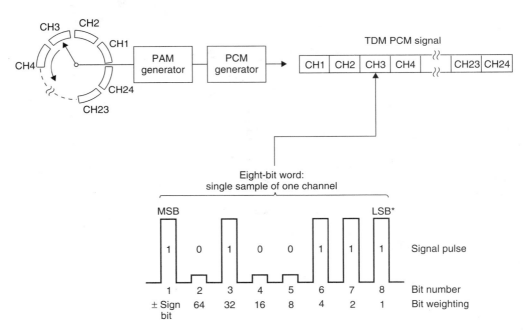

Figure 8-37 Coding for the T1 carrier PCM signal.

T1 carrier frames are transmitted in groups of 12. Twelve frames make up a *superframe*. The PCM receiver synchronizes to an alternating pattern of ones and zeros (101010) transmitted as framing bits in *odd*-numbered frames within the superframe. Even-numbered frames within the superframe have a framing bit pattern of 001110. The format of the T1 carrier frame and its relationship to the superframe are shown in Figure 8-38.

Signaling information, such as on-hook/off-hook condition or ringing, must also be transmitted between digital channel banks. This is accomplished by *robbing* the least significant bit position of each channel's eight-bit word. To keep distortion to a minimum, this is performed *only* in the sixth and twelfth frames, as shown in Figure 8-38. In all other frames, the LSB position conveys the normal voice data.

The preferred framing format for all new designs of DS-1 rate terminals or equipment utilizes the *extended superframe* (ESF) format.[†] ESF extends the DS-1 superframe structure from 12 to 24 frames (4632 bits) and divides the 8-kbps, 193rd bit position pattern (F-bits) previously used for basic frame and robbed bit signaling synchronization into a number of subchannels.

8.6.2 Digital Channel Banks

The PSTN uses *digital channel banks* to perform the sampling, encoding, and multiplexing of 24 voice channels for the basic DS-1 signal. Digital channel banks are designated D1 through D5, in accordance with the respective technology used over the years. D1 channel banks, the first of the channel banks, employed discrete components. A seven-bit data word (sign plus six-bit data) was originally used to represent the sampled voice. The eighth bit was used exclusively for signaling. As technology advanced, new specifications were written to meet the improved performance resulting from integrated-circuit technology. D3 channel banks incorporated SSI (small-scale integration) technology. Most of today's channel banks are D4 and D5, which use custom-designed LSI (large-scale integration) technology. Eight-bit data words are used to represent voice, thus improving on the SNR. For the most part, D1 channel banks have become obsolete. D2 and D3 channel banks are slowly being replaced with more recent equipment as well.

8.6.3 Pulse Train for the T1 Carrier

The PCM signal discussed up to now has been presented as a *unipolar pulse train;* that is, the pulses are positive. Although these pulses are regenerated, occasionally an error is made. These errors have been known to produce clicks in

---

[†]Bell Communications Research, Inc., 1985, *The Extended Superframe Format Interface Specification.*

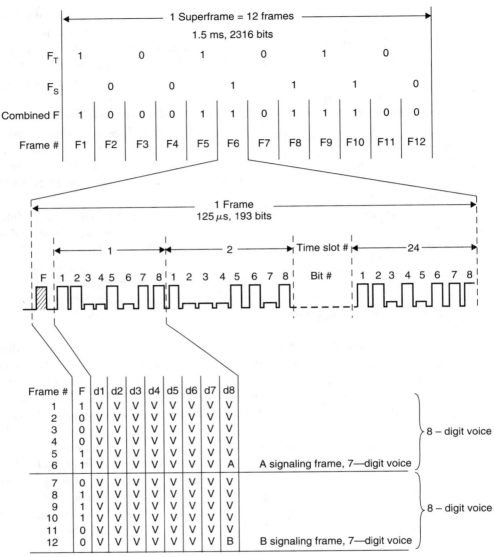

Figure 8-38 Frame structure for a T1 carrier frame and superframe. (From Bell System Technical Reference Publication 62411: *High Capacity Digital Service Channel Specification*, Sept. 1983.)

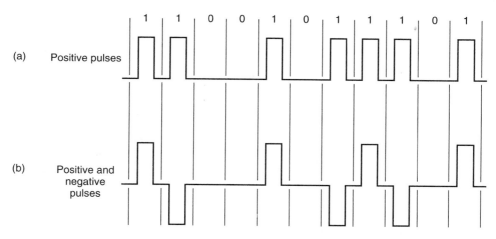

Figure 8-39 (a) Unipolar pulse train; (b) bipolar pulse train.

the received telephone signal. Fewer errors are produced if the PCM signal is converted from a unipolar to a *bipolar pulse train,* in which the polarity of each consecutive pulse must be reversed (Figure 8-39). The bipolar DS-1 pulse train is the actual signal presented to the T1 carrier line. Notice that the bipolar pulse train must maintain a 50% duty cycle whenever a pulse is present. If a pulse is present within a time slot, positive or negative, it represents a 1-bit. If there is no pulse present within a time slot, it represents a 0-bit.

8.6.4 B8ZS Coding

For receivers to maintain synchronization to the DS-1 bipolar signal, there must be an adequate amount of energy in the signal at any given time. This is true for many synchronous serial transmission links. The problem arises when an excessive number of 0's are transmitted. For the DS-1 bipolar pulse train shown in Figure 8-39b, suppose the line is quiet (all 0's). No alternating pulses (representing 1's) would be present for the receiver to synchronize to. To circumvent this problem within the T1 network, common carriers require a minimum number of 1-bits within a given amount of time or within a given number of bits transmitted. One technique used for ensuring compliance with pulse density requirements for synchronization is called **B8ZS** (*bipolar with 8 zero substitution)* **coding.** B8ZS coding ensures adequate signal energy by substituting any string of eight consecutive zeros within the serial bit stream with a specific eight-bit code. There are two eight-bit replacement codes. They are inserted as follows:

1. If the pulse preceding the string of eight consecutive 0's is a positive one pulse, the inserted eight-bit code is $0\ 0\ 0 + -\ 0 - +$. For example,

	1	2	3	4	5	6	7	8
Original word:	0	0	0	0	0	0	0	0
Substituted word:	0	0	0	+1	−1	0	−1	+1

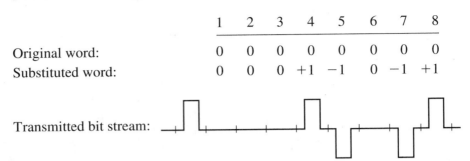

Transmitted bit stream:

2. If the pulse preceding the string of eight consecutive 0's is a negative 1 pulse, the inserted eight-bit code is: $0\ 0\ 0 - +\ 0 + -$. For example,

	1	2	3	4	5	6	7	8
Original word:	0	0	0	0	0	0	0	0
Substituted word:	0	0	0	−1	+1	0	+1	−1

Transmitted bit stream:

The receiver, upon receiving either of the two substituted bit streams above, simply replaces the B8ZS code with the original string of eight zeros. The obvious question now arises: How does the receiver distinguish between the B8ZS code and eight bits of data that happen to be the same sequence? A close look at the *transmitted bit stream* for either of the B8ZS codes above will reveal that *bipolar violations* occur in the **fourth** and **seventh** bit positions. (See preceding pulses to these bit positions.) A bipolar violation occurs in a T1 signal whenever two consecutive pulses occur with the *same* polarity, regardless of the number of zero bit times that separate the two pulses. Each pulse, or 1-bit, must be alternating in polarity, as shown in Figure 8-39b. The receiver now has a mechanism for distinguishing between eight bits of data that should be replaced with zeros and eight bits that should be left alone as part of the message stream.

If 16 or more zeros occur in the bit stream, two or more consecutive B8ZS codes are inserted into the transmitted serial bit stream. This is possible since the eighth bits of both B8ZS codes are positive and negative logic 1 pulses that will precede the next B8ZS code under the same rules. Figure 8-40 depicts the pulse characteristics of the bipolar DS-1 signal.

The Telephone Network   **Chap. 8**

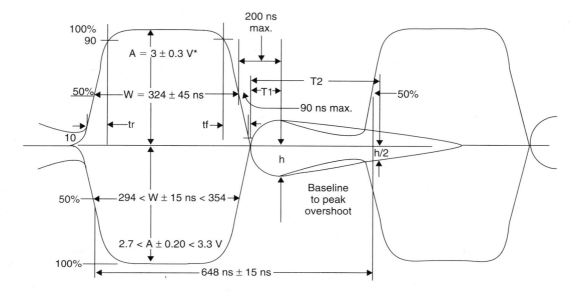

tr, tf < 100 ns

*At 60°F (2.40 to 3.00 V with surge
protection)

Figure 8-40 Output pulse characteristic for the DS-1 signal. (Bell System Technical
Reference Publication 62411: *Addendum 1 M24-Digital Data Throughput,* June 1985.)

PROBLEMS

1. What types of trunks are used to interconnect the following?
   (a) Central offices
   (b) Tandem office to a central office
   (c) Central office to a toll center
   (d) Toll center to toll center

2. How many classes of switching exchanges are used in North America, and what are
   their names?

3. Name five advantages that a private leased line has over a switched or dial-up line.

4. What is the name of the device used by switching offices to convert the two-wire
   subscriber loop to a four-wire circuit?

5. Compute the minimum round-trip delay time for a signal to propagate between two
   telephones separated by 2900 miles. Assume the speed of light in free space. Would
   an echo suppressor or an echo canceller be necessary?

6. What is the name of the process for compressing a signal at the transmitter and expanding it at the receiver?

7. What major parameter is improved as a result of the process in Problem 6?

8. Name five types of transmission mediums used for trunk circuits.
9. Explain what happens when a call is blocked.
10. Name four customer services that an ESS system provides.
11. What are the three classifications of multiplexing? Briefly define their differences.
12. Compute the quantization range for a 10-bit A/D converter.
13. The upper cutoff frequency of a telephone line is 3400 Hz. Based on Nyquist's sampling theorem, what is the minimum sampling rate necessary to recover the highest signal component that would pass through the lines?
14. Compute the signal-to-quantization noise ratio for an eight-bit A/D converter.
15. Compute the following:
    (a) Bit time for the T1 carrier system
    (b) Frame rate for the T1 carrier system
    (c) Length of time to transmit a superframe
16. Derive the bit rate for the T1 carrier system.
17. What special bit pattern is used in the T1 carrier system for frame synchronization?
18. What is the primary difference between a *superframe* and an *extended superframe*?
19. Refer to Figure 8-39. Draw both a unipolar pulse train and a bipolar pulse train for the serial bit stream 1011101011.
20. What does the acronym *B8ZS* stand for?
21. Draw the B8ZS transmitted bit stream for a string of eight consecutive 0's assuming that the pulse preceding the eight 0's is a negative 1 pulse.
22. Repeat Problem 21 assuming that the pulse preceding the eight 0's is a positive 1 pulse.

9

Modems

It is often necessary for computers, like people, to communicate with each other beyond rooms and buildings. Normally, a direct connection would suffice. For longer distances it becomes economical to use the existing PSTN. Since telephone circuits are analog carriers and computers use digital signals, a device is necessary to interface the two. This device is called a *modem*.

The term *modem* is a contraction of the words *mo*dulator and *dem*odulator. These are the two principal functions that the modem performs. A computer's digital serial bit stream is used by the modem to modulate a carrier tone suitable for transmission over the phone lines. The modem at the receiving end demodulates the carrier tone, thus restoring the signal to its original digital form.

In this chapter, various types of modems and their signal characteristics are discussed. Some of the standard techniques used to enhance communications via the modem interface are also discussed. This includes line conditioning, data compression, and error control.

9.1 MODEM FEATURES

To meet the demands of the communications field, today's modems have become extremely sophisticated. Figure 9-1 illustrates some of U.S. Robotics' high-speed modems. Consider some of the features offered by these high-speed modems:

Figure 9-1(a) U.S. Robotics Sportster 28,800 bps V.FAST fax modem with V.32bis. (Courtesy of U.S. Robotics.)

Figure 9-1(b) U.S. Robotics Courier Dual Standard with V.34 Ready Data/fax with V.FC and V.32bis. (Courtesy of U.S. Robotics.)

— Autodial, Autoanswer, and Autoredial

— Synchronous and asynchronous operation

— 9600 bps CCITT Group III send/receive FAX

— Error detection and correction

— Adaptive phase equalization

— Data compression

Figure 9-1(c) WorldPort 14,400 Fax/Data Modem, with 14,400-bps Group III fax capabilitites, V.32*bis* data transmission, and V.42/V.42*bis*. (Courtesy of U.S. Robotics.)

— Phone number storage

— Speed conversion from 300 to 28,800 bps

— Self-test

— Analog and digital loopback test

— Protocol management

— On-screen help menu

— RS-232 pins 2 and 3 reversibility

— ASL (Adaptive Speed Leveling)†

— Battery-operated portability

9.2 INTERFACE TECHNIQUES

Modems are connected to the phone lines either through an *indirect connection* or a *direct connection.* An indirect connection is made by *acoustically coupling* the modem to the line. A direct connection to the line involves a hard-wired connection directly to the line, usually through a switch, internal to the modem.

---

†ASL is a registered trademark of U.S. Robotics Inc.

9.2.1 Acoustically Coupled Modems

An acoustically coupled modem provides a method of coupling a modem's transmitted and received data to the phone lines through the use of transducers. The modem's transducers convert sound energy to electrical energy, and vice versa. Figure 9-2 illustrates the acoustically coupled modem. Two rubber cup-sized sockets on the modem are used to mount the handpiece of the telephone set. Inside one of the cups is the modem's speaker, and in the other is the modem's microphone. The spacing of each rubber cup is such that the mic and speaker of the handset can be seated snugly into the appropriate socket; that is, the modem's mic is seated against the telephone handset's speaker. Conversely, the modem's speaker is seated against the telephone handset's mic. The modulated carrier tones representing the transmitted data are made audible through the audio amplifier section of the modem. These audible tones can be heard at the modem's speaker. They are transduced from sound energy to electrical energy in the same manner as voice by the microphone of the telephone set. Received data are acoustically coupled from the telephone handset's speaker to the microphone of the modem.

With an acoustically coupled modem, data can be transmitted and received from any location with a standard telephone set. To transmit or receive data, a call is manually dialed in the usual manner. The user listens for ringing and the

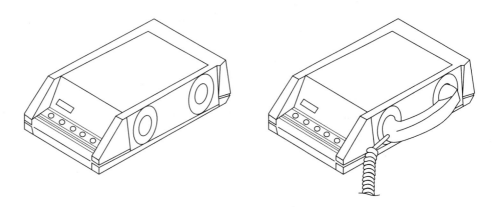

Another common form of connection to the switched network is the acoustically coupled modem, or acoustic coupler. Acoustic couplers were first introduced in 1967 from a design originating at the Stanford Research Institute. Three types of couplers are available. Two are 103- and 202-compatible devices, the third is the VA3400-compatible coupler operating at 1200 bps full duplex.

Figure 9-2   Acoustically coupled modem. (Reprinted with permission from Racal-Vadic.)

off-hook condition. When an answer tone is heard, acknowledging connection, the telephone is placed in the modem's rubber fittings; carrier signals are exchanged, and data communication begins.

Transmission speeds of acoustically coupled modems are typically limited to less than 1200 bps due to the limitations of bandwidth and noise immunity. Even with the modem's rubber fittings tightly sealed to the telephone handpiece, ambient noise can leak through. Following approval by the FCC to permit modems to be directly connected to the PSTN, acoustically coupled modems have become virtually extinct.

9.2.2 Direct-Connect Modems

In the past, telephone installations required a service call to the local telephone company in order to connect a phone to an existing line. After divestiture of the Bell System, modular telephones and connections have become standardized, making it possible for anyone to install a phone. Likewise, modems registered (FCC Part 68) and approved (FCC Part 15) by the FCC are permitted to be connected to the line directly. A standardized cable is used to interconnect today's telephones and modems to the telephone network. One of the most common connectors used is the RJ11C voice jack module shown in Figure 9-3. Other common voice and data jacks used with today's modems are the RJ41S and RJ45S.

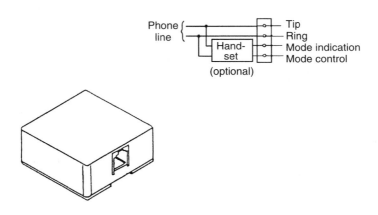

Modems registered as "permissive" devices are most commonly connected to an RJ11C voice jack. This connection is used to connect a standard telephone to the telephone line. Permissive devices must limit the amplitude of the signal presented to the telephone line to a maximum of −9 dBm.

Figure 9-3 Common modem voice and data jacks. (Reprinted with permission from Racal-Vadic.)

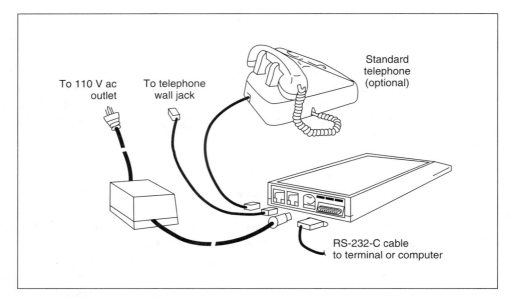

Figure 9-4 Typical direct-connection installation. (Reprinted with permission from Racal-Vadic.)

Modems are designed to accept the same cable and connector used for the telephone set, thus allowing direct connection to existing telephone jacks. Switching between normal telephone use and the modem is controlled either by commands generated by a user from DTE or by a manual switch located on the modem. Figure 9-4 illustrates the direct connection.

9.3 MODULATION TECHNIQUES

Since the frequency spectrum of a pulse train consists of an infinite number of related harmonics, the limited bandwidth of the telephone network, approximately 3 kHz, does not lend itself to transmitting a computer's digital pulses. The filtering loss of high-frequency components inherent in the pulse train causes serious degradation to the pulse train's shape. A modem is necessary to convert the digital signal into analog form, which is suited for the telephone lines. To perform this task, four basic modulation techniques are employed by modems. They are illustrated in Figure 9-5. The technique used depends on the data transmission rate. Generally, the higher the transmission rate, the more sophisticated the modulation technique must be to meet the passband requirements of the phone lines. The basic modulation techniques are described next.

Frequency Shift Keying (FSK). FSK is used in low-speed asynchronous transmission from 300 to 1800 bps. The carrier frequency of the modem is shifted between two discrete frequencies in accordance with the logic levels of

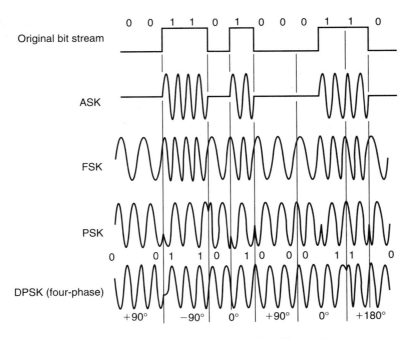

Figure 9-5 Modulation techniques employed by modems.

the digital signal. The higher of the two frequencies represents a mark (logic 1) and the lower frequency represents a space (logic 0). Mark and space frequencies lie within the 300- to 3400-Hz passband of the PSTN. Bell 103/113-type modems utilize FSK.

Amplitude Shift Keying (ASK).   ASK is a form of AM in which the amplitude of the carrier frequency is turned on and off in accordance with the digital data. ASK is also referred to as OOK, which stands for *on–off keying*. The reverse channel of Bell 202-type modems use ASK for error control. Secondary RS-232-C and RS-449 circuits are used as the driver and receiver interface. Transmission speeds for this application are generally limited to 5 bps due to noise problems and bandwidth constraints.

Phase Shift Keying (PSK).   PSK is a modulation technique in which the carrier frequency remains the same and its phase is shifted in accordance with the digital data. When only two phases are used, this is referred to as *binary phase shift keying* (BPSK).

Differential Phase Shift Keying (DPSK).   DPSK is a variation of PSK that is employed by modems with data transmission rates of 1200 bps and above. Multiple phases are used to encode groups of bits called *dibits, tribits, quadbits,* and so on. This reduces the number of phase changes that the carrier frequency undergoes, thus allowing the data transmission rate to be increased. Figure 9-5 illustrates four-phase DPSK. Two bits, or *dibits,* are encoded into one of four phase changes. The advantage of DPSK over PSK is that an absolute phase reference is not necessary for demodulation of the data. Phase shift is *differential,* meaning that it is referenced to the phase of the carrier during the previously encoded interval. Although DPSK is more costly to implement, it is the most efficient modulation technique used by modems in terms of bandwidth utilization. Several bits can be encoded into a single phase change using DPSK. The Bell 212A employs four-phase DPSK to achieve 1200 bps. Bell 208-type modems use eight-phase PSK to achieve 4800 bps.

Quadrature Amplitude Modulation (QAM). QAM is a combination of ASK and PSK. It is used to achieve higher transmission rates than PSK alone. The Bell 209A modem, for example, operates at a transmission rate of 9600 bps. The QAM signal is comprised of 16 carrier states and three different amplitudes. Data rates of 14,400, 16,800, 19,200, and 28,800 bps are possible using the QAM technique. As many as 64 carrier states are required at these speeds.

9.4 MODEM TRANSMISSION MODES

Three transmission modes of operation are used by modems: *simplex, half-duplex,* and *full-duplex.* They are defined next.

Simplex. When a modem is configured for simplex operation, data are sent or received in one direction only.

Half-Duplex. In half-duplex mode, the communications channel is shared between sending and receiving stations. The transmission and reception of data are performed in an alternating manner. Before a change in transmission direction can occur, the transmitter at one end must be turned off while the other end is turned on. The time that it takes for transmission directions to change between two stations is referred to as *modem turnaround time.*

Full-Duplex. Modems capable of operating in full-duplex can transmit and receive data simultaneously. In two-wire telephone circuits, this is performed by using frequency-division multiplexing (FDM) whereby two separate frequency channels, *low band* and *high band,* are allocated within the passband of the telephone lines. One modem transmits on the low band and receives on the high band. The other modem transmits on the high band and receives on the low band, thus allowing full-duplex operation. In private or leased lines, four-wire circuits are often used, whereby the transmit and receive circuits are physically separate from each other. The entire frequency band can be utilized by each modem in this case since the channel is not shared. Another technique is called *echo cancelling*. Echo cancelling allows both modems to transmit simultaneously on the same frequency.

When operating in full-duplex mode over a single channel, an agreement between the two stations is typically made beforehand as to which bands will be used to transmit and receive data on. This eliminates the problem of both stations transmitting and receiving on the same frequency band. Communications cannot occur in this case unless echo cancelling is used. Historically, full-duplex modems have been designated to operate in one of two modes:

1. Originate mode
2. Answer mode

A modem configured for the *originate mode* is the station that originates the call. In the originate mode, transmission occurs on the low band of frequencies and reception occurs on the high band of frequencies. The station that answers the call must be configured for the *answer mode*. In the answer mode, transmission occurs on the high band of frequencies and reception occurs on the low band of frequencies. When a host computer services several remote terminals, the interfacing modem to the host computer is set to the answer mode; that is, it answers the calls placed by the remote terminal's modems that have been configured for the originate mode.

9.5 THE BELL FAMILY OF MODEMS

The Bell System, historically, has dominated the modem market. Operating specifications of the various types of Bell modems have become de facto standards and have evolved into international standards set forth by CCITT. Some of the most common types of modems will now be discussed.

9.5.1 Bell 103/113 Modem

The Bell 103/113 modem is an asynchronous modem designed to operate full-duplex over switched or leased lines. Transmission speed is limited to 300 bps. To operate in the full-duplex mode, the Bell 103/113 modem employs FDM within the 300- to 3400-Hz bandwidth of the switched phone lines. See Figure 9-6. FSK is the modulation technique used. The 3-kHz passband has been divided into two separate frequency channels: a low band and a high band. The low band includes the FSK mark and space frequencies of 1270 and 1070 Hz, respectively. Mark and space frequencies for the high band are 2225 and 2025 Hz, respectively. The Bell 113A/D operates in the originate mode only. The Bell 113B/C operates in the answer mode only.

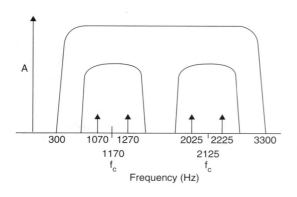

Specifications
Data:
Serial, binary, asynchronous, full duplex
Data Transfer Rate:
0 to 300 bps
Modulation:
Frequency shift-keyed (FSK) FM
Frequency Assignment:

		Origin
Transmit	1070-Hz SPACE 1270-Hz MARK	2025-Hz SPACE 2225-Hz MARK
Receive	2025-Hz SPACE 2225-Hz MARK	1070-Hz SPACE 1270-Hz MARK

Transmit Level:
0 to −12 dBm
Receive Level:
0 to −50 dBm simultaneous with adjacent channel transmitter at as much as 0 dBm.

Specifications and channel assignments for the full-duplex 300-bps asynchronous Bell 103/113 modem are shown in this illustration.

Figure 9-6   Bell 103/113 frequency assignment.   (Courtesy of Racal-Vadic.)

9.5.2 Bell 202 Modem

A disadvantage with the Bell 103-type modem is having to divide the existing bandwidth of the phone lines into one-half in order to obtain two frequency channels for full-duplex operation. Transmission speed is sacrificed in this case. The Bell 202 modem operates in the half-duplex mode. Since transmission occurs in only one direction at a time between modems, the entire bandwidth can be utilized by the transmitted signal. A higher transmission rate is achieved. For switched lines, the transmission rate is 1200 bps, whereas for leased lines with C2 conditioning, 1800 bps is used. FSK is the modulation technique employed. A mark is represented by 1200 Hz and a space is represented by 2200 Hz. Figure 9-7 illustrates the frequency assignment for the Bell 202 modem.

Handshaking is performed by the Bell 202 modem through the use of RS-232-C interface control circuits: RTS and CTS. When data are to be transmitted by DTE, RTS is activated. RTS causes DCE to turn its carrier frequency ON and inhibit its receiver section. The remote DCE detects the carrier and turns its CD line ON. When DCE is ready to transmit data onto the phone lines, CTS is sent to DTE and DTE transmits its data to DCE. When all of DTE's data have been transmitted, it turns RTS OFF. This instructs DCE to complete the transmission of the remaining data onto the communications channel and turn its carrier OFF. DCE inhibits CTS. The channel is quiet at this time. Once the remote DCE detects that the carrier is gone, CD is turned OFF and the remote station begins its transmission using the same procedure.

The process of changing the direction of transmission in half-duplex operation is called *modem turnaround time.* For long-distance communications, echo

Specifications

Data:

Serial, binary, asynchronous, half duplex on two-wire lines

Data Transfer Rate:

0 to 1200 bps — switched network

0 to 1800 bps — leased lines with C2 conditioning

Optional 5-bps AM reverse channel transmitter and receiver available for switched-network units

Modulation:

Frequency-shift keyed (FSK) FM

Frequency Assignment:

MARK 1200 Hz; SPACE 2200 Hz

Transmit Level:

0 to −12 dBm

Receive Level:

0 to −50 dBm switched network

0 to −40 dBm leased network

Specifications and channel assignments for the half-duplex 1200-bps asynchronous Bell 202 modem are shown in this illustration.

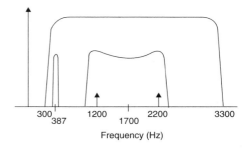

Frequency (Hz)

A significant factor to consider when using a 202 modem is line turnaround time required for switching from transmit mode to the receive mode. Echo suppressors in the telephone equipment that are required for voice transmission on long-distance calls must be turned off by the modem to transmit digital data. The modem must provide a 200-ms signal to the line to turn off the echo suppressors every time it goes from transmit to receive mode; hence, if short records are being transmitted, the turnaround time can slow the throughput considerably. In addition, many terminals and computers do not have the capability to control the Request To Send lead on the Bell 202 interface and can only support full-duplex 103-line discipline.

Figure 9-7 Bell 202 frequency assignment. (Courtesy of Racal-Vadic.)

suppressors in the line must have sufficient time to *turn around* once the local carrier is turned off and the remote carrier is turned on. The turnaround time of an echo suppressor can be as high as 100 ms. Additional time is also necessary for echoes to subside and for the local receiver to turn on its CD signal upon receiving the return carrier frequency. A 150- to 200-ms delay is inserted between RTS and CTS for this purpose. Modem turnaround time is a significant factor to consider in the transfer of files back and forth. Short files requiring numerous turnarounds can reduce the *throughput* of data considerably. (See Sections 9.12 and 9.14 for a definition of throughput.)

9.5.2.1 Pseudo full-duplex operation. Although the Bell 202 is generally considered a half-duplex modem, *pseudo full-duplex* operation is possible through a FDM low-speed, ASK, 5-bps *reverse channel*. The reverse channel is supported by RS-232-C secondary interface circuits, as depicted in Figure 9-8. A 387-Hz carrier frequency is amplitude shift keyed (ASK) at a low enough rate to prevent the resulting AM sidebands from spilling into the passband of the main channel. The reverse channel is used primarily to indicate that a remote station is connected to the interface. It is also used as a feedback signal requesting the retransmission of data in the event of a detected error.

9.5.2.2 Disabling the echo suppressors. For full-duplex operation, echo suppressors must be disabled in the long-distance trunk facilities. The PSTN has set provisions for disabling echo suppressors by applying a tone within the range

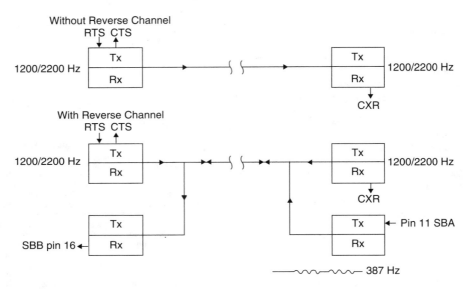

Figure 9-8 Bell 202 reverse channel operation. (Courtesy of Racal-Vadic.)

of 2010 and 2240 Hz for 350 ± 50 ms. The tone must be presented at a time when the channel is quiet. The tone may be generated from either direction. Typically, the answer modem will apply this tone during a call setup procedure before shifting to its idle frequency. The echo suppressor will remain disabled as long as one of the interconnecting modems continues to assert a signal within the range 300 to 3400 Hz. They will become enabled again if there is any period as long as 100 ms without any signal on the line. In relation to the Bell 202 modem, another purpose of its reverse channel is to hold echo suppressors disabled in pseudo full-duplex operation.

There are several versions of the Bell 202 modem. Later versions have upgraded the reverse channel transmission rate to 75 and 150 bps through the use of FSK. Mark and space frequencies for the reverse channel FSK signal are 390 and 490 Hz, respectively. CCITT's V.23 international standard is very similar to the BELL 202 specification. The forward channel rate is the same, at 1200 bps. The reverse channel FSK rate is 75 bps with mark and space frequencies of 390 and 450 Hz.

9.5.3 Bell 212A Modem

The Bell 212A is a two-speed modem that operates full-duplex and supports asynchronous or synchronous transmission modes over the switched lines. The low-speed asynchronous mode operates in accordance with the BELL 103 specification with a 300-bps data transfer rate and FSK employed as the modulation technique. Transmit frequencies for the originate mode are 1070 Hz (space) and

1270 Hz (mark). Receive frequencies for the originate mode are 2025 Hz (space) and 2225 Hz (mark). In the answer mode, the opposite frequencies are used to permit full-duplex operation.

In the high-speed mode, characters can be transmitted synchronously or asynchronously at 1200 bps. Four-phase DPSK is used to phase shift a 1200-Hz tone for the originate mode and a 2400-Hz tone for the answer mode. Figure 9-9 illustrates the Bell 212A frequency assignment for the high-speed mode. A mark and a space are not represented by two discrete frequencies, as is the case with FSK. Instead, each consecutive two bits of the serial binary data sent to the 212A modem are encoded into a single phase change of the carrier frequency. The encoded *two* bits are called *dibits*. Since a dibit represents two bits, there are four possible phase changes that the carrier frequency can undergo. This is shown in the phasor diagram of Figure 9-10. Notice that the encoded bits are Gray coded. Four-phase DPSK is also referred to as *quadrature PSK* (QPSK).

Specifications*
    Data: Serial, binary, asynchronous, full duplex
    Data transfer rate: 0 to 1200 bps
    Modulation: Differential phase shift keying (DPSK)
    Originate frequency: 1200 Hz
    Answer frequency: 2400 Hz
*Also see Bell 103 specification (Figure 9-6) for Bell 212A
  low-speed mode.

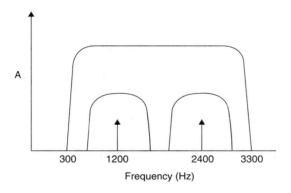

Figure 9-9 Bell 212A frequency assignment for the high-speed mode. Frequency assignment for the low-speed mode is identical to the Bell 103 specification (see Figure 9-6).

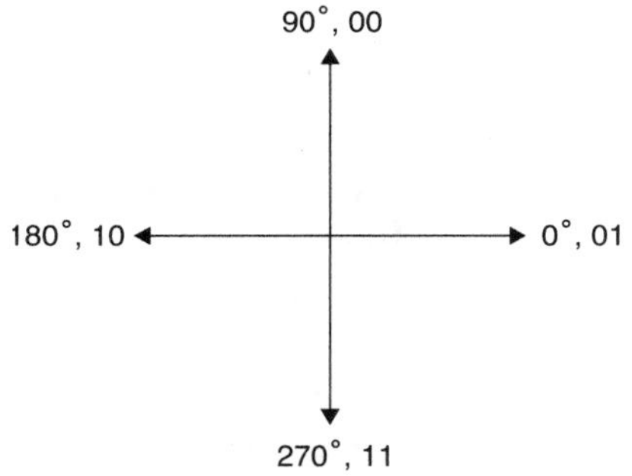

Figure 9-10 Phasor diagram for the Bell 212A modem.

9.5.3.1 M-ary. When PSK or QAM is used as a modulation technique, the term *M-ary*, derived from bi-nary, is used to denote the number of encoded bits used to modulate the carrier frequency. M-ary is governed by the equation

$$n = \log_2 M \qquad (9\text{-}1)$$

where n is the number of encoded bits used to represent a carrier state, M-ary, and M is the number of state changes that the carrier can undergo represented by n bits.

Example 1

The Bell 208 modem employs eight-phase DPSK as its modulation technique. Compute the M-ary.

Solution: $\quad n = \log_2 M$
$\qquad\qquad = \log_2 8$
$\qquad\qquad = 3$

Therefore, M-ary $= 3$ bits.

9.5.3.2 Baud rate versus bit rate. When more complex modulation techniques are employed by modems to achieve higher data transfer rates, a distinction must be made between the transmitted signal's *bit rate* and its *baud rate*. They are not always equal. Bit rate and baud rate are defined as follows:

Baud rate: A signal's baud rate is defined as the rate at which the signal changes per unit time. For modems it is the actual *modulation rate* of the carrier frequency as it is transmitted or received via the communications channel. Baud rate is also referred to as *signaling rate.*

Bit rate: A signal's bit rate is the actual number of binary bits transmitted per second (bps) onto the communications channel.

Bell 212A modems have a modulation rate of 600 baud when operating in the high-speed mode. This is equal to one-half of its bit rate of 1200 bps since its carrier frequency is phase shifted at dibit intervals ($1200 = 2 \times 600$). In its low-speed mode, however, the baud rate and bit rate (300) are equal; that is, the carrier is frequency shifted at the same rate as the binary serial data stream.

9.5.4 Bell 201B/C Modem

The Bell 201 family of modems is designed to operate at a fixed data transfer rate of 2400 bps over the basic, unconditioned, 3002-type line or two- or four-wire private line. The Bell 201A is an obsolete 2000-bps modem, the Bell 201B is for private or leased line applications, and the Bell 201C is for switched or leased line applications. Each modem is designed for half-duplex operation over the switched line or full-duplex operation over the four-wire private line. Four-phase

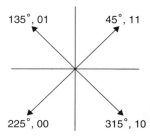

135°, 01   45°, 11

225°, 00   315°, 10

Figure 9-11 Phasor diagram for the Bell 201B/C modem.

DPSK is the modulation technique employed to achieve 2400 bps. The phasor diagram for the Bell 201B/C is shown in Figure 9-11.

9.5.5 Bell 208A and 208B Modems

Bell 208A and 208B modems are designed for synchronous transmission and reception of data at 4800 bps over four-wire private leased lines and switched lines, respectively. Eight-phase DPSK is employed on a 1800-Hz carrier frequency. Each consecutive *three* bits, called *tribits,* of the binary serial input data are encoded into a single phase change of the carrier frequency. The encoding of three bits into a tribit allows the representation eight possible phase changes of the carrier frequency. A transmission rate of three times the baud rate of 1600 is achieved ($4800 = 1600 \times 3$). Figure 9-12 depicts the phasor diagram for the Bell 208A modem. Notice, here again, that a Gray code is used.

9.5.6 Bell 209A Modem

Data transfer rates can be increased further by encoding additional bits into a greater number of phase changes. *Four* bits, or a *quadbit,* for example, can be encoded into 16 possible phase changes (M-ary = 4). The phase differential between adjacent phasors would amount to 22.5 degrees (360 degrees/16 = 22.5 degrees). The problem here, however, is that any phase jitter in excess of 11.25

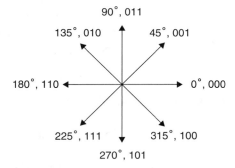

90°, 011

135°, 010   45°, 001

180°, 110   0°, 000

225°, 111   315°, 100

270°, 101

Figure 9-12 Phasor diagram for the Bell 208A modem.

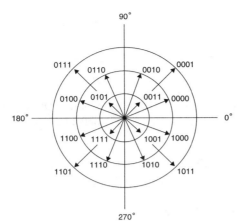

Figure 9-13 Phasor diagram for the Bell 209A modem depicting 9600-bps QAM.

degrees would result in the detection of erroneous data. This amount of phase jitter is not uncommon in long-haul networks that utilize regenerative repeaters and digital multiplexers. For this reason, 16-phase PSK is generally not used.

To avoid the problem of phase jitter, the Bell 209A modem employs a combination of ASK and PSK called *quadrature amplitude modulation* (QAM). QAM, pronounced "Kwamm," is a modulation technique that uses 12 different phases and three different amplitudes to represent 16 possible carrier states. Susceptibility to phase jitter is effectively reduced by increasing the separation between adjacent phasors. The phasor diagram for the Bell 209A QAM signal is shown in Figure 9-13. Each state is represented by one of 16 possible quadbits. By employing QAM, 9600-bps full-duplex synchronous communications is achieved using a baud rate of 2400 ($9600 = 4 \times 2400$) over private four-wire lines with D1 conditioning. The Bell 209A also has provisions for multiplexing multiples of 2400 bps into 9600 bps.

9.6 CCITT MODEMS AND RECOMMENDATIONS

Data transmission standards outside the United States are set by CCITT, which is part of the International Telecommunications Union, headquartered in Geneva, Switzerland. Standards V.21, V.23, and V.26 describe modems similar to the Bell 103, 202, and 201, respectively. V.22 describes modems similar to the Bell 212A, and V.29 is similar to the Bell 209A specification. Table 9-1 lists CCITT modems and recommendations.

9.6.1 CCITT Modem Recommendation V.22bis

CCITT's V.22bis (*bis* means *second revision* in French and *ter* means *third revision*) specification provides for 1200- and 2400-bps synchronous full-duplex communication over switched and two-wire leased lines. Four-phase DPSK is employed as the modulation technique for modem operation at 1200 bps. QAM is used to achieve a data transfer rate of 2400 bps. Modulation rate is specified at 600 baud for both operating speeds. Full-duplex operation is achieved by phase and amplitude shift keying a low-channel carrier frequency of 1200 Hz and a high-channel carrier frequency of 2400 Hz.

Phase and amplitude assignment for dibit (1200 bps) and quadbit (2400 bps) encoding is depicted by CCITT as a *16-point signal constellation* rather than a phasor diagram. They are essentially the same. Figure 9-14 illustrates the signal constellation for the V.22bis specification. For 1200-bps operation, the data stream is divided into groups of two consecutive bits, or dibits. Each dibit is encoded into a quadrant phase change relative to the preceding phase of the carrier frequency. This is shown in Table 9-2. For 2400-bps operation, the data stream is divided into groups of four consecutive bits, or quadbits. The two least significant bits of the quad bit are encoded into a quadrant phase change in the same manner as for 1200-bps operation. The most significant two bits of the quadbit define one of four signaling elements associated with the new quadrant.

TABLE 9-1 CCITT Modems and Recommendations

Recommendations	Description
V.21	0 to 200 (300) bps (similar to Bell 103). Defined for FDX switched network operation.
V.22	1200-bps, FDX, switched, and leased line network operation.
V.22bis	1200/2400-bps, FDX, switched, and leased line network operation.
V.23	600/1200-bps (similar to Bell 202). Defined for HDX switched network operation. Optional 75-bps reverse channel.
V.24	Definition of interchange circuits (similar to EIA RS-232-C).
V.25	Automatic calling units (similar to Bell 801).
V.25bis	Serial interface autocalling.
V.26	2400 bps (identical to Bell 201B). Defined for four-wire leased circuits.
V.26bis	2400/1200 bps (similar to Bell 201C). Defined for switched network operation.
V.26ter	2400 bps over the switched network using echo cancelling.
V.27	4800 bps (similar to Bell 208A). Defined for leased circuits using manual equalizers.
V.27bis	4800/2400 bps with autoequalizers for leased lines.
V.27ter	4800/2400 bps for use on switched lines.
V.28	Electrical characteristics for interchange circuits (similar to RS-232-C).
V.29	9600-bps FDX (similar to Bell 209). Defined for leased circuits.
V.32	9600-bps FDX for switched or leased-line circuits using echo cancelling.
V.32bis	14,400-bps FDX for switched or leased-line circuits using echo cancelling.
V.33	14,400-bps FDX for leased lines.
V.42	Error-correction procedures for DCEs using asynchronous to synchronous conversion.
V.42bis	An extension of V.42 that defines data compression for use with V.42.
V.34	(V.FAST) ITU-TS 28,800 bps standard.

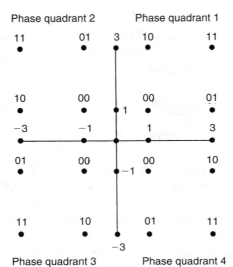

Figure 9-14 V.22bis 16-point signal constellation.

TABLE 9-2 V.22bis Line Encoding

First two bits in quadbit (2400 bps) or dibit values (1200 bps)	Phase quadrant change
00	$1 \rightarrow 2$ $2 \rightarrow 3$ $3 \rightarrow 4$ $4 \rightarrow 1$ } 90°
01	$1 \rightarrow 1$ $2 \rightarrow 2$ $3 \rightarrow 3$ $4 \rightarrow 4$ } 0°
11	$1 \rightarrow 4$ $2 \rightarrow 1$ $3 \rightarrow 2$ $4 \rightarrow 3$ } 270°
10	$1 \rightarrow 3$ $2 \rightarrow 4$ $3 \rightarrow 1$ $4 \rightarrow 2$ } 180°

9.6.2 CCITT Modem Recommendation V.29

CCITT's V.29 specification is the first internationally recognized standard for 9600-bps communications. This standard provides for synchronous data transmission over four-wire leased lines. The same 16-point QAM signal constellation used for V.22bis is used for the V.29 specification. The higher data transfer rate is made possible by using a single carrier frequency of 1700 Hz and increasing the baud rate from V.22bis's 600 baud to 2400 baud. The entire bandwidth is utilized. Some modem manufacturers have elected to use V.29-compatible modems in half-duplex mode over the switched lines. Pseudo-full-duplex operation over the switched lines is also performed by one of two techniques: *ping-pong* or *statistical duplexing*. These are described next.

Ping-Pong. Ping-pong is a method of simulating full-duplex operation between two modems. Data sent to each modem by DTE are buffered and automatically exchanged over the link. By rapidly turning carriers on and off in a successive fashion (hence the name *ping-pong*) through flow control procedures, full-duplex operation is simulated.

Statistical Duplexing. Statistical duplexing uses a low-speed, 300-bps reverse channel, similar to the manner in which the Bell 202 operates in pseudo full-duplex. The intent of the reverse channel is to allow keyboard data entry from an operator while at the same time receiving a file from a remote station. By monitoring the modem's data buffers, the direction of the data transaction can be sensed and high and low-speed channels can be reversed to suit the desired condition.

9.6.3 CCITT Modem Recommendation V.32

CCITT's V.32 recommendation is intended for the use of 9600-bps synchronous modems on connections to switched and leased lines. The recommendation also specifies signaling rates of 2400 bps (based on V.26ter) and 4800 bps. QAM is the modulation technique employed on a carrier frequency of 1800 Hz. V.32 is very similar to V.29 except that an optional encoding technique called *trellis encoding* is specified.[†] Trellis encoding divides the data stream to be transmitted into groups of five consecutive bits or *quintbits*. This unique encoding technique results in superior signal-to-noise ratios. A 32-point signal constellation is achieved (M-ary = 5). Figure 9-15 depicts the 32-point trellis signal constellation.

9.6.3.1 Echo cancellation. One principal characteristic outlined in the V.32 standard is to provide for a 9600-bps modem with true full-duplex operation over the switched lines. Through the advanced technology of digital signal processors (DSPs), full-duplex operation is achieved by a technique called *echo cancellation*. Echo cancellation is performed by adding an *inverted* replica of the

[†]For further details on trellis encoding, refer to CCITT Recommendation V.32, 1984.

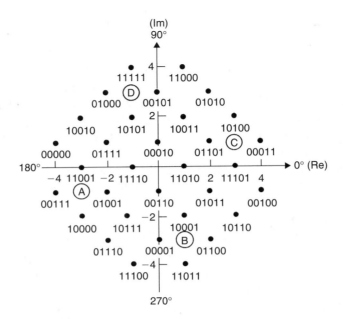

32-point signal structure with trellis coding for 9600 bps and states A B C D used at 4800 bps and for training

Figure 9-15 V.32 trellis 32-point signal constellation.

transmitted signal to the received data stream. This permits the transmitted data from each modem to use the same carrier frequency and modulation technique simultaneously. The two clashing signals are separated by the receiver section of each modem.

9.6.4 CCITT Modem Recommendation V.32bis and V.32 terbo

In 1991, the *V.32bis* standard created a new benchmark in the industry, allowing modem speeds of 14,400 bits per second (bps), 50% faster than the V.32 9600 bps modem standard at the time. Instead of using a 16-point signal constellation (four bits per baud) as in the V.32 standard, V.32*bis* uses a 64-point signal constellation (six bits per baud) and maintains the same 2400 baud rate as V.32. Thus, 14,400 bps (6 × 2400 = 14,400), full-duplex operation over the two-wire switched lines is achieved. Another improvement of V.32*bis* over V.32 is the inclusion of *automatic fall forward*, the ability to return to a higher transmission speed when the line quality improves. V.32*bis* modems can also *fall back* or quickly slow down to 12,200 bps, 9600, or 4800 bps if line quality degrades. Most V.32*bis* modems also support Group III fax. Group III fax is a standard for fax communication that specifies the connection procedure between two fax machines or fax modems and the data compression procedure that will be followed during the transmission.

In August 1993, U.S. Robotics Inc. announced a major evolution to its product line: the **V.32 terbo** protocol with its proprietary *Adaptive Speed Leveling*[†] technology, which boosts modem speeds to 21.6 kbps. These new features fall into three new categories: increased data rates, FAX enhancements, and high-end features.[‡] V.32 terbo is the new 19.2-kbps data transmission rate developed by AT&T. It is designed to deliver a 33% increase in speed over the 14,400-V.32bis standard.

9.6.5 CCITT Modem Recommendation V.33

The CCITT V.33 Recommendation is designed for modems that will operate over point-to-point four-wire leased lines. It is similar to V.32 except that it encodes a redundant bit and six information bits to produce a transmission rate of 14,400 bps at 2400 baud. The carrier frequency is also 1800 Hz. The V.33 128-point signal constellation is shown in Figure 9-16.

9.6.6 CCITT Modem Recommendation V.42

A relatively new modem protocol adopted in 1988 by CCITT is the V.42 standard: *Error Correcting Procedures for DCEs.* The V.42 standard is designed to address asynchronous-to-synchronous conversions, error detection and correction, and modems that do not have such protocols. V.42's main impetus revolves around a new protocol called *Link Access Procedure for Modems* (LAP M). LAP M is similar to the packet switching protocol used in the X.25 standard. An alternative procedure developed by Microcom Inc. has also been adopted by the V.42 standard. This procedure is called *Microcom Networking Protocol* (MNP). Both MNP and LAP M are discussed later in this chapter.

9.6.7 CCITT Modem Recommendation V.42bis

In an effort to enhance the performance of error-correcting modems that implement the V.42 standard, CCITT adopted the V.42bis standard, which addresses *data compression for DCEs using error-correcting procedures.* Modems employing data compression have significant throughput performance over their predecessors. The V.42bis standard can achieve 3:1 to 4:1 data compression ratios for ASCII text. The algorithm specified by the standard is British Telecom's *BTLZ* technique. In 1990, the CCITT Study Group XII voted to include Microcom's *MNP 5* and *MNP 7*. These two new revisions to the standard are

[†]*ASL* is a trademark of U.S. Robotics.
[‡]*U.S. Robotics News,* U.S. Robotics, Skokie, Ill.

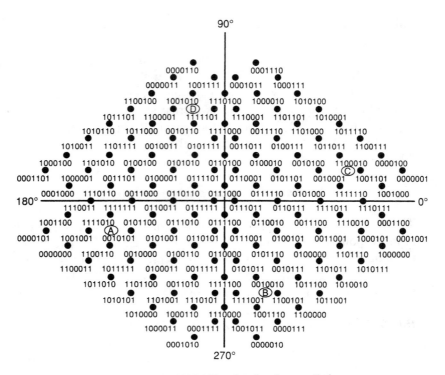

Figure 9-16   V.33 128-point signal constellation.

data compression and advanced data compression algorithms, respectively. Throughput rates up to 56,600 bps can be achieved by today's modems that employ V.42bis data compression.

9.6.8 CCITT Modem Recommendation V.34 (V.fast)

The *International Standards Union–Telecommunications Standardization Sector* (ITU-TS and formerly CCITT) established the TR–30 Group (formerly SGXVII) to work on a recommendation for the next generation of modem speed: *V.fast*. V.fast was officially adopted in June of 1994 and has been designated V.34. The new standard's speed is 28,800 bps *without* data compression. With data compression, the new modems will be able to send asynchronous data two to three times as fast, thus dramatically reducing telephone costs. V.34 will automatically adapt to line conditions and adjust its speed up or down to ensure data integrity.

Increasing both complexity and speed doesn't come easy, however. There are definite boundaries for both and V.34 pushes these limits. A number of innovations enable V.34 to go faster.[†]

[†]Dale Walsh, U.S. Robotics Vice President, Advanced Products, "V.fast: The modem's next generation," *U.S. Robotics The Intelligent Choice in Data Communications*, 1993.

- **Nonlinear coding** combats the effects of nonlinearities, such as harmonic distortion and amplitude proportional noise.
- **Multidimensional coding** and **constellation shaping** gives the data greater immunity to noise in the channel.
- **Reduced complexity decoding** makes it possible to use more complex codes. A 64-state, 4-dimensional code is now feasible with this technology. Without it, this code would increase the computations 40 percent.
- **Precoding** allows more of the available channel bandwidth to be used, which means more symbols can be sent faster. With older standards the symbols were sent closer to the middle of the bandwidth. The outer limits of the bandwidth is where amplitude attenuation and phase distortion are most severe.
- **Line probing** is a scheme that quickly looks during the training sequence at the line for impairments that might be encountered. It then attempts to select the best solution to counteract the circuit problems.

Inside the V.34 modems, look for more sophisticated hardware elements. For example, high resolution sigma-delta converters lower self-generated receiver front end noise. V.34 algorithms chew up memory; so they'll need larger and faster SRAMs. Finally, a doubling of modem-to-modem speed means that the modem-to-computer speed will also escalate, at the most to 115,200 bps.

9.7 ANALOG LOOPBACK TEST (ALB)

Among the standard features of today's modems are self-test, remote test, and loopback test capabilities. These test capabilities permit fault isolation down to the modem, customer facilities, or the telephone facilities. An *analog loopback test* (ALB) is a test feature that verifies the operation of the local modem and its connection to the DTE. The modem's transmitter section, normally connected to the telephone channel, is *looped back* to its own receiver section, as illustrated in Figure 9-17. A character typed from a terminal is therefore modulated onto the carrier frequency of the modem in the normal process. The analog signal that has been looped back is demodulated by the receiver section of the modem and sent back to the terminal for display. If the character typed is not displayed, a faulty modem should be considered.

9.8 DIGITAL LOOPBACK TEST (DLB)

The *digital loopback test* (DLB) permits data generated from a terminal to be sent between two modem interfaces, thus testing the transmit and receive sections of both modems, the interconnecting telephone lines, the central office, and the terminal itself.

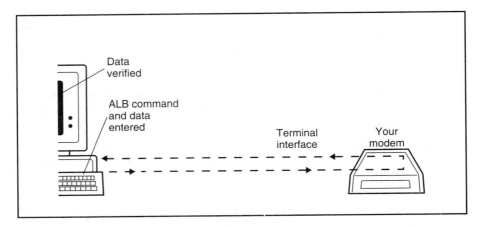

Figure 9-17 Analog loopback test setup. (Reprinted with permission from Racal-Vadic.)

A *local* DLB uses data from a remote terminal to test each modem and the interconnecting telephone lines. This is shown in Figure 9-18a. The data are said to be looped back *locally*. A command must be initiated and sent by the local terminal to the locally attached modem to begin the test. An operator at the remote station is necessary to send and verify that characters typed at the remote terminal are looped back by the local modem, back through the telephone facilities, and displayed by the remote terminal.

A *remote* DLB, as shown in Figure 9-18b, performs the same test as a local DLB, except that the *remote* operator initiates the DLB test and the local modem loops back the data typed by an operator at the remote terminal. No local operator is required for this test. In either test, if the data do not match what has been typed (assuming that the terminal and its interface to the modem is functioning), a problem exists either in the telephone lines or one of the modems. An ALB test should then be performed on each modem. If the ALB test is successful for both modems, this indicates that the telephone facilities have failed.

9.9 MODEM SYNCHRONIZATION

Modem receivers are classified into two categories: *coherent* and *noncoherent*. Coherent receivers extract and recover the carrier frequency of the received signal. The recovered frequency is phase locked to the original carrier frequency and used in the demodulation process. High-speed modems that utilize multiphase PSK or QAM contain a coherent receiver section. In contrast, noncoherent receivers do not require carrier synchronization for the purpose of demodulation. The carrier frequencies of the transmitting and receiving modems are independent of each other. Data are recovered by detecting the *shape* of the carrier frequency. Envelope detectors are often used in noncoherent receivers.

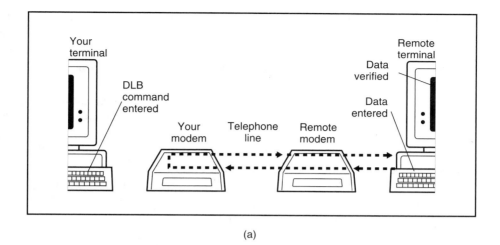

(a)

(b)

Figure 9-18   (a) Local DLB test setup; (b) remote DLB test setup. (Reprinted with permission from Racal-Vadic.)

9.10 SCRAMBLING AND DESCRAMBLING

To extract the carrier frequency in modems with coherent receivers, phase-locked loop (PLL) technology is employed. A PLL circuit is a feedback-controlled circuit. Its basic components include a voltage-controlled oscillator (VCO), a low-pass filter, and a phase comparator. The output of the VCO is phase compared with the incoming signal. Any difference between the two produces an error voltage which is used to raise or lower the VCO's frequency to maintain phase lock. The frequency spectrum of the digital signal usually contains spectral energy at the carrier frequency (or symmetrical about) in which the PLL can achieve phase lock. If a bit pattern of all 1's or 0's is transmitted by the modem for a prolonged period of time, carrier synchronization can be lost and demodulation cannot occur. This is due to the lack of carrier spectral energy in the received signal necessary to maintain phase lock. To prevent this, high-speed modems typically scramble their data with a *scrambling* circuit to produce an optimum bit pattern for the receiving modem to achieve phase lock to. The data stream is scrambled in accordance with a defined mathematical algorithm prior to modulation of the carrier frequency. At the receiving end, the detected data stream is unscrambled by a *descrambler* circuit.

CCITT's V.22bis recommendation specifies self-synchronizing scrambler and descrambler circuits to be included in the transmitter and receiver sections of the modem. The transmitted data stream is scrambled by dividing it by the polynomial given by

$$1 + X^{-14} + X^{-17} \tag{9-2}$$

The coefficients of the quotients of this division, taken in descending order, form the data sequence at the output of the scrambler. This sequence is equal to

$$D_s = D_i \oplus D_s * X^{-14} \oplus D_s * X^{-17} \tag{9-3}$$

The demodulated data stream is descrambled by multiplying it by the polynomial given by equation (9-2). The coefficients of the recovered data sequence, taken in descending order, form the output data sequence given by

$$D_o = D_R (1 \oplus X^{-14} \oplus X^{-17}) \tag{9-4}$$

where D_s = data sequence at the output of the scrambler
$\quad D_i$ = data sequence applied to the scrambler
$\quad D_R$ = data sequence applied to the descrambler
$\quad D_o$ = data sequence at the output of the descrambler
$\quad \oplus$ = modulo 2 addition
$\quad *$ = binary multiplication

9.11 MODEM FILE TRANSFER PROTOCOLS

Historically, file transfers over the PSTN were error prone. Errors occurring in text messages were tolerable since the messages could still be interpreted by users. Recently, however, there has been growing concern over the integrity of interactive asynchronous data communications over the PSTN. Some studies estimate that over 90% of users communicate asynchronously in an interactive mode with a personal computer (PC) or mainframe linked to various bulletin board services (BBS) and other mainframes and PCs. Many of these applications make speed and error control a necessity rather than a luxury. Software developers have hastened to fill the need for efficient file transfer protocols. As a result, an array of protocols are available to suit our needs for uploading and downloading files to and from other terminals. A brief discussion of the most widely used file transfer protocols offered by BBS systems will now be given.

9.11.1 Xmodem

When the first computer bulletin board services went on line, there was an obvious need for an error-controlling mechanism that would allow program files to be reliably exchanged among users of PCs. Xmodem was one of the first software data communications protocols designed to meet this need. Xmodem is sometimes referred to as the *Christensen protocol* after its designer Ward Christensen. It has become a de facto standard since the late 1970s for modem file transfer verification between PCs.

Xmodem is an ACK/NAK alternating protocol that operates in half-duplex mode. This means that when a packet of data is transmitted and received without error, a positive acknowledgment (ACK) control packet is sent back to the transmitter. In the event of an error, a negative acknowledgment (NAK) control packet is returned to the transmitter. The packet is retransmitted in this case. Packets must be acknowledged before communications can resume. This form of error control is referred to as *automatic repeat request* (ARQ). As shown in Figure 9-19, 128 bytes are used in the data field of the Xmodem packet. An eight-bit checksum is appended to the data field as the block check character (BCC).

A variation of Xmodem is Xmodem/CRC. Xmodem/CRC uses a 16-bit cyclic redundancy check (CRC) of the binary data within the packet instead of an eight-bit checksum. The 16-bit CRC used as a BCC is much more reliable in terms of detecting errors.

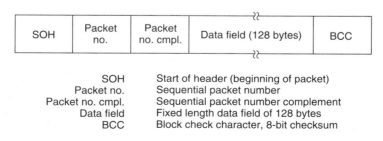

| SOH | Packet no. | Packet no. cmpl. | Data field (128 bytes) | BCC |

SOH	Start of header (beginning of packet)
Packet no.	Sequential packet number
Packet no. cmpl.	Sequential packet number complement
Data field	Fixed length data field of 128 bytes
BCC	Block check character, 8-bit checksum

Figure 9-19 XMODEM packet format.

9.11.2 1K-Xmodem

This file transfer protocol is identical to Xmodem described above, except that the block size has been increased from 128 bytes in the data field to 1024 (1K), hence the name *1K-Xmodem*. For relatively quiet telephone lines, 1K-Xmodem is faster than its counterpart. This is true since a smaller number of blocks are necessary to transfer the same amount of data, and therefore fewer blocks need to be checked for errors.

9.11.3 1K-Xmodem/G

Another variation of Xmodem is *1K-Xmodem/G*. This protocol also uses a 1-kbyte data field. The "/G" denotes the use of MNP (Microcom networking protocol), a protocol that resides in the modem's hardware (see the detailed discussion of MNP below). For modems that support MNP, 1K-Xmodem/G offers a very fast and efficient file transfer protocol. Unlike many protocols that perform a block-by-block handshake for error control, the use of MNP allows multiple frames to be transmitted before an acknowledgment is required. This ability increases the overall throughput.

9.11.4 Ymodem

The unique advantage of *Ymodem* is that it is a *multiple file transfer protocol*. It was developed by Chuck Forsberg of Omen Technology Inc. Up to a maximum of 99 files can be uploaded (transmitted from the local terminal to the remote terminal) or downloaded (transmitted from the remote terminal to the local terminal) in succession. Block sizes are typically 1K bytes in length and the BCC is CRC-16, a 16-bit CRC algorithm standard.

9.11.5 Ymodem/G

A very fast and efficient protocol that supports multiple file transfers is *Ymodem/G*. This protocol, like 1K-Xmodem/G, requires a modem with MNP residing in its hardware. Most of the latest 9600-bps modems and above utilize MNP. Data compression is also performed in the modem's hardware. Both error control and data compression, residing in the modem's hardware, adhere to CCITT (Consultative Committee for International Telephony and Telegraphy) V.42 and V.42bis standards that will be discussed later. For communication sessions requiring large file transfers at high data rates (9600 bps, 14,400 bps, 19,200 bps, 38,400 bps), Ymodem/G, or *Zmodem*, discussed below, is among the recommended choices.

9.11.6 Zmodem

One of the most widely used file transfer protocols is *Zmodem*. Some statistical surveys indicate that Zmodem is utilized in over 70% of BBS file transfers. This high-performance and high-reliability protocol was also developed by Chuck Forsberg. Zmodem uses a 32-bit CRC to reduce the number of undetected errors. Error checking of this degree is accurate to 99.9999%. Multiple file transfers and variable block sizes of data are supported. MNP hardware is not necessary with Zmodem and if connecting links do not support Zmodem, it can step down to Ymodem.

Zmodem does not wait for acknowledgments from the receiving device. Instead, it assumes that the data have been received without error unless a repeat request is sent for a specific block. Zmodem also has the unique capability to resume file transfers that have been aborted for some reason and thus only partially completed. This is called *crash recovery*.[†] For overall performance and reliability, Zmodem is recommended as one of the best file transfer protocols to use.

9.11.7 Kermit

Another commonly used protocol is Kermit, developed at Columbia University. It is used extensively with PCs, minicomputers, and mainframes. Like Xmodem, Kermit is an ACK/NAK alternating protocol that uses ARQ for error correction. Figure 9-20 illustrates the format of the Kermit packet. Although the data field of the packet is shorter than that of Xmodem, a length field (LEN) offers the flexibility of varying the packet size up to a maximum of 94 bytes. Small buffers can be accommodated, and the final packet in a transmission is not restricted to a fixed number of bytes.

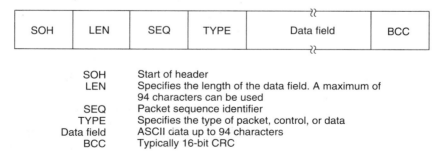

SOH	LEN	SEQ	TYPE	Data field	BCC

SOH	Start of header
LEN	Specifies the length of the data field. A maximum of 94 characters can be used
SEQ	Packet sequence identifier
TYPE	Specifies the type of packet, control, or data
Data field	ASCII data up to 94 characters
BCC	Typically 16-bit CRC

Figure 9-20   Kermit packet format.

[†]This information was downloaded from the *Wildcat BBS*. A more detailed description of Zmodem can be downloaded from many BBS services under the file name of *ZMODEM8.ZIP*.

9.11.8 Microcom Networking Protocol (MNP)

Unfortunately, the move to higher data transfer rates over the switched telephone lines involves the trade-off of increased errors. To maintain the same error rate, a V.22bis compatible modem operating at 2400 bps requires a signal-to-noise ratio (SNR) several decibels higher than a Bell 212A-compatible modem operating at 1200 bps. A Massachusetts-based company called Microcom, Inc., has developed a method of controlling these errors. This method is MNP, an acronym for *Microcom networking protocol;* it offers far more capability than the protocols discussed thus far.

MNP resides in the modem's hardware. This relieves the user's DTE from the burden of controlling errors via software-based protocols such as Xmodem. The CPU is left to perform other tasks, thus increasing system productivity. MNP formats ASCII data sent from DTE into an SDLC frame, similar to that used on larger mainframe systems. The basic frame structure is shown in Figure 9-21. By using a synchronous format, MNP is capable of reducing the number of bits in the user's data. Once the data have been converted, MNP will calculate a check sum or CRC for the number of bits in the data. This CRC is then sent with the data. At the receiving modem, the CRC is removed from the frame and a new CRC is computed. This new CRC is then compared with the CRC originally received from the transmitting modem. If the new check sum matches the original CRC computed by the transmitting modem, the data are passed on to the user's DTE in its original asynchronous format. If the two do not match, the frame is in error.

An advantage of MNP over other ACK/NAK protocols is that multiple frames can be transmitted without having to wait for an acknowledgment from the receiver for each individual frame. Throughput is increased. This is possible through a technique called *Go Back N*. Should any frame be in error using this technique, an ARQ control frame is sent back to the transmitting modem. The transmitting modem must then go back and retransmit the frame in error, including all other frames that followed.

"GO BACK N"

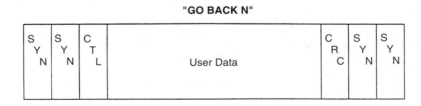

Basic MNP frame

Figure 9-21   MNP frame structure.

Figure 9-22 illustrates the Go Back N technique. A typical number of frames that may be transmitted in succession by a modem that features MNP may be five. Modem A has transmitted five frames of data. Frame number 3 contains an inaccurate CRC calculation. Modem B requests a retransmission of frames 3 through 5. This scheme results in a total number of eight frames that have been transmitted in order to send five frames of data.

MNP has been disseminated within the data communications industry through licensing agreements with Microcom, Inc. It has become a de facto standard for a number of manufacturers and service providers: Racal-Vadic and GTE Telenet to name just two.[†] To date, there are nine designated classes of MNP. Each is designed to reduce data errors and optimize throughput over varying line conditions. The nine MNP classes are listed in Table 9-3. Classes 1 through 4 have become part of the CCITT V.42 recommendation.

9.11.9 Link Access Procedure (LAP)

Another widely used data communications protocol currently being recognized by CCITT's V.42 (Error Correction Procedures for DCEs) standard is link access procedure (LAP). Several variations of LAP have been written to meet specific applications of the X.25 standard. LAP B, for example, is the link-level protocol used for the X.25 public data networks. LAP D is the link-level protocol for use in ISDN (Integrated Services Digital Network). For modems, LAP M has been designed specifically as the link access protocol. LAP M is an error controller such as MNP. AT&T, Hayes, the United Kingdom (led by British Telecom), and Japan support LAP M. This raises the issue of compatibility between modems

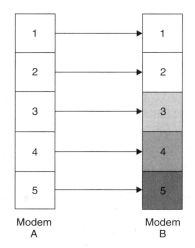

"Go Back N" MNP Scheme

Should any frame be errored the transmitting modem must retransmit the bad frame and all frames that follow.

Example:
Modem A has transmitted 5 frames of data. Frame 3 contains an inaccurate CRC calculation. Modem B requests a retransmission of frames 3 through 5. This scheme results in a total of 8 frames that have been transmitted in order to send 5 frames of data.

Figure 9-22 MNP's "Go Back *N*" technique. (Courtesy of Racal-Vadic.)

[†]*A Primer on MNP*, from Racal-Vadic.

TABLE 9-3 The Nine Designated Classes of MNP

Class 1	Half-duplex asynchronous character-based error control (Bell 202 types of modems). Frame size is a maximum 260 octets (bytes).
Class 2	Full-duplex asynchronous character-based error control (Bell 212A types of modems). Frame size is a maximum 64 octets.
Class 3	Full-duplex synchronous block-based error control (Bell 212A or CCITT V.22bis types of modems). Frame size is a maximum of 64 octets. Frame size is determined during initial call setup negotiations.
Class 4	Enhancement to class 3, adaptive packet lengths. Increases throughput to 120%. Frame size is extended to a maximum of 256 octets (V.22bis modems at 2400 bps).
Class 5	Data compression. Uses a real-time adaptive algorithm to compress data. The algorithm continually analyzes the data and adjusts its compressing parameters. Increased throughput up to 200% (2400-bps modems and above).
Class 6	Universal link negotiation. This procedure links at 2400 bps and then looks for a higher-speed protocol such as CCITT V.29 or V.32.
Class 7	Enhanced data compression. Utilizes an adaptive real-time algorithm like class 5 to compress data. Also, a predictive algorithm is used to predict the probability of characters in the data stream. Increased throughput up to 300% (2400-bps modems and above).
Class 8	Enhanced data compression with V.29 fast train, HDX.
Class 9	Enhanced data compression with V.32 modem engine, FDX.

employing MNP versus LAP M. Currently, modems compatible with the V.42 CCITT standard will query the other modem and use LAP M only if it is compatible. If one of the modems is not equipped with LAP M but has MNP capability, both modems will default to communications using MNP.

9.12 IMPROVING MODEM PERFORMANCE

Increasing the data transmission speeds of modems has brought about extensive hardware and software options that have seriously affected the performance of today's modems. Several standards have been written to utilize this performance and achieve uniformity, particularly among high-speed modems.

Three parameters must be controlled by the high-speed modem: *error detection and correction, data compression,* and *throughput.* They are defined as follows:

Error detection and correction: The mechanism used to reduce the number of errors that may occur between a transmitting and receiving modem.

Data compression: The mechanism used to reduce the number of data bits that is normally used to express a given amount of information.

Throughput: A measure of transmission rate based on the time that it takes to successfully transmit and receive a maximum number of bits, characters, or blocks per unit time. This includes the added time for acknowledgments, framing, and error control. Throughput is measured in one direction.

In data communication systems, these three parameters are of major concern to the user. Consider, for example, an electronic banking transaction. A sudden burst of noise induced into the communications channel from an external source can cause errors in the data bit stream. These errors *must* be detected and corrected, which takes additional system time. If the error is not corrected or goes undetected, the results could be catastrophic. The customer therefore expects prompt and error-free communications. We now consider some of the basic principles regarding these three parameters.

9.13 MODEM ERROR CONTROL

Several techniques are used to detect if there are errors in a data stream. Typically, the transmitted data stream is formatted in accordance with a communications protocol that inserts additional bits into the data stream for error control purposes. In asynchronous transmission, the parity bit is used for error detection purposes. For synchronous transmission, blocks of data can include several hundred contiguous characters sent in succession. The block is framed with redundant information called the *preamble* and *postamble*. The postamble of the block typically includes a block check character (BCC) used for error detection. The BCC is the result of some mathematical or logical operation performed on the transmitted data bits and inserted in the postamble of the block during transmission. The receiver performs the same mathematical or logical operation on the received data block. A comparison is made to the BCC received from the transmitter. If there is a difference, an error has been detected. Generally, the receiving station requests a retransmission of the block in error. If communications is simplex (one way), a request for retransmission is impossible. Special techniques must be employed in the receiver to correct the detected errors. In Chapter 12, error detection and correction will be covered in depth.

9.14 THROUGHPUT

The *throughput* of a communications channel is of major concern in applications requiring intensive data transfer rates. A user being charged by the minute for downloading information from a data bank would no doubt benefit from the increased throughput of a 9600-bps modem over a 1200-bps modem. Many defense- and medical-related situations often require immediate demand for data.

Several factors govern the throughput of a system: the bit rate; the electrical characteristics of the channel, such as bandwidth and phase and amplitude distortion; the occurrence of errors; block size; communications protocol; and data compression.

Throughput is not to be confused with the absolute maximum instantaneous bit transfer rate. An asynchronously transmitted character from a terminal, for example, may have a bit time associated with a terminal setting of 9600 bps. However, the duration between the transmitted character and when the next character is typed is wasted in terms of channel utilization. Throughput, in this case, is dependent primarily on the speed of the typist, which may be slower than a few characters per second in this mode of operation. Even if the asynchronous characters are transmitted in succession, throughput is lowered by the overhead included in each character for framing and error control (start and stop bits and parity). Ideally, maximum throughput is obtained by transmitting as many characters as possible in a given period of time without having to include redundancy for framing, error control, and other overhead requirements.

For synchronous transmission, throughput is reduced in the preamble and postamble of the transmitted block. The smaller the block size, the greater the amount of time that is spent transmitting redundant information relative to the actual data within the block. Block size should therefore be maximized to increase throughput. Maximum block size is governed by the synchronous communications protocol used.

Several methods are commonly used to compute system throughput. One such formula commonly used by system designers is TRIB, an acronym for *transmission rate of information bits*. TRIB is given by the equation

$$\text{TRIB} = \frac{B(L - C)(1 - P)}{(BL/R) + T} \quad \text{bps} \tag{9-5}$$

where B = number of information bits per character
L = total number of characters (or bytes) in the block
C = average number of noninformation characters in the block
P = probability of an error occurring in the block
R = modem transmission speed (bps)
T = interval of time between blocks

Example 2

A 2400-bps modem transmits five contiguous blocks of HDLC information frames with no idle time between blocks. Each block contains 256 bytes of raw data. The format of each block is given in Figure 9-23. Prior tests indicate that an average of three blocks in every 100 contains an error. Compute the TRIB.

Solution: $\text{TRIB} = \dfrac{8(262 - 6)(1 - 3/100)}{[(8 \times 262)/2400] + 0} = 2275 \text{ bps}$

Opening flag (8)	Address (8)	Control (8)	Data field (256 bytes)	Frame check sequence (16)	Closing flag (8)

Figure 9-23 HDLC frame for computing the throughput of a communications system.

This example illustrates how throughput is reduced by protocol overhead and errors. The parameter T was assumed to be 0 for the full-duplex HDLC link. If a half-duplex protocol such as BISYNC were used, T must be considered in the equation for acknowledgments of individual blocks.

Throughput is clearly improved by increasing the block size and transmission rate of the modem. However, at some point the percentage of errors will diminish this effect. An optimum block size can therefore be established for any given communication system. System designers must conduct random tests to ascertain this information. Modeling by computer simulation is often used. Many programs are available for this.

9.15 DATA COMPRESSION

One method of drastically increasing system throughput is through the process of *data compression.* Data compression can increase system throughput by a consistent 4:1 margin. It is employed in many of today's high-speed modems. A 2400-bps modem, for example, can now communicate at an effective 9600 bps without any modifications to the modem's transmitter and receiver sections.

The term *data compression* refers to the ability of a data communication system to remove redundant bits from a transmitted data stream, thus reducing the total number of transmitted bits necessary to convey the *same* information. The bandwidth constraints of the telephone lines are indirectly avoided. To accomplish this, data compression takes advantage of the fact that characters that are transmitted within a block of data do not occur with equal probability of use. Some vowels, for example, are used much more frequently than some of the consonants in the alphabet. Many ASCII and EBCDIC punctuation and control characters are rarely used in text files, yet they are still represented by eight bits. This is also true with files containing graphic and numeric data. Spreadsheets and graphs typically include numerous repeated characters, such as lines, dots, and spaces. In addition, the probability of groups of characters statistically dependent on each other exists in many cases. The letter q, for example, is almost always followed by the letter u in the English language. Or a t is often followed by the letter h. For these reasons and more, it is well known that for any specific application, a more efficient encoding scheme is desirable.

9.15.1 Huffman Encoding

Huffman encoding is one of the oldest data compression techniques. It has been used for more than 20 years and has served as a predecessor for more advanced techniques. The theory behind Huffman encoding is to reduce the number of bits representing those characters that have a high frequency of occurrence.

For example, the letter *e* is the most common letter in the English text and the letter *z* is the least common. Huffman encoding takes this probability of occurrence into account, encoding the most probable characters with fewer than eight bits and the least probable characters with more than eight bits. The average word length after compression is much less than eight bits.

In the previous paragraph, there are 38 lowercase *e*'s and one lowercase *z*. It is easy to see the benefit of reencoding the letter *e* as a three-bit word and the letter *z* as a 16-bit word. This is precisely the strategy behind a Huffman-based compression technique. Figure 9-24 is a graph of the frequency of occurrence of each character in the preceding paragraph.

Table 9-4 illustrates an example of how a Huffman encoding table might look. The table is used to represent 256 possible characters of an alphanumeric character set, such as ASCII or EBCDIC. Three *fields* are used to divide the

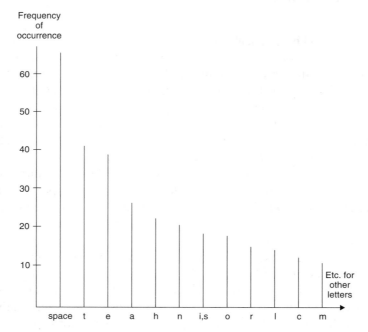

Figure 9-24   Graph depicting the frequency of occurrence of the characters used in the boldface paragraph above. (Reprinted with permission from Concord Data Systems, *All About Adaptive Data Compression for Asynchronous Applications*.)

TABLE 9-4 Huffman Table[a]

HUFFMAN ANTICIPATES CHARACTERS			
3 BIT BOUNDARIES 7		15	
CHAR CODE			
CHAR CODE			
CHAR CODE			7 CHARS
CHAR CODE			
CHAR CODE			Number of characters in table segment
"NO-CHAR"	CHAR CODE		
"NO-CHAR"	CHAR CODE		
"NO-CHAR"	CHAR CODE		15 CHARS
"NO-CHAR"	CHAR CODE		
"NO-CHAR"	CHAR CODE		
"NO-CHAR"	"NO-CHAR"	CHAR CODE	
"NO-CHAR"	"NO-CHAR"	CHAR CODE	
"NO-CHAR"	"NO-CHAR"	CHAR CODE	234 CHARS
"NO-CHAR"	"NO-CHAR"	CHAR CODE	
"NO-CHAR"	"NO-CHAR"	CHAR CODE	
"NO-CHAR"	"NO-CHAR"	CHAR CODE	

[a] **The Huffman table** is organized from top to bottom in descending order of frequency. Each entry could consist of, say, 15 bits divided into three fields of three, four, and eight bits. Seven characters can be represented in the field containing three bits (plus the "noncharacter" code). Fifteen characters can be represented in the field containing four bits (plus the noncharacter code).

Source: Reprinted with permission from Telcor Systems Corporation.

character set into a code based on each character's frequency of use. The right of the table shows how the 256 characters are grouped into one of three categories that are represented by a three-bit code, a seven-bit code, or a 15-bit code.

Seven of the most frequently used characters, space and vowels, for example, are represented at the top of the table by a three-bit *character code* (CHAR CODE). The 15 next most frequently used characters are encoded into a seven-bit code, where the first three bits are a special noncharacter (NO-CHAR) code, and the remaining four bits represent the 15 characters in descending order of frequency. The three-bit NO CHAR code indicates to software to look in the second CHAR CODE field for the actual character code. Seven bits are still an improvement over the original encoded eight-bit character.

If the character to be transmitted is not one of the 22 (7 + 15) most frequently used characters, it is represented by a 15-bit code. This includes the remaining 234 entries (256 − 15 − 7), with the least frequently used character occupying the 234th entry. Although 15 bits used to represent these characters are greater than the original eight bits, the average transmitted word length after Huffman compression is much less than eight bits.

Huffman tables may be either static or dynamic.[†] A static table is one in which the frequency of each character in the message is assumed, so the order of the character in the table is determined ahead of time. That order is fixed so that even if the actual frequency changes, the software will continue using the same table. Obviously, both the sender and the receiver will have created identical tables prior to transmission to enable proper interpretation of the message codes.

In a dynamic Huffman model, a frequency algorithm determines which characters are represented at which levels in the table. Every time a character is used, its position in the table is exchanged for the position of the character immediately above it. The bit patterns in the table themselves do not actually change. What changes is the assignment of the bit patterns within a table entry to represent a particular character. An exchange is always made after the code currently assigned is sent across the line. This ensures that both sender and receiver can update their respective copies of the table in sync.

For example, let's suppose that a data stream shifted from using uppercase and lowercase letters to all uppercase. In the first instance, the lowercase *e* might be the most frequently used character and be at the top of the table. Uppercase *E*, on the other hand, could very well be at the bottom. Once the shift to all uppercase occurred, every time *E* was used, its table assignment would be swapped to the next-higher level until it reached the top. Along the way, it would cross the boundary between 15- and 7-bit representation and then the boundary between 7- and 3-bit representation.

Typical performance for Huffman-based data compression is about a 2:1 ratio, depending on the statistics of the data being sent. The primary advantage of the Huffman encoding technique is its simplicity and the fact that it does not require a large amount of memory to implement.

More advanced data compression techniques are currently being used by modem manufacturers today. Huffman encoding is based on the *statistical independence* of characters. In other words, a character's probability of occurrence does not take into account what the preceding character is. More advanced data compression techniques do. They are also *adaptive* to the data being sent. In the English language, for example, a *q* is most likely to be followed by a *u,* or a *t* is likely to be followed by an *h*. In addition, characters used in spreadsheets and graphs often repeat themselves (e.g., spaces, dots, dashes, etc.). Rather than sending a repeated string of these characters, the data can be encoded into a word indicating the number of repeated characters that follow. This is referred to as *run-length encoding.* Several sophisticated algorithms employ these techniques, achieving as much as a 4:1 compression ratio. The disadvantage is that this added complexity requires dedicated processors and extra money. Within time, however, the added cost will certainly decline.

[†]Francis Bacon (Telcor Systems Corporation), "How to Quadruple Dial-Up Communications Efficiency," *Mini-Micro Systems,* Feb. 1988: *Technology Forum.*

PROBLEMS

1. What does the term *modem* stand for?

2. Explain the difference between a direct-connect modem and an acoustically coupled modem.

3. Why are acoustically coupled modems limited to speeds of less than 1200 bps?

4. If the bit stream shown in Figure 9-5 were changed to 101110100010, draw the waveforms resulting from the following modulation techniques: ASK, FSK, PSK, and DPSK.

5. What is a *dibit*?

6. What Bell modem standard utilizes tribit encoding as a modulation technique?

7. What is the transmitting MARK frequency for a Bell 103 modem operating in the answer mode?

8. What is the transmitting originate frequency for a Bell 212A modem?

9. Explain the differences between simplex, half-duplex, and full-duplex transmission modes for a modem.

10. What range of frequencies is used by modems to disable echo suppressors? How long must the tone be present?

11. A 4800-bps modem uses tribit encoding to represent eight phases of a DPSK signal. Compute its baud rate.

12. A 9600-bps modem uses QAM for its modulation technique. Compute its baud rate.

13. Compute the M-ary for the 9600-bps modem in Problem 12.

14. What do *bis* and *ter* mean?

15. Explain the difference between an ALB test and a DLB test.

16. What is the number of data bytes used in the data field of Xmodem?

17. Explain the difference between Xmodem and 1K-Xmodem/G.

18. How accurate, in terms of error detection, is Zmodem's 32-bit CRC?

19. What error-controlling technique is used to allow multiple frames to be transmitted without having to wait for an acknowledgment for each frame?

20. Refer to Figure 9-23. A 4800-bps modem transmits four contiguous HDLC blocks with the format shown in Figure 9-23. There is no idle time between blocks. An average of 2.5 blocks out of every 100 sent are received in error. Compute the transmission rate of information bits (TRIB) using equation (9-5).

21. Repeat Problem 20 for a 9600-bps modem.

22. What is *data compression*?

23. What is a typical compression ratio that can be achieved with a Huffman-based data compression technique?

10

Protocols

In data communications, a *protocol* is defined as a set of rules and procedures that has been developed for purposes of communicating between devices. A multitude of protocols has been documented. They range from specifying the type of connector and voltage levels to be used by a device, EIA–RS–449 for example, to the management of software used to control the signaling sequence necessary for communications.

Rapid technological advances have forced the standardization of protocols in recent years to become more revolutionary and anticipatory rather than evolutionary and documentary as in the past. As a result, there has been widespread concern among manufacturers and buyers as to whether the more recent standards will adapt themselves as *accepted* standards in the industry. Manufacturers and buyers both benefit in this case in terms of costs, compatibility, and application. On this premise, the discussion of protocols and their classification will be limited to that which has been universally observed as an accepted standard.

10.1 STANDARDS ORGANIZATIONS FOR DATA COMMUNICATIONS

A consortium of standards organizations, vendors, and users of data-communicating devices meets on a regular basis to establish guidelines and standards for communications between two or more devices. A brief look at some of the major organizations that have contributed to the success of the telecommunications industry will be given here. These organizations continue to develop and refine standards concurrent with the technological trends in communications.

ISO: The International Standards Organization is *the* Standards Organization. ISO creates the Open Systems Interconnection (OSI) protocols and standards for graphics, document exchange, and related technologies. ISO endorses and coordinates work with other groups, such as CCITT, ANSI, and IEEE.

CCITT: The Consultative Committee for International Telephony and Telegraphy (CCITT) was founded over 100 years ago. It is now an agency of the United Nations. CCITT develops the recommended standards and protocols for telecommunications. The group consists of many government authorities and representatives of the PSTN. CCITT has developed the V series specifications for modem interface, the X series for data communications, and the I and Q series for Integrated Services Digital Network (ISDN). CCITT is now known as ITU–TS, International Standards Union–Telecommunication Standardization Sector.

ANSI: The American National Standards Institute is the official U.S. agency and voting representative for ISO. ANSI has developed information exchange standards above 50 Mbps. The institute is involved with coordinating manufacturers through the Computer and Business Equipment Manufacturers Association.

IEEE: The Institute of Electrical and Electronic Engineers is a professional organization of U.S. electronics, communication, and computer engineers.

EIA: The Electronic Industries Association is a U.S. organization of manufacturers that establishes and recommends industrial standards. EIA has developed the RS *(recommended standard)* series for data and telecommunications.

ITU–TS: See CCITT.

10.2 ISO-OPEN SYSTEMS INTERCONNECT (OSI) SEVEN-LAYER MODEL

In previous chapters we focused our attention on fundamental interfacing techniques and devices used for serial communications. The electrical, mechanical, and functional aspects of serial interfacing were considered. In summary we have been dealing with the elements of data communications on a *hardware* level. This is by no means the whole of data communications but rather, a crucial aspect of the overall requirement for a system or network of systems. Consider, for example, the software necessary for handling error conditions as a result of the transmission loss of data. Also, consider how data may be interpreted by a computer on a bus in determining whether or not they are data that should be attended to or passed on to another device on the bus. For informational resources to be shared among users who are either centrally located, as in a business building, or remotely located throughout the world, the necessary hardware and software must be adapted to standardized guidelines. Worldwide networks that currently exchange enormous amounts of data on a 24-hour basis all follow stringent sets of defined and documented protocols. All the aforementioned standards organizations have collaborated in the development of these standards.

With the ever-increasing need for standards and dependence on standards, the combined efforts of ANSI, CCITT, EIA, IEEE, ISO, and others have led to the development of a hierarchy of protocols called the *Open Systems Interconnect* (OSI) model. The model encourages an open system by serving as a structural guideline for exchanging information between computers, terminals, and networks. Figure 10-1 depicts the OSI model. The OSI model categorizes data communications protocols into seven levels. The hierarchy of each level is based on a *layered* concept. Each layer serves a defined function in the network. Each layer depends on the lower adjacent layer's functional interaction with the network. If level 1, the physical layer, for example, were to experience complications, all layers above would be affected. On the other hand, since each layer serves a defined function, that function may be implemented in more than one way. In other words, more than one protocol can serve the function of a layer, thus offering the advantage of flexibility.

Physical Layer. The lowest layer of the OSI model defines the electrical and mechanical rules governing how data are transmitted and received from one point to another. Definitions such as maximum and minimum voltage and current levels are made on this level. Circuit impedances are also defined in the physical layer. An example would be the RS–232–C serial interface specification.

Data Link Layer. This layer defines the mechanism in which data are transported between stations in order to achieve error-free communications. This includes error control, formatting, framing, and sequencing of the data. IBM's BISYNC and SDLC are examples that fall into the data link layer.

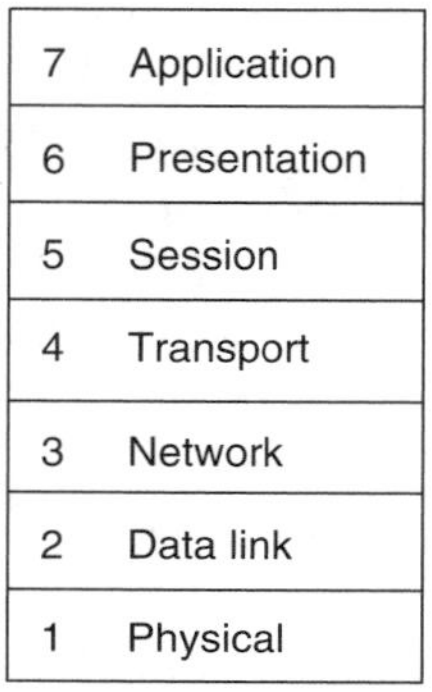

7	Application
6	Presentation
5	Session
4	Transport
3	Network
2	Data link
1	Physical

Figure 10-1　Seven-layer OSI model.

Network Layer. This layer defines the mechanism in which messages are broken into data packets and routed from a sending node to a receiving node within a communications network. This mechanism is referred to as packet switching. Individual packets representing the original message may take various routes throughout the network in order to arrive at their final destinations. The order in which they are received may or may not be the same order in which they were sent.

Transport Layer. The transport layer of the OSI model ensures the reliable and efficient end-to-end transportation of data within a network. It is the highest layer in terms of communications. Layers above the transport layer are no longer concerned about the technological aspect of the network. The upper three layers address the software aspects of the network, whereas the lower three layers address the hardware. The functions served by the transport layer are ensuring the most simplified and efficient service, error detection and recovery, and multiplexing of end-user information onto the network.

Session Layer. When a user interacts with a computer within a network, it is often referred to as a session. A session is initiated and terminated by a user during log-in and log-out procedures. The session layer concerns itself with the management of a session. This includes the recognition of a use's request to use the network for communications, as well as terminating the user's session. If a break occurs during the session, this layer addresses the full recovery of the session without any loss of data.

Presentation Layer. The services provided on this level address any code or syntax conversions necessary to present the data to the network in a common format for communications. This includes alphanumeric code sets (ASCII and EBCDIC), data encryption, data compression, file formats, and so on.

Applications Layer. The applications layer is the highest ISO/OSI layered protocol. At this level the specific applications program that performs the end-user task is defined. This includes, for example, database management programs, word processing, spreadsheets, banking, and electronic mail.

10.3 BIT-ORIENTED PROTOCOLS VERSUS BYTE-ORIENTED PROTOCOLS

We have learned that protocols are a set of rules that govern the orderly flow of information between two parties. In serial communications, the information can be transmitted in either synchronous or asynchronous format. As the communications system becomes more complex, the need for speed and efficiency rises.

Synchronous transmission is used. A closer look at synchronous transmission methods and protocols that serve the function of the data link layer (level 2) is necessary to understand the development and maintenance of data communication systems.

Synchronous data link protocols can be subdivided into two categories: *bit-oriented* protocols and *byte-oriented* protocols.

Bit-Oriented Protocols. In bit-oriented protocols, special groups of uniquely defined bit patterns are used to control the framing, error checking, and flow of data between devices. The data in bit-oriented protocols may be of any content and may not necessarily represent an encoded character set such as ASCII or EBCDIC. An example would be raw data from an A/D converter. Special transmission schemes must be employed so that the receiver can distinguish between the actual control characters for framing and error control versus the raw data patterns from the A/D converter that may coincidentally take on the same bit pattern as the control character.

Byte-Oriented Protocols. In byte-oriented protocols, the transmission of data blocks is controlled by ASCII or EBCDIC control characters such as SYN, SOH, and ETX. Like the data characters, control characters are uniquely defined as part of the ASCII or EBCDIC character set. They are placed at the beginning and end of the transmitted data block for purposes of framing and error control. Since the actual data within the block are typically ASCII or EBCDIC, the receiving device can distinguish between data and control.

10.4 BISYNC

BISYNC stands for *binary synchronous communications protocol.* BISYNC is also referred to as BSC. Developed in 1964 by IBM, BISYNC was one of the most widely used synchronous protocols until recent protocols have made it relatively slow and inefficient in comparison. Many systems that have been installed with BISYNC hardware and software are still being used to this date. The BISYNC data link protocol has served as a predecessor for more recent protocols to improve on.

BISYNC is a byte-oriented synchronous serial communications protocol. It is designed for half-duplex operation between two or more stations connected in a *point-to-point* or *multipoint* configuration. Figure 10-2a illustrates two computers configured for point-to-point operation. Point-to-point can also be a direct connection between computers or computer and terminal without the use of a modem. Figure 10-2b illustrates a computer connected to several terminals in a multipoint (or multidrop) configuration. Several terminals are shown here sharing a private line through the use of modems.

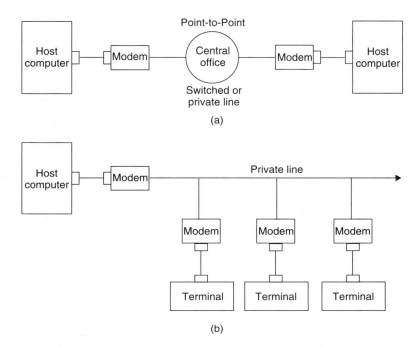

Figure 10-2 (a) Computers connected point-to-point; (b) computer connected to several terminals in a multipoint (or multidrop) configuration.

10.4.1 BISYNC Message Block Format

Figure 10-3 illustrates the format of a BISYNC message block. Notice that the block contains *data link control* characters. Data link control characters are used for framing and management of the data between devices. A complete list of BISYNC control characters is given in Table 10-1. These control characters are

TABLE 10-1 BISYNC Data Link Control Characters

ACK	Affirmative acknowledgment
DLE	Data link escape
ENQ	Enquiry
EOT	End of transmission
ETB	End of transmission block
ETX	End of text
ITB	End of intermediate transmission block
NAK	Negative acknowledgment
SOH	Start of header
STX	Start of text
SYN	Synchronous idle
WACK	Wait before transmit positive acknowledgment

Figure 10-3 BISYNC message block format.

included in the ASCII and EBCDIC code sets. The components of the BISYNC message block are defined as follows.

SYN: The message block begins with the framing control character, SYN. Two SYN characters precede the message. The USART in Chapter 6 can be programmed to enter the SYNC HUNT mode in a search for these characters. Once found, the remaining fields of the block can be interpreted in their respective order.

SOH: The start of header is the control character used to introduce the header field to the receiving device.

HEADER: The header is a variable-length field that is typically used for addressing, that is, selecting a device or polling a device. It is an optional field.

STX: Start of text is a framing control character used to inform the receiving device that text will immediately follow.

TEXT: This is a variable-length field that generally includes ASCII or EBCDIC characters.

ETX or *ETB:* End of text or end of text block is a framing control character that identifies the end of the text field.

BCC: The block check character immediately follows ETX or ETB. It is computed by the transmitting device and inserted at the end of a message block that includes a TEXT field. All characters following STX are computed in the BCC. The length of the BCC is typically 8 to 16 bits. It can be a simple LRC character or a 16-bit CRC character.

10.4.2 Transparent Text Mode

A problem is encountered in BISYNC when the message block contains data that are normally restricted within the TEXT field. These would be data that do not conform to the given standard code set, EBCDIC or ASCII. Take, for example, a block of A/D converter data. Several binary bit patterns within the block's TEXT field are likely to be equivalent to the control characters mentioned above. To prevent the receiver from interpreting the raw binary data within the TEXT field as control characters, the transmitting device must somehow inform the receiver that the following data are *transparent.* Any bit patterns that happen to be equivalent to control characters are to be disregarded by the receiver in terms of its normal control procedures. BISYNC uses the data link control character DLE (data link escape) followed by STX (start of text) to enter the *transparent text mode.* To exit the transparent text mode, DLE followed by ETX (end of text) or ETB (end of text block) is used. Figure 10-4 illustrates the format of a BISYNC

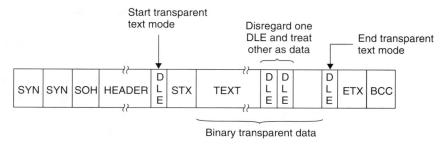

Figure 10-4   BISYNC transparent text mode format.

transparent message block. Upon receiving DLE followed by STX (start of text), the receiver interprets the following data within the block as *transparent data.*

This raises an obvious question: How does the receiver know when the end of the block occurs? The problem is solved by the transmitter inserting a second DLE whenever an equivalent DLE bit pattern is to be sent within the transmitted TEXT field. The receiving device, in turn, disregards one of the two DLE bit patterns in a detected *pair* of DLEs as shown in Figure 10-4. When it receives only one DLE after the beginning of a defined transparent block (by the first DLE), it returns to the normal mode of operation and the next character is to be interpreted as a control function, for example, ETX. Any data within the TEXT field can now be issued by the transmitting device once the transparent text mode has been started. Table 10-2 lists BISYNC DLE control sequences used in the transparent text mode.

TABLE 10-2   BISYNC DLE Control Sequences Used in the Transparent Text Mode

DLE STX	Data link control characters for beginning the transparent text mode.
DLE DLE	When a bit pattern equivalent to DLE occurs within the transparent data, a second DLE is sent in succession. The receiver disregards one DLE and the other is treated as data.
DLE ETX	Terminates the transparent text mode, returns the data link to normal mode, and calls for a reply.
DLE ETB	Terminates a block of transparent text, returns the data link to normal mode, and calls for a reply.
DLE SYN	Used to maintain SYNC or as time-fill sequence for transparent mode.
DLE ENQ	Indicates "disregard this block of transparent data" and returns the data link to the normal mode.
DLE ITB	Terminates an intermediate block of transparent data, returns the data link to normal mode, and does not call for a reply. BCC follows DLE ITB. Transparent intermediate blocks may have a particular fixed length for a given system. If the next intermediate block is transparent, it must start with DLE STX.

10.4.3 BISYNC Point-to-Point Line Control

Computers often need to transfer large files of data between each other. These files are broken into blocks of data using the format we have just discussed. Take, for example, two computers that are configured point-to-point as illustrated in Figure 10-2a. To ensure the integrity of the data transferred between computers, they are broken into blocks, framed, and sent with a block check character. The size of the block can vary depending on several factors:

— Transmission media   — Overhead

— Message content   — Baud rate

— System noise   — Probability of error

— Transmission distance   — Retransmission time

If the block size is large, the probability of an error increases and retransmission time must be considered. However, the data are transferred in the most expedient manner if no errors occur. If the block size is small, the overhead increases and the relative time to transmit data increases. If an error occurs in the transmission of a small block, retransmission time is shorter. There are certainly trade-offs involved. A nominal block size of 256 bytes is commonly used in BISYNC and other protocols.

10.4.3.1 Handshaking blocks of data. BISYNC uses a system of *handshaking* blocks of data from one computer to another. The following BISYNC data link control characters are used for this purpose.

ACK 0 and *ACK 1* (acknowledgment): ACK is an acknowledgment that a block has been received successfully. ACK 1 always acknowledges the reception of the first message block of data that was received and all odd-numbered blocks thereafter. ACK 0 is used to acknowledge all even-numbered blocks received.

ENQ (enquiry): ENQ is used by a device to gain control of the line for transmitting data. It is also used to enquire with the receiving device as to why there was no acknowledgment of the previously transmitted block.

NAK (negative acknowledgment): NAK indicates to the transmitting device that the previous block was received in error (checksum error). A retransmission is typically performed. NAK is also used as a "not ready" indicator.

WACK (wait before transmit positive acknowledgment): WACK is used as a temporary "not ready" indicator to the transmitting device. It can be sent as a response to a test or heading block, line bid, or an identification (ID) line bid sequence.

The method of handshaking blocks of data is called ACK ALTERNATING, whereby the data link control characters ACK 0 and ACK 1 are sequentially used to acknowledge even- and odd-numbered blocks that are received successfully.

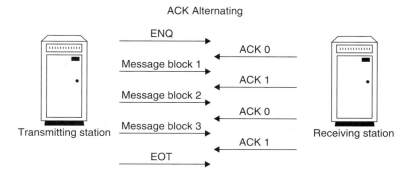

Figure 10-5 BISYNC point-to-point line control sequence.

Figure 10-5 depicts a typical BISYNC, point-to-point, half-duplex line control sequence using ACK ALTERNATING. Assuming that both stations shown here are equal contenders for use of the line, the transmitting station is the station that *bids* for the line first. This is achieved by asserting ENQ. If both stations assert ENQ at the same time, a *collision* occurs: two transmitters are transmitting at the same time on the half-duplex link. An agreement is made beforehand as to which system has priority, in which case the other system backs down and acts as the receiving station. This process is referred to as *line contention*. A closer look at line contention is presented in Chapter 11.

Framing control characters have been left off for purposes of clarification. Notice the receiving station's acknowledgment, ACK 0, of the transmitting station's bid for the line first. ACK 1 is an acknowledgment to the first message block and all odd-numbered blocks thereafter. ACK 0 is an acknowledgment to all even-numbered blocks received without any errors. The block check character BCC is not used unless a text field is included in the message block.

10.4.3.2 Encountering a BCC error.   Figure 10-6 illustrates an example of a block that gets garbled en route as a result of impulse noise. The receiving station receives the block. However, the computed BCC does not agree with the BCC that was sent. The receiving station issues a NAK to the transmitting station. NAK is a request for retransmission of the second block. The transmitting station, upon receiving NAK (instead of ACK 0), retransmits the previous even-numbered message block, and the handshaking process continues.

10.4.3.3 A time-out error.   When two devices are handshaking data between one another, a limited amount of time must be imposed on the receiving device to acknowledge the reception of the transmitted data. If this were not the case, and the receiving device failed to acknowledge the received data, all further transmissions would cease, pending the acknowledgment. Both systems would "hang" until an operator attends to the problem. Computers incorporate, in either

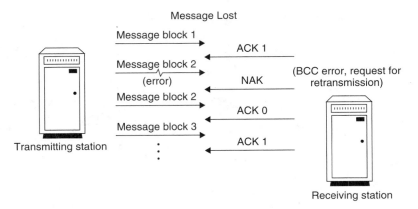

Figure 10-6 An error condition encountered causes the receiving station to request a retransmission.

hardware or software, what is referred to as a *time-out error.* A time-out error occurs when a device fails to respond to a message within a given period of time. It may be a few microseconds to several seconds.

In BISYNC, if the transmitting station does not receive ACK within an agreed period of time, a time-out error will occur (Figure 10-7). Presumably, the last message block was received but the receiving station's ACK was lost. The transmitting station then issues an ENQ. ENQ, in this case (when not bidding for the line), is an enquiry to the receiving station to see if it received the last message block. An ACK 0 or NAK in return will allow the transmitting station to determine whether or not the last message was received. In Figure 10-7, ACK 0 is the response to ENQ. It is therefore presumed that Message block 2 was received and the previous ACK 0 was lost.

10.4.4 BISYNC Multipoint Line Control

Thus far we have discussed the format of a BISYNC message block and the method of handshaking data blocks in a point-to-point configuration. In a point-to-point configuration, it is certain where the data are going to and coming from since there are only two devices in question. In multipoint operation, the process is similar in terms of transmitting messages between individual stations. The difference is that a method of addressing is needed in multipoint operation to determine which station is transmitting and receiving the data blocks. We now discuss this addressing method.

10.4.4.1 Device polling and selecting. There is no bidding for the use of the line in multipoint operation. One station is designated as the *host computer* and the remaining stations on the line are designated as *tributaries.* See Figure

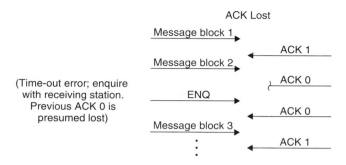

Figure 10-7   Acknowledgment, ACK, lost en route to the transmitting station.

10-8. The host computer supervises all activity on the line by a method called *device polling* and *device selecting*.

Device Polling.   Each tributary on the line has a unique *polling address*. The host computer sequentially *polls* (addresses) each tributary with its address followed by ENQ to enquire if it has any data to be sent. If the tributary has data for the host computer, it is handshaked in the same manner as discussed under point-to-point operation. If no data are available, a corresponding EOT message is sent to the host and the next device in sequence is polled. Device polling is used when data are sent from the tributary to the host computer.

Device Selecting.   *Device selection* is used when data are to flow from host computer to tributary. Each tributary on the line has a unique *select address*. A tributary's select address is different from its polling address. This allows the device to send or receive data from the host upon receiving one of its two addresses. A tributary is selected by the host computer by asserting its *select address* followed by ENQ. If the device is ready to receive data that the computer has for it, it responds with ACK 0. Data are then handshaked between the host and tributary in the same manner as point-to-point operation. If a tributary is not ready for data from the host, it asserts WACK in response to being *selected*. The host responds to WACK with EOT and checks for readiness at a later time.

10.4.4.2 BISYNC multipoint communications between host computer and several terminals.   Figure 10-9 depicts several terminals that have been configured for multipoint operation. In this mode, internal and external switch settings, or the terminal's software are configured for block mode operation. When an operator is ready to transmit his or her buffer of data, the ENTER key is depressed. Data are not sent immediately to the host computer. The operator's terminal must wait for its turn to be polled before it can begin to transfer its blocks of data.

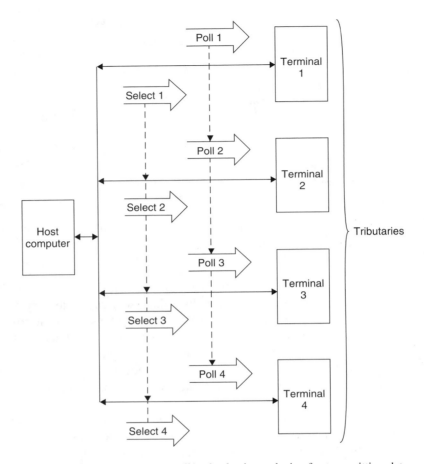

Figure 10-8 Host computer poll and selecting a device for transmitting data in a multipoint setup.

Each terminal in this example has been assigned a unique polling address and select address. The host computer begins by polling terminal 1, address P1, for possible message blocks. Terminal 1 responds with three message blocks that are acknowledged by the host. After the data are interpreted, the host sends a one-block message to the terminal as a response. The terminal is first selected with its select address, S1. The select enquiry is acknowledged by ACK 0, indicating its readiness for the message block. The block is then transferred to terminal 1 and acknowledged. EOT is sent to terminal 1, thus terminating the exchange.

Terminal 2 is then polled for any possible messages. An EOT is sent to the host, indicating that there are no pending data at this time. The host computer, however, has a message block to send to terminal 2. Its select address, S2, along with ENQ, is asserted by the host. Terminal 2, however, is not ready to receive the message block, so it responds to the selection with WACK. The host com-

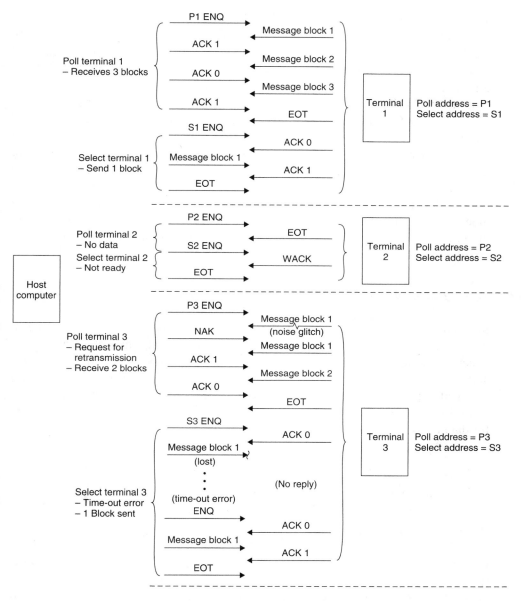

Figure 10-9 Diagram depicting BISYNC multipoint operation between a host computer and several terminals.

puter terminates the handshaking process with terminal 2 by sending EOT, and its readiness to receive a message will be checked at a later time.

When terminal 3 is polled at its polling address, P3, its returned messaged block is garbled en route to the host computer. The computed BCC is in conflict with the BCC sent with the data. As a result, the host computer sends NAK to terminal 3; a request for retransmission. The message block, as shown in Figure 10-9, is transmitted again by terminal 3. The host computer this time successfully receives the message and consequently responds with ACK 1, acknowledging the first message block. The second block is then transferred to the host and the transaction is terminated with EOT. The computer at this time has a message for terminal 3. S3 ENQ is asserted to test for terminal 3's readiness. Upon receiving ACK 0 (ready for data), message block 1 is transmitted. During the transmission, the message block is lost due to an intermittent connector contact. After a 3-s time-out, there is still no responding acknowledgment to the message block sent to terminal 3. The host computer enquires at this time by asserting ENQ. ACK 0 from terminal 3 is an indicator to the computer that message block 1 was lost rather than the acknowledgment, ACK 1, from the receiving device. Message block 1 is therefore retransmitted by the host computer. The message is received this time and acknowledged. The host computer has no further messages for terminal 3. The transaction is terminated with EOT.

10.5 TELECOMMUNICATION SWITCHING FACILITIES

In Section 10.4 we discussed the BISYNC protocol and its *point-to-point* and *multipoint* or *multidrop* interface connection. Many of the first connections between computers and terminals were interfaces of this type. As computer systems and telecommunication switching facilities grew, the need for new and more sophisticated methods of ensuring data integrity and end-to-end connectivity grew. Traditionally, there have been three switching technologies utilized by the PSTN for data transmission:

1. Circuit switching
2. Message switching
3. Packet switching

10.5.1 Circuit Switching

Circuit switching is a method of allowing data terminal equipment (DTE) to establish an immediate full-duplex connection to another data station on a temporary basis. Once the connection is established, exclusive use of the channel and its available bandwidth is provided by the switching network until the channel is relinquished by one of the stations (e.g., the standard PSTN voice-grade switched line service). Figure 10-10 illustrates an example of a circuit-switched network. Circuit switching has its advantages over message and packet switching services. The switched connection is dedicated for the entire communication session and no time buffering is necessary; thus it is highly efficient in cases when relatively high data transfer rates occur throughout the session. Addressing information

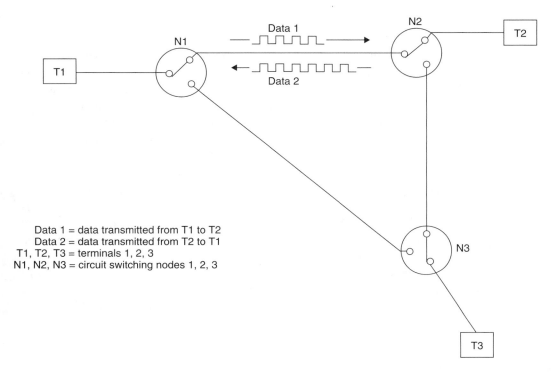

Data 1 = data transmitted from T1 to T2
Data 2 = data transmitted from T2 to T1
T1, T2, T3 = terminals 1, 2, 3
N1, N2, N3 = circuit switching nodes 1, 2, 3

Figure 10-10 Circuit switching network.

occurs only once during the call setup procedure by the network. A circuit-switched channel can also be full-duplex and provide interactive communication between stations.

A subcategory of circuit switching is called *channel switching.*[†] Channel switching provides the ability to modify multiplexers and network configurations remotely in response to outages or time-of-day traffic variations, or to accommodate growth and organizational changes.

The problem arises with the circuit-switched network when multiple stations require the use of the switching facilities. The switch must support a wide range of transmission rates from all users simultaneously. Typically, these user rates are "bursty"; that is, peak periods of transmission are followed by periods in which no data at all are transmitted. This tends to be very inefficient in cases when the switch is overloaded. Other users could be transmitting at maximum rates during these idle periods. *Message switching* and *packet switching* systems have been devised to overcome this problem.

[†]Joseph Pecar, Roger O'Conner, and David Garbin, *Telecommunications Factbook* (New York: McGraw-Hill, 1993), p. 225.

10.5.2 Message Switching

Many telephone companies offer automatic message switching services to their customers. Message switching (Figure 10-11) is a method of transferring messages between DTE by temporarily buffering or "storing" the message at the switching exchange and "forwarding" it to the next switching exchange acting as a successor. This occurs when traffic on the communications channel is favorable for transmission. The next successor, if necessary, repeats the process until the message is routed to its final destination. This message switching technique is known as **store and forward**. Since immediate connection between DTEs are not necessary, the telecommunications facilities can be shared among several users on a per-message basis. Efficiency of the switching network is increased as a result of this technique. An example of message switching that has been around for many years is the international *Telex* network: a *Western Union* worldwide Teletype exchange service that uses the PSTN.

A message switching system operating at 100% of design capacity will have substantial delays, regardless of message length.[†] However, a message switching system operating at 80% capacity or less will have short delays pro-

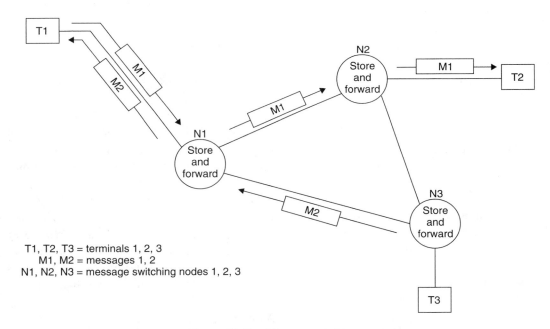

T1, T2, T3 = terminals 1, 2, 3
M1, M2 = messages 1, 2
N1, N2, N3 = message switching nodes 1, 2, 3

Figure 10-11 Message switching network.

[†]John E. McNamara, *Technical Aspects of Data Communications,* 2nd ed. (Bedford, Mass.: Digital Press, 1982), p. 196.

vided that there are no long messages in the system that might block certain routing paths and delay other messages. Long messages should be broken up into "packets" and appended with address and control information; hence *packet switching* should be considered.

10.5.3 Packet Switching

Until recently, packet switching technology has been the most advanced and established data communications switching technology used for *wide area networks* (WANs). Packet switching is a method of segmenting a user's message into discrete variable-length units called **packets** (Figure 10-12). By limiting the length of the packets, other users can effectively share the use of the channel. Packets are assembled and disassembled by a *packet assembler–disassembler* (PAD). Individual packets are appended with control information for routing, sequencing, and error detection. Packets are then routed to various switching nodes throughout the network, depending on availability of the channel. Individual packets may not necessarily take the same switching route as other packets derived from the same message. In addition, they may not arrive at the terminating switching node in the same order in which they were transmitted. Eventually, all packets arrive at their destination node, where they are disassembled by a PAD. Data are placed into its original message form and transported to its final destination.

Packet switching networks can be enormously large and complex, spanning the world and permitting many users to share the use of the communications facilities. For these reasons, the user's data in a packet-switched network are not highly time critical. Typical applications involve low-to-moderate data transfer rates of files containing banking information, electronic mail, airline reservation, ticket information, and so on. The packet-switched network can deliver this type of information in fractions of a second, or it can take up to a few seconds in cases where system loading, reliability, and propagation delays due to distance spans become a factor. *Telenet,* now called *Sprint Net,* and the *ARPANET* (Advanced Research Projects Agency Network) are examples of packet-switched networks.

The establishment of packet-switched networks throughout the world has created a need to produce a standard protocol that facilitates international internetworking. The CCITT recommended protocol for packet switching over public data networks is the *X.25 Protocol.* The X.25 protocol was written in Geneva in 1976 and amended in 1980. It addresses the interface between data terminal equipment (DTE) and data circuit terminating equipment (DCE) operating in the packet mode on public data networks. With respect to the ISO-OSI seven-layer model discussed in Section 10.2, the X.25 protocol is a three-layer process: layers 1, 2, and 3, with most of its protocol complexity defined in layer 2. In the next section we concentrate on the X.25 *Link Access Procedures* (LAP) for data ex-

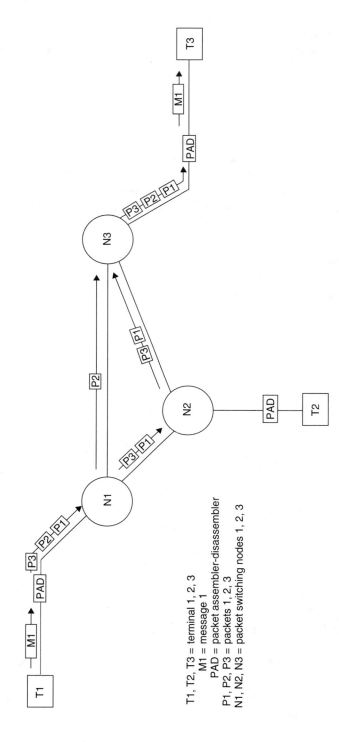

T1, T2, T3 = terminal 1, 2, 3
M1 = message 1
PAD = packet assembler-disassembler
P1, P2, P3 = packets 1, 2, 3
N1, N2, N3 = packet switching nodes 1, 2, 3

Figure 10-12 Packet switching network.

change between DTE and DCE as recommended by CCITT and specified by the International Standards Organization (ISO).

10.6 HIGH-LEVEL DATA LINK CONTROL (HDLC)

High-level data link control (HDLC) is the standard communications link protocol proposed by the International Standards Organization (ISO). The HDLC protocol has been accepted internationally and used to implement the X.25 packet switching network. *Synchronous data link control* (SDLC) is a variation and predecessor to HDLC. SDLC was developed in 1974 by IBM. Unlike BISYNC, these two bit-oriented protocols are ideally suited for full-duplex communications. The two protocols are functionally identical. Since HDLC encompasses SDLC and HDLC has been internationally recognized as the proclaimed standard, our discussion will focus on the latter. Any differences between the two protocols will be pointed out as needed.

10.6.1 HDLC Frame Format

The frame format for HDLC is shown in Figure 10-13. In HDLC, the term *frame* is synonymous with the term *block* used in BISYNC. For bit-oriented protocols, including HDLC, the framing structure is the same for all messages. There are no framing characters within the block, such as SYN, STX, and EOT, as in BISYNC, a character-oriented protocol. Framing is achieved by placing the unique bit pattern **7EH** (01111110) at the beginning and end of the HDLC frame. All fields within the frame consist of *bytes* (eight bits) or multiples thereof, with the exception of the information transfer field. In HDLC a byte is also referred to as an *octet*.

10.6.1.1 Flag byte and bit order of transmission. The beginning and end of all frames are enclosed with the unique flag byte 7EH, or 01111110 in binary. The flag byte is the only framing character used in HDLC. Figure 10-14 depicts the bit structure of the flag as well as the remaining HDLC fields.

The order of bit transmission is such that the low-order (LSB) bit, bit 1 (as defined in the X.25 protocol), is transmitted first for address, commands, responses, and sequence numbers. The order of bit transmission is not specified for

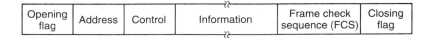

Figure 10-13 HDLC frame format.

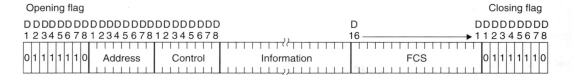

Figure 10-14   HDLC framing pattern and bit order of transmission.

the information field. The *frame check sequence* (FCS) field is transmitted to the line, commencing with the coefficient of the highest term of the 16-bit CRC character. Figure 10-14 depicts this order.

10.6.1.2 Address byte.   The address byte[†] indicates the address of the secondary station that the frame corresponds to. In multipoint operation the address byte within the frame represents which secondary station the frame is either going to or coming from. There is no address that represents the primary station since it is controlling the communications.

In HDLC, the address byte can be extended to include multiple address bytes. The receiving device must therefore be able to discern between a single-address byte field and a multiple-address byte field. In HDLC, the LSB of the address byte is set to a logic 0 if the following byte is to be interpreted as an extension of the address field. The following byte conforms to the same rule. If the LSB of the address byte is a logic 1, that byte is the last byte of the address field (Figure 10-15).

10.6.1.3 Control byte.   The control byte[‡] serves many of the same functions as data link control characters do in character-oriented protocols, for example, ACK, NAK, and EOT. It also serves to identify the *type* of HDLC frame that is being sent. There are three types, as defined by the low-order bits, D1 and D2, of the control byte. For each type, the control byte has a different format. The three types of HDLC frames are *information transfer, supervisory,* and *unnumbered.* Figure 10-16 illustrates the structure of the CONTROL BYTE for each type of HDLC frame.

Definition of Control Byte Terms

NR (Receive Sequence Number).   NR acknowledges the number of frames that have been received successfully. It is included in information transfer and supervisory types of frames. NR is incremented by 1 with each frame received

---

[†]The address byte for SDLC does not offer extended addressing capability.

[‡]The control byte for HDLC can be extended in the same manner as the address byte. It is limited to one extension byte only. SDLC does not offer extension of the control byte.

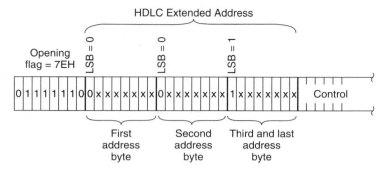

Figure 10-15 Extending the HDLC ADDRESS BYTE field.

with no errors detected. The maximum number of frames that can be received without an acknowledgment is 7. The value of NR is an indicator of what the *next* received frame's NS value should equal. The value of NR can also be thought of as an indicator that the DTE or DCE sending the NR has correctly received all information frames numbered up to and including NR1.

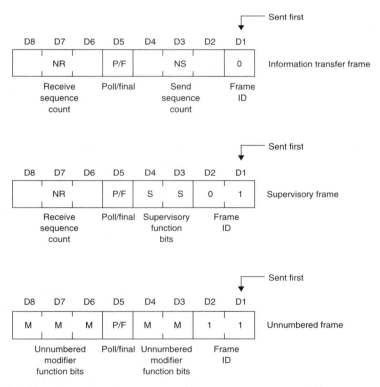

Figure 10-16 Format of CONTROL BYTE for information transfer, supervisory, and unnumbered HDLC frames.

P/F (Poll/Final Bit). The P/F bit serves a dual control function. When set by the primary station, P (poll) is used to poll a secondary station for a response. The secondary station, designated by the address field of the given frame, is required to respond to the poll. When set by a secondary station, F (Final) is used to indicate the final frame that the secondary has for the primary. The primary is required to acknowledge the reception of the secondary's frames upon receiving a frame with the F (Final) bit set. When the P/F bit is 0, P = 0, in a frame sent by the primary station, a secondary station shall not respond since it is not a poll. When the P/F bit is 0, F = 0, and sent in a frame from a secondary station to the primary station, the frame sent from the secondary station is not the last frame. The primary station in this case shall not acknowledge the reception of the secondary's frames.

NS (Send Sequence Number). The NS field is unique to information transfer frames. NS is initialized to 0 for the *first* information transfer frame sent by a station and incremented by 1 with each information transfer frame sent thereafter. NS and NR work in conjunction with each other in terms of control and acknowledgment of the number of information transfer frames sent and received. The maximum number of frames that can be sent without an acknowledgment is 7.

Frame ID (Frame Identifier). The least significant bits of the control byte identify the HDLC frame type as follows:

D2	D1	
(NS)	0	Information transfer frame
0	1	Supervisory frame
1	1	Unnumbered frame

S (Supervisory Function Bits). The supervisory function bits (S) are used to further classify the supervisory frame into one of three different types: receiver ready (RR), receiver not ready (RNR), and reject (REJ). Bits D4 and D3 determine the supervisory frame type as follows:

D4	D3	
0	0	Receiver ready (RR): Used to indicate a primary or secondary station's readiness to receive an information frame. Used as an acknowledgment to receiving a frame.
0	1	Receiver not ready (RNR): Used for acknowledgments and to indicate that a station is busy.
1	0	Reject (REJ): Used for error control and requesting retransmission of a frame starting with the frame numbered NR.

TABLE 10-3 HDLC Unnumbered Frame Definition

Name	Binary format $D_8 \cdots D_1$	Description
DISC	010 P 0011	Disconnect
DM	000 F 1111	Disconnect Mode
FRMR	100 F 0111	Frame Reject
RD	010 F 0011	Request Disconnect
RIM	000 F 0111	Request Initialization Mode
SIM	000 P 0111	Set Initialization Mode
SNRM	100 P 0011	Set Normal Response Mode
UA	011 P/F 0011	Unnumbered Acknowledgment
UI	000 P/F 0011	Unnumbered Information

M (Unnumbered Modifier Function Bits). There are five bits within the control field of the unnumbered frame that serve as the unnumbered modifier (M) function bits: D8, D7, D6, D4, and D3. The state of these five bits determines the function of the unnumbered frame (Table 10-3).

Definition of HDLC Frame Types

Information Transfer Frame. The information transfer frame carries data in the information field between the primary and secondary stations. It includes control for polling and data acknowledgment through the use of the P/F bit. This type of frame is identified by a logic 0 for the LSB, D1, of the control byte (Figure 10-17). Stations transmitting information transfer frames use the NS count to indicate the number of information transfer frames sent. The NR count is used to acknowledge the number of information transfer frames received.

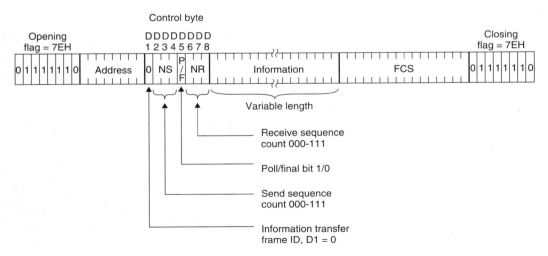

Figure 10-17 HDLC information transfer frame structure.

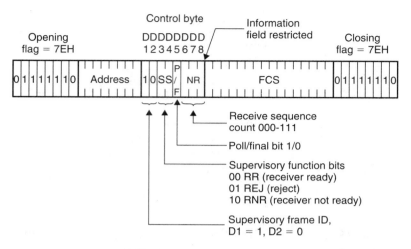

Figure 10-18 HDLC supervisory frame structure.

Supervisory Frame. The supervisory frame is used for polling, data acknowledgment, and control. The control byte of this type of frame has a unique two-bit supervisory function (S) field that is used to indicate a station's readiness: RR (receiver ready) and RNR (receiver not ready). The two-bit field is also used for requesting the retransmission of an information frame with REJ (reject). Figure 10-18 illustrates the format of the supervisory frame. Note that a supervisory frame cannot have an information field. The least significant two bits of the control byte identify the supervisory frame. D1 and D2 are a logic 1 and a logic 0, respectively.

Unnumbered Frame. This type of frame is identified by the lower two bits of the control byte. D1 and D2 both equal a logic 1 (Figure 10-19). The unnumbered frame is used for polling, testing between stations, initialization, and control of stations. The information field may or may not be present in this type of frame. Table 10-3 lists the various types of unnumbered frames with a description of each.

10.6.1.4 Information field. The HDLC information field is a variable-length field that contains any number of bits. The vast number of applications, however, tend to use multiples of eight bits. This field is always present in information transfer-type frames and may be present in unnumbered-type frames. In supervisory-type frames, the information field does not exist. The code within this field is transparent to the receiver station. The size of the field is entirely left to the system designer. Here, again, the protocol overhead, system noise, distance between stations, and various other factors must be considered. Typically 256 bytes are used.

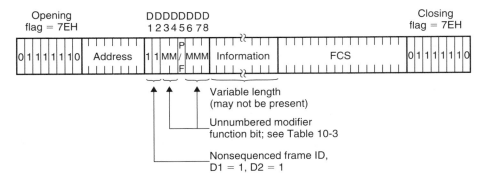

Figure 10-19 HDLC unnumbered frame structure.

10.6.1.5 Frame check sequence (FCS). The FCS is a 16-bit CRC character computed by the transmitting device and checked by the receiving device. All bits within the frame, excluding the flag pattern, are used to compute the FCS character. Inserted *zero bits* (see below) used to attain transparency are not included in the calculation. The mathematical polynomial used to compute the FCS has been documented by CCITT's V.41 specification.

10.6.2 Zero-Bit Insertion and Deletion

The information field in HDLC is said to be *transparent* to the receiver section. As is the case with character-oriented protocols, a method of distinguishing between data and control within the frame is needed. HDLC is a bit-oriented protocol. Bit-oriented protocols do not identify groups as *characters*. In BISYNC, a character-oriented protocol, dedicated characters are used for framing and control as well as resolving the transparency problem. Transparency for bit-oriented protocols must be achieved in a different manner due to the absence of these dedicated characters.

Since 7EH is the *only* framing character used in HDLC, the transmitting device must ensure that this delimiting pattern does not exist *within* the boundaries of the frame itself. The receiving device would otherwise misinterpret the end of the frame. For purposes of discussion, we will refer to this portion of the frame as the *data field* (Figure 10-20). This includes all HDLC fields with the exception of the flags. The technique used to achieve transparency within the data field in HDLC (and other bit-oriented protocols) is referred to as *zero-bit insertion* or *bit stuffing*. Zero-bit insertion occurs within the range of the data field as depicted in Figure 10-20.

Notice that the eight-bit FLAG pattern, 01111110 (7EH), marking the beginning and end of the frame, includes a series of *six* consecutive 1-bits between 0-bits. In zero-bit stuffing, the transmitting device, *after* sending the opening flag, automatically inserts, or "stuffs," a binary zero-bit after any succession of *five*

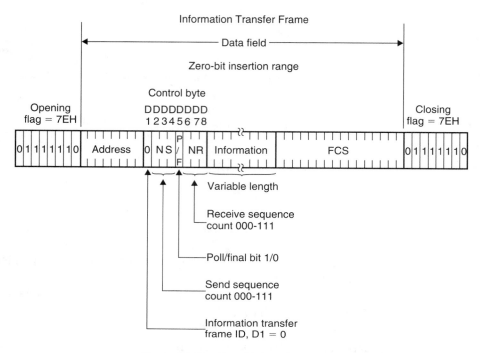

Figure 10-20   Zero-bit insertion range.

consecutive 1's. This will ensure that no pattern of the flag, 01111110, is ever transmitted within the data field. It is important to recognize that bit stuffing occurs *only* in the data field.

Figure 10-21 shows how the bit-stuffing technique works. Once the receiver detects the opening flag, it monitors for five consecutive 1's. If five consecutive 1's are received after the opening flag, the sixth bit is automatically deleted *if* it is a binary 0, thus resulting in the original bit pattern. If the six consecutive 1-bits are detected after the opening flag, then it is assumed to be the closing flag or an *abort character.* The 16 bits preceding it are taken to be the frame check sequence (FCS) character. Inserted zeros are not included in the calculation of the FCS character.

10.6.3 HDLC Abort and Idle Condition

Another requirement for the receiving device is the detection of two other bit patterns: the *abort pattern* and *idle pattern*. The HDLC abort pattern[†] is *7 to 14* consecutive 1-bits without zero-bit insertion (one bit more than the closing flag). The idle pattern is 15 or more consecutive 1-bits without zero-bit insertion.

[†]IBM's SDLC Abort Pattern is eight consecutive 1-bits without zero-bit insertion. Seven contiguous 1-bits in SDLC are used for polling purposes.

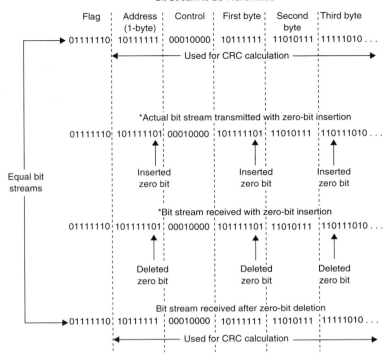

Figure 10-21 HDLC zero-bit insertion, bit-stuffing technique.

When the abort pattern has been detected by the receiver, the current frame is prematurely terminated. That portion of the frame already received is disregarded. An example of an abort condition may be a computer's loss of its modem carrier frequency or the loss of one of its control signals, such as CTS. The computer in this case may be in the process of transferring data to other computers or terminals. If the loss of its modem's carrier frequency or modem control lines occurs during a frame transfer, the transmitting device sends seven consecutive 1's without zero-bit insertion to the station that the frame is being transmitted to. The receiving station detects the abort patten and invalidates the current frame.

A channel is defined in HDLC as being in an idle condition when the receiver detects 15 or more contiguous 1's on the line. The specific action to take upon detecting the idle condition is up to the system designer.

10.6.4 Control and Information Exchange

In Figure 10-22, examples of HDLC frame sequences are shown between a host computer and three terminals. The same multipoint setup used earlier for BISYNC line control will be used here. Although HDLC has virtually twice the

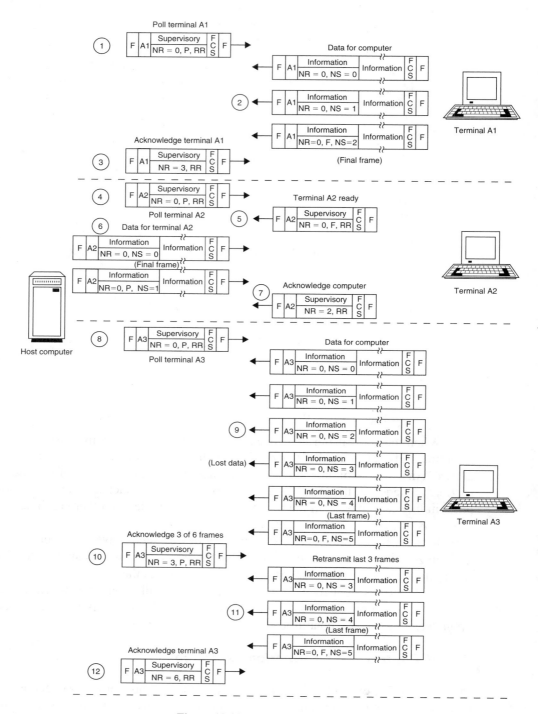

Figure 10-22 HDLC frame sequencing.

informational transfer rate of BISYNC (since it is suited for full-duplex operation), a half-duplex control sequence is illustrated here for simplicity. The formats of the frames are depicted from left to right with their respective types, information, supervisory, or unnumbered, identified within each frame's control field.

Step 1. Initially, terminal A1 is polled by the host computer for possible information. See step 1 of Figure 10-22. NR, the number of frames received by the host, has been initialized to 0. P/F has been set to 1 to indicate that terminal A1 is being polled (P). The host computer is ready to receive data, RR.

Step 2. Terminal A1 responds to the poll by transmitting three information frames (NS = 0, 1, 2). NS, the number of frames sent, is incremented by 1 with each frame sent to the host starting with frame 0, NS = 0. Notice the third and final (F) frame, P/F, is set to 1, indicating the last frame has been sent. The FCS character is sent with each of the three information frames, thus allowing the host to perform error checking on each individual frame.

Step 3. The host computer acknowledges the reception of Terminal A1's three frames. The acknowledgment is a supervisory frame. NR in this frame is set to a value of 3 as an indicator to terminal A1 that no errors were detected in the three frames received. If an error were detected in one of the received frames, frame number 2, for example, then the value of NR would be set to 1, notifying terminal A1 that only one frame was *successfully* received. Terminal A1 in turn would retransmit the last two frames with NS = 1 and NS = 2, respectively.

Step 4. A new sequence is initiated in step 4. Terminal A2 is polled (P), P/F = 1, for information and readiness to receive two frames that the host computer has for it.

Step 5. Terminal A2 has no information frames at this time for the host, as indicated by the final (F) bit in this supervisory frame. P/F is set to 1. Terminal A2, however, is ready to receive data, RR.

Step 6. Two information frames are sent to terminal A2. In the first frame sent, NS = 0. In the second and final (P) frame sent, NS is incremented by 1 and P/F is set to 1 in order to poll terminal A2 for acknowledgment of the transmitted data.

Step 7. The two information frames sent by the host computer are successfully received, NR = 2, and acknowledged by terminal A2.

Step 8. The host computer polls terminal A3 (P), P/F = 1, for information.

Step 9. In step 9, terminal A3 responds to being polled by transmitting six consecutive frames (NS = 0 through NS = 5) to the host computer. The value of

NS is incremented by 1 with each frame sent. A portion of the data within the fourth frame, however, is lost en route to the host due to an intermittence in the transmission line. The host computer will have to catch this error through the CRC check and request a retransmission of the fourth frame and all subsequent frames. The sixth and final (F) frame is identified by P/F = 1.

Step 10. As a result of an error detected in the fourth frame received from terminal A3, the host computer sets NR to a value of 3, since it successfully received three frames. It polls (P) terminal A3 with a supervisory frame. NR = 3 and P/F = 1 act as a request for retransmission of the last three information frames: NS = 3, NS = 4, and NS = 5. Notice that the supervisory frame's NR value in this case serves the dual purpose of acknowledging the reception of three frames and, at the same time, requesting the retransmission of frames starting with NS = 3. The NR value sent by the host is also the value that it is anticipating in the NS field of the next frame sent from terminal A3. Since the host computer knows the total number of frames sent based on the NS value received, it can further anticipate the number of frames that must be retransmitted.

Step 11. Terminal A3 retransmits its last three frames, NS = 3, NS = 4, and NS = 5. The final (F) frame again is identified by P/F = 1.

Step 12. The host computer receives the three retransmitted frames from terminal A3. NR is incremented from 3 to 6 with the reception of each frame. A supervisory frame is sent to terminal A3 to acknowledge with NR = 6 that all six frames have been received without errors.

10.7 EMERGING TECHNOLOGIES

As scientists continue their research and development (R&D) of new technologies and refine existing technologies, the general public and private sectors reap the benefiting results of their labor. There are adverse effects to R&D, however. The technological growth that it fosters continues to spawn new technologies, new standards, and new equipment that must adhere to these standards. No sooner does one technology achieve global acceptance before another more advanced technology arrives on the scene with the potential to displace it. This global competition in the marketplace has left many industry leaders uncertain and cautious about emerging technologies. Furthermore, with technology changing so rapidly, working people must enhance their skills continuously to remain productive in the workforce. This will remain a way of life for many of us.

Over the past decade, several emerging technological standards in telecommunications have gained the acceptance of equipment manufacturers serving the needs of the community. Each of these telecommunication standards has its unique advantages and disadvantages over one another, depending primarily on the user's application. As stated above, there is no long-term assurance for any standard to prevail. Among the more widely recognized communication standards are the following:

— ATM
— ISDN
— SONET
— FDDI
— Fibre channel
— Frame relay
— SMDS

Together, these standards form an infrastructure that experts believe will provide a flexible, unifying set of communication protocols capable of carrying multimedia information at a wide range of speeds across a variety of communications platforms. These include *local area networks* (LANs) and *wide area networks* (WANs). An introduction to each of these standards, with emphasis on ATM, ISDN, and SONET, will now be given.

10.7.1 ATM

Most experts believe that in the forthcoming years, a new technology called *Asynchronous Transfer Mode* (ATM) will profoundly change the communication industry as we know it today. In 1992, this new telecommunications technology exploded from virtual obscurity to one that will likely change the basic structure of the PSTN. Standards are currently being developed by CCITT (Study Group XVIII) that are enabling equipment manufacturers and local exchange carriers (LECs) to unleash several new ATM products and switching services.

10.7.1.1 The ATM network. ATM is a high-speed form of packet switching developed as part of the *Broadband Integrated Services Digital Network* (BISDN). It is intended to be carried on the *Synchronous Optical Network* (SONET) and serve the needs that revolve around corporate private networking. Figure 10-23 illustrates a paradigm of an ATM network. Experts believe the ATM network will gradually displace private leased T1 facilities and customer switching equipment. Traditional analog or digital TDM switches used to route traffic through the network use a central processor (ESS) to establish the proper path through the switch. In contrast, ATM switches will be built using self-routing procedures where individual *cells* consisting of user data will find their own way through the ATM switching network on the fly using their own address, rather than having an external process establish a path.

ATM uses the concept of *virtual channels* (VCs) and *virtual paths* (VPs) to accomplish the routing of cells. A VC is a connection between communicating

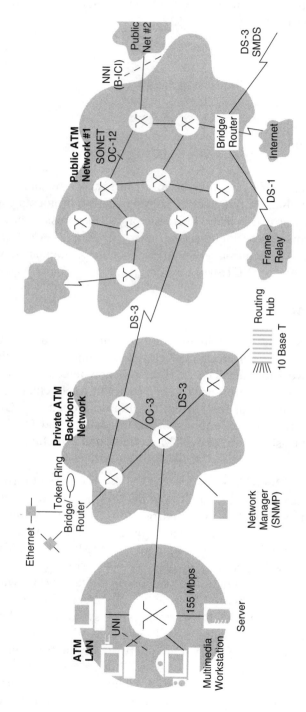

Figure 10-23 Paradigm of an ATM network. (Courtesy of Agile Networks Inc.)

312

entities. It may consist of several ATM links between local exchange carriers (LECs). All communications occur on the VC, which preserves cell sequence. In contrast, a VP is a group of VCs carried between two points and may consist of several ATM links. Further details of ATM network operations and switching is far beyond the scope of this book. We will, however, introduce the format of the ATM cell for purposes of comparing structure and definition with other protocols discussed.

10.7.1.2 The ATM cell header.

ATM transmits short fixed-length packets called *cells*. As shown in Figure 10-24, each cell is 53 bytes in length and is comprised of two parts: a five-byte *Header Field* containing address and control information and a 48-byte *Information Field* containing the user's data. The Information Field can represent digitized voice, data, image, and video. Individual cells can be mixed together and routed to their destination via the telecommunications network.

ATM and its high transmission speeds will provide a boost to fiber-optic installation. Thus far, the transmission speeds agreed upon by the *ATM Forum*[†] are 45, 100, and 155 Mbps, which matches the *Synchronous Optical Network* (SONET) OC-3 rate.[†] Transmission rates are currently used up to 622 Mbps and will continue to rise as computer speeds increase.

The heart of the ATM communications process lies in the content and structure of the cell. The ATM cell contains all of the network information for relaying individual cells from node to node over a pre-established ATM connection. The *Header Field* of the cell, which includes the first five bytes of the cell, has been designed for addressing and flow control. Its role is for networking purposes only. The format of the Header Field is illustrated in Figure 10-25 and its various fields are defined as follows.[‡]

Generic Flow Control Field (GFC): The first four bits of the first byte controls the flow of traffic across the user–network interface (UNI) and into the network.

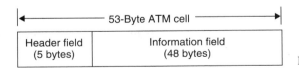

Figure 10-24   ATM cell structure.

[†]The *ATM Forum* was founded in 1991 by Adaptive, Cisco, Northern Telecom, and Sprint. Its membership currently includes over 120 firms.

[‡]Michael Fahey, "ATM: The Choice of the Future?" *Fiber Optic Product News*, May 1993, p. 21.

[‡]The following cell information was compiled with permission from Jim Lane, *Asynchronous Transfer Mode: Bandwidth for the Future* (Norwood, Mass.: Telco Systems, 1992), pp. 18–20.

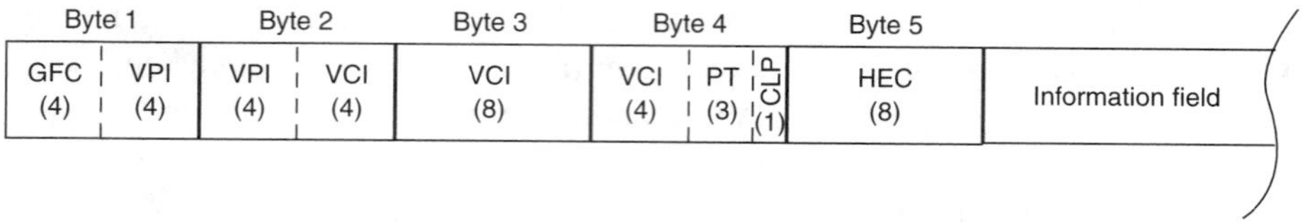

Byte 1		Byte 2		Byte 3	Byte 4			Byte 5	
GFC (4)	VPI (4)	VPI (4)	VCI (4)	VCI (8)	VCI (4)	PT (3)	CLP (1)	HEC (8)	Information field

Figure 10-25 ATM five-byte header field structure.

Virtual Path Identifier (VPI)/Virtual Channel Identifier (VCI):   The next 24 bits, which include the second half of the first byte, the second and third bytes, and the first half of the fourth byte, make up the ATM address.

Payload Type (PT):   The first three bits of the second half of the fourth byte specify the type of message in the payload. The three-bit code is encoded into one of eight types ($2^3 = 8$) of payloads. Types 0 to 3 are reserved at this time for identifying the type of user data. Types 4 and 5 indicate management information and 6 and 7 are reserved for future definition.

Cell Loss Priority (CLP):   The last bit of the fourth byte indicates the eligibility of the cell for discard by the network under congested conditions. The CLP bit is set by the user. If set, the network may discard the cell depending on traffic conditions.

Header Error Control (HEC):   The fifth and final byte of the Header Field is designed for error control. The value of the HEC is computed based on the four previous bytes of the Header Field. The HEC is designed to detect header errors and correct single bit errors within the Header Field only. This provides protection against the misdelivery of cells to the wrong address. The HEC does not serve as an entire cell check character.

10.7.1.3 The ATM information field.   The 48-byte Information Field of the ATM cell contains the user's data. Insertion of the user's data into the cell is accomplished by the upper half of layer 2 of the ISO–OSI seven-layer model. This layer is specifically referred to as the **ATM Adaptation Layer** (AAL). It is the AAL that gives ATM the versatility to carry many different types of services from continuous processes like voice to the highly bursty messages generated by LANs, all within the same format. Since most users' data require more than 48 bytes of information, the AAL divides this information into 48-byte segments suitable for packaging into a series of cells to be transmitted between endpoints on the network. There are five *types* of AALs:

Type 1: Constant Bit Rate (CBR) Services: This AAL handles T-Carrier traffic like DS0s, DS1s, and DS3s. This permits the ATM network to emulate voice or DSn services.

Type 2: Variable Bit Rate (VBR) Timing-Sensitive Services: This AAL is currently undefined but reserved for data services requiring transfer of timing between endpoints as well as data (e.g., packet video).

Type 3: Connection-Oriented VBR Data Transfer: This AAL transfers VBR data (i.e., bursty data generated at irregular intervals) between two users over a pre-established connection. The connection is established by network signaling similar to that used by the PSTN. This class of service is intended for large, long-period data transfer, such as file transfers or backup and provides error detection at the AAL level.

Type 4: Connectionless VBR Data Transfer: This AAL provides for the transmission of VBR data without pre-established connections. It is intended for short, highly bursty transmission as might be generated by LANs.

Type 5: Simple and Efficient Adaption Layer (SEAL): This new and well-defined AAL offers improved efficiency over type 3. It serves the same purpose and assumes that the higher layer process will provide error recovery. The SEAL format simplifies sublayers of the AAL by packing all 48 bytes of the Information Field with data.

10.7.1.4 The future of ATM. ATM is already beginning to find its way into LECs throughout the world. With most of the major players in the computer networking industry working on ATM products, the evolution of ATM looks promising but not without its obstacles. ATM technology as it stands today faces many of the bottlenecks that any new technology faces. These include the high initial costs of ATM equipment, installation, training, and more. Many analysts believe that the expense of converting to ATM technology will not be justified unless multimedia catches fire throughout the general public and private sectors. However, most of these obstacles are only because ATM is new. Prices are expected to come down dramatically in the next few years, which will speed the development of this revolutionary technology.

10.7.2 ISDN

ISDN stands for *Integrated Services Digital Network.* ISDN was developed in the 1970s by Bellcore (Bell Communications Research, Livingston, NJ) and marketed by the seven Bell operating companies (BOC). It was progressively standardized by CCITT in 1984. Its purpose is to provide a set of standardized interfaces and signaling protocols for delivering integrated voice and data over a standard telephone line. By ordering ISDN services through the local telephone company, custom ISDN telephones, computers, FAX machines, and more can be used to send and receive digital signals simultaneously upon dial-up to any accessible location in the world via the telephone network. As users of the PSTN become more and more committed to nonvoice communications (i.e., computer data, facsimile, video, and graphics information), the all-digital network will soon prevail. This is the basic premise behind the major thrust toward ISDN.

10.7.2.1 Architectural overview. Figure 10-26 compares the traditional data connection to the PSTN versus the ISDN interface. In the traditional connection, data originating from a terminal or host computer (DTE) must be transformed to an analog signal through the use of a modem. The analog signal is then transported to another end user whose modem reconverts the analog signal to digital data for computer and terminal use. A more efficient process is the ISDN connection shown. Here, the signals originating from digital sources remain digital throughout the network. Terminals and computers connect directly to an ISDN digital line. Even the analog voice signal is digitized by the ISDN telephone prior to being placed on the network.

10.7.2.2 Equipping subscriber loops with ISDN. Most major PBX (Private Branch Exchange) vendors offer ISDN interfaces that are used in business facilities today. However, the deployment of a standard ISDN interface to residential customers over the existing two-wire facilities has been slow due primarily to the limited bandwidth of the telephone lines. These lines, by and large, have been designed to accommodate the analog voice signal only. ISDN signals are digital and require a greater bandwidth than the 3.1 kHz (300 to 3400 Hz) available over the standard switched voice-grade lines. Unfortunately, the loading coils used in the subscriber loop that have been designed to minimize losses within this frequency range have an abrupt increase in loss above 3400 Hz. This loss plays havoc with phase and amplitude characteristics of the ISDN signal. To provide ISDN basic rate services to the home, existing subscriber loops must be modified in the following manner:

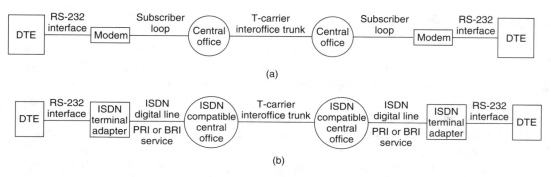

Figure 10-26 (a) Traditional and (b) ISDN data connection to the PSTN.

1. Loading coils must be removed in the loop.

2. Digital echo cancellers must be used on both ends of the loop.

3. High-grade twisted-pair telephone cables must be used.

4. Central office amplifiers and repeater amplifiers should be designed for ISDN signals.

10.7.2.3 BRI versus PRI. Figure 10-27 illustrates how the ISDN system provides end-to-end digital connectivity to a myriad of user services. There are two standard ISDN interfaces: the *Basic Rate Interface* (BRI) and the *Primary Rate Interface* (PRI). The BRI delivers ISDN services to the subscriber over a standard twisted-pair telephone line.[†] The BRI carries two 64-kbps B (Bearer)-channels for voice, data, or video and one 16-kbps packet-switched D (Delta)-channel that carries user packet data and call-control messaging. B-channels always operate at 64 kbps and carry end user voice, audio, video, and data. The primary purpose of the D-channel is to exchange signaling messages necessary to request services on the B-channel. The D-channel is also used for packet switching and low-speed telemetry at times when no signaling is waiting.[†] For user services requiring additional bandwidth for high-speed transmission, *H-channels* are used. H-channels carry multiple B-channels at rates of 384 kbps (H0: 6 B-channels), 1536 kbps (H11: 24 B-channels), and 1920 kbps (H12: 30 B-channels). They are used for audio, video and graphics conferencing, business television, and other high-speed applications. Table 10-4 lists the various ISDN channel types, their transmission rates, and applications.

The PRI is a CCITT-defined ISDN trunking technology that delivers ISDN services to digital PBXs, host computers, and LANs. PRI interconnections illustrated in Figure 10-27 are made up of digital "pipelines" or "pipes" capable of carrying multiple trunk lines. A PRI has the bandwidth of a T1 link, carrying 23 B-channels, each at 64 kbps, and one D-channel for messaging, also at 64 kbps. The overall data rate is 1.544 Mbps. PRI can serve *customer premises equipment* (CPE), such as PBXs, LAN gateways, and host computers. PRI can also serve as an interoffice trunk (IOT). ISDN experts are hopeful that BRI and PRI services combined will some day extend beyond the PBX and into most homes.

10.7.2.4 ISDN customer premises. The customer premises segment of ISDN is defined by CCITT in terms of *functional groupings* and standard *reference points*. There are five functional groupings and four reference points designed to provide interface specifications and facilitate access to the ISDN network. Figure 10-28 depicts the customer premises interface. ISDN functional groupings and reference points are defined as follows:

[†]David Moody, *ISDN*, Issue 2, Northern Telecom, 1991, p. vii.

[‡]William Stalling, *Data and Computer Communications* (New York, Macmillan, 1988), p. 598.

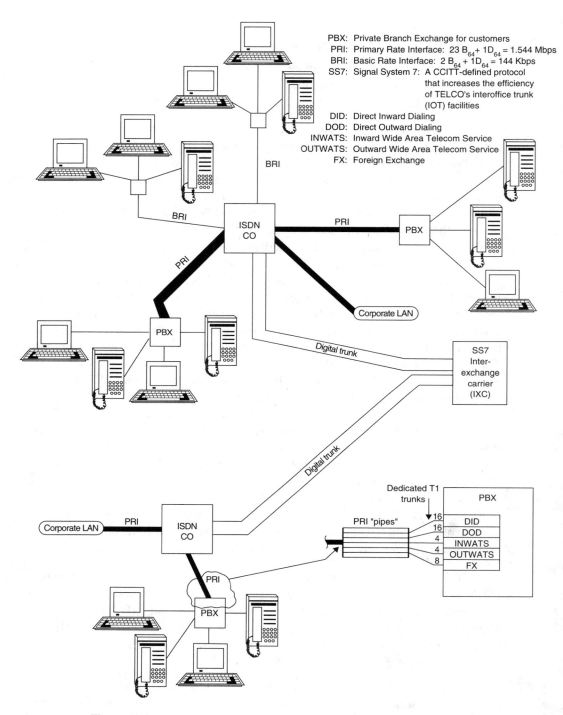

PBX: Private Branch Exchange for customers
PRI: Primary Rate Interface: $23 B_{64} + 1D_{64} = 1.544$ Mbps
BRI: Basic Rate Interface: $2 B_{64} + 1D_{64} = 144$ Kbps
SS7: Signal System 7: A CCITT-defined protocol
　　　 that increases the efficiency
　　　 of TELCO's interoffice trunk
　　　 (IOT) facilities
DID: Direct Inward Dialing
DOD: Direct Outward Dialing
INWATS: Inward Wide Area Telecom Service
OUTWATS: Outward Wide Area Telecom Service
FX: Foreign Exchange

Figure 10-27　ISDN connection depicting BRI (Basic Rate Interface) and PRI (Primary Rate Interface) services.

ISDN functional groupings:

NT1   Network Termination 1
NT2   Network Termination 2
TA   Terminal Adapter
TE1   Terminal Equipment Type 1
TE2   Terminal Equipment Type 2

TABLE 10-4   ISDN Channel Types

Channel type	Transmission rate	Switching technology	Applications
B	Up to 64 kbps	Circuit switched, packet switched	Digital voice, FAX, E-mail, graphics, bulk data, interactive datacom, slow-scan video
D	16 kpbs (BRI), 64 kbps (PRI)	Packet switched (LAP-D)	Telemetry, remote alarms, energy management, E-mail, interactive datacom
H0	384 kbps	Channel switched	High-quality audio, high-speed digital information
H11	1536 kbps	Channel switched	Video/teleconferencing, high-speed digital information
H12	1920 kbps	Channel switched	Video/teleconferencing, high-speed digital information
H4	Up to 150 Mbps (approximate rate)	Channel switched	HDTV, interactive video

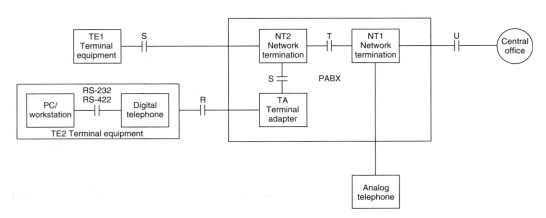

Figure 10-28 ISDN functional groupings and reference points for customer premises. (From TE&M, Feb. 15, 1991, pp. 42–43.)

ISDN reference points:

R   The R reference point is between TA and TE2. It provides a non-ISDN interface between non-ISDN customer premises equipment (CPE) and adapter equipment. It is recommended that this interface complies with CCITT's V. and X. series.

S   The S reference point is between a TA and NT2 or TE1 and NT2. It separates terminal equipment from network communications.

T   The T reference point is between NT1 and NT2. It defines the boundary between the CPE and the network provider's segment of the network.

U   The U reference point separates NT1 of the customer premises from the PSTN transmission line.

10.7.2.5 Broadband Integrated Services Digital Network (BISDN). As new bandwidth-intensive services such as *high-definition television* (HDTV) and ultrahigh-speed computer links are developed, *local area network* (LAN), *metropolitan area network* (MAN), and *wide area network* (WAN) services must fulfill the demand for high-speed data delivery. Standards organizations are now developing a *Broadband Integrated Services Digital Network* (BISDN) to address this need. Whereas ISDN applies mainly to the narrowband telephony world, the BISDN network will use fiber-optic transmission systems capable of transmitting at speeds ranging from 150 Mbps to 2.5 Gbps. BISDN will use the new multiplexing technique employed in ATM. It will also utilize SONET transport networks designed for BISDN facilities. The future of BISDN, like other new technologies, will depend on its ability to offer users the same or better services, features, and cost benefits.

10.7.3 SONET

The culminating efforts of standards organizations have resulted in a worldwide standard for optical communications. This standard, called *Synchronous Optical Network* (SONET), was published by the American National Standards Institute (ANSI) in 1988 (ANSI T1.105). The SONET standard has also been incorporated into the *Synchronous Digital Hierarchy* (SDH) recommendations of CCITT. It has since emerged as a forerunner in today's leading telecommunications technologies. The major goal of SONET is to standardize fiber-optic network equipment and allow the internetworking of optical communication systems from different vendors. Several telephone carriers have been aggressively deploying SONET equipment with hopes that it will reach its greater potential as a broadband switching network. There is sufficient flexibility designed into the SONET payload such that it will be used as the underlying transport layer for BISDN and ATM cells.

10.7.3.1 SONET signal hierarchy and line rates. In brief, SONET defines optical carrier levels and their equivalent electrical *synchronous transport signals* (STS) for a fiber optic–based transmission hierarchy. Table 10-5 lists the standard transmission rates for SONET. Those rates that are marked with an asterisk are the most widely supported by network vendors and providers.

The STS-1 rate of 51.84 Mbps is the basic building block for SONET optical line rates. Higher rates are integer multiples of this rate. For STS-N, only certain values of N are permitted. These values of N are N = 1, 3, 9, 12, 18, 24, 36, and 48. Higher rates than STS-48 may be allowed in future revisions with a maximum of 255 under the current standard.

TABLE 10-5   SONET Optical Transmission Rates

Optical carrier	Synchronous transport signal	Line rate (Mbps)
OC-1*	STS-1	51.84
OC-3*	STS-3	155.52
OC-9	STS-9	466.56
OC-12*	STS-12	622.08
OC-18	STS-18	933.12
OC-24	STS-24	1244.16
OC-36	STS-36	1866.24
OC-48*	STS-48	2488.32

*Popular SONET physical layer interface.

10.7.3.2 The SONET frame format. The transport system adopted in SONET utilizes a synchronous digital bit stream comprised of groups of bytes organized into a frame structure in which the user's data is filled. The basic STS-1 SONET frame format is illustrated in Figure 10-29. The frame format is typically depicted as a 9-row by 90-column matrix representing individual bytes of the synchronous signal. The frame is transmitted byte by byte starting with byte one and scanning from left to right and top to bottom for a total of 810 bytes (9 × 90) or 6480 bits (810 × 8). The entire frame is transmitted in 125 μs, which works out to be the STS-1 line rate of 51.84 Mbps (6480 bits/125 μs = 51.84 Mbps).

STS-1 frames are divided into two main areas: the *transport overhead* (TOH) and the *synchronous payload envelope* (SPE). The first three columns of the 90-column matrix is occupied by the TOH. The remaining 87 columns make up the SPE. The first column of the SPE is the *STS Path Overhead*, which contains payload specific data. These data do not change as the SPE traverses the network. The remaining 86 columns of the SPE are used to carry the payload or network traffic. The TOH and SPE are defined as follows.

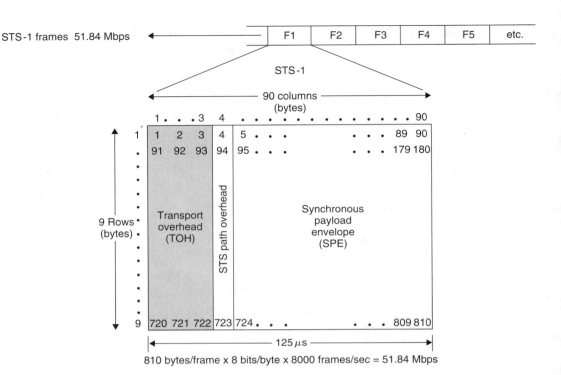

Figure 10-29 SONET STS-1 frame format.

TOH. The **transport overhead** portion of the STS-1 frame is for alarm monitoring, bit error monitoring, and data communications overhead necessary to ensure the reliable transmission of the synchronous payload envelope (SPE) between nodes in the synchronous network.

SPE. The **synchronous payload envelope** is designed to transport a tributary's signal across the synchronous network from end to end. The SPE is assembled and disassembled only once even though it may be transferred from one transport system to another on its route through the network. In most cases the SPE is assembled at the point of entry to the synchronous network and disassembled at the point of exit from the network.[†]

10.7.3.3 STS-1 payload. The *payload* can be thought of as the revenue-producing traffic being transported and routed over the SONET network. Once the payload is assembled (or multiplexed as discussed below) into the SPE, it can be routed through the SONET network to its destination. The STS-1 payload has the capacity to transport the circuits (or equivalent of) listed in Table 10-6.

[†]*Introduction to SONET Networks and Tests* (Palo Alto, Calif.: Hewlett-Packard, 1992), pp. 36–38.

TABLE 10-6 STS-1 Payload Capacity

Capacity	Signal type	Signal rate (Mbps)	Voice circuits	T1's	DS3's
28	DS1	1.544	24	1	—
21	CEPT1	2.048	30	—	—
14	DS1C	3.152	48	2	—
7	DS2	6.312	96	4	—
1	DS3	44.736	672	28	1

The 86 columns of the SPE that are designed to carry the payload are arranged in accordance with standard "mappings" or rules that are a function of the data services carried (e.g., DS1, DS2, etc.). The specific arrangement used is called a *virtual tributary* (VT). SONET specifies different sizes of VTs. For example, a VT1.5 is a frame consisting of 27 bytes structured as three columns of nine bytes. With a 125-μs frame time (8 kHz), these bytes provide a transport capacity of 1.728 Mbps and will accommodate the mapping of a DS1 signal at 1.544 Mbps. Twenty-eight VT1.5's may be multiplexed into the STS-1 SPE.

The SPE may not need a VT frame structure if the service occupies the entire SPE. The DS3 signal in Table 10-6 is an example. SONET's STS-1 SPE has been specifically designed to provide transport for a DS3 tributary signal. A DS3 signal occupies the entire SPE and does not need a VT frame structure.

10.7.3.4 SONET multiplexing. Higher levels of synchronous transport signals (STS) are created by a technique called *byte-interleaved multiplexing*. STS-N is constructed by byte-interleave multiplexing N STS-1 signals together. STS-3, for example, is three STS-1 signals multiplexed together as illustrated in Figure 10-30. The STS-3 frame therefore has 90 $\times$ 3 or 270 columns and nine rows for a total byte count of 2430 (270 $\times$ 9). Since the STS-3 (or STS-N) frame is also transmitted in 125 μs, the transmission line rate is three times that of the STS-1 rate (3 $\times$ 51.84 Mbps), or 155.52 Mbps. Note that byte-interleaved process results in a matrix of multiplexed columns also. Thus the first nine columns of the STS-3 frame are occupied by the TOHs of the three STS-1 signals. The remaining 261 columns are occupied by the SPEs of the three STS-1 signals.

10.7.4 FDDI

As our demand for information increases, the telecommunication system becomes increasingly more complex. First-generation network standards such as Ethernet and Token Ring cannot handle the enormous volumes of data that users are demanding today, particularly over longer-distance spans. The lack of bandwidth on existing transmission media is the bottleneck. With the evolution of

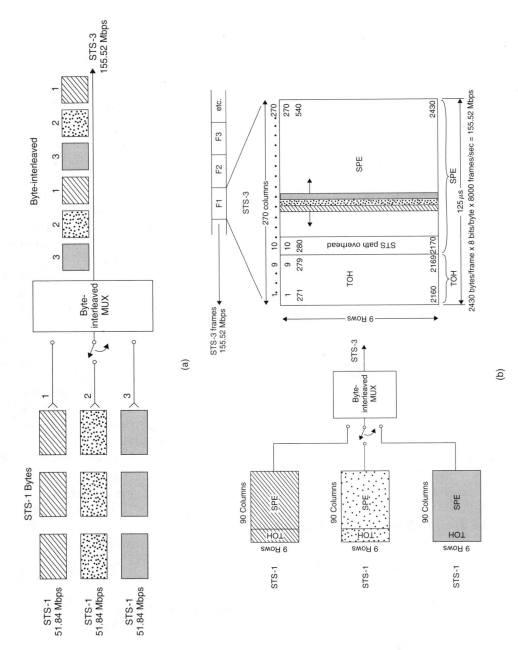

Figure 10-30 Byte-interleaved multiplexing of an STS-3 frame: (a) Byte-interleave process; (b) STS-3 frame format.

fiber-optics technology, bandwidth is virtually unlimited; hence the *Fiber Distributed Data Interface* (FDDI) standard was conceived.

The *American National Standards Institute* (ANSI) committee has developed the FDDI standard (X3T9.5) to support users who require the flexibility, reliability, and speed provided by fiber-optic technology. The standard specifies a 100-Mbps transmission signaling rate with up to 2 km between stations. A dual counter-rotating ring structure is used for the high-speed token-passing network.

The relatively high cost of fiber-optic transmission media for FDDI has kept the 100-Mbps standard from dominating the PC and workstation LAN market. The ANSI X3T9.5 committee continues to work on an alternative to the standard to lower the cost. The alternative is to attain 100 Mbps over copper wire, both *shielded twisted pair* (STP) and *unshielded twisted pair* (UTP). In addition, several vendors have combined their efforts in an attempt to unite the benefits of FDDI with those of copper: a transmission medium that is less expensive than fiber and is likely to be installed in the user's facility already.

10.7.5 Fibre Channel

Although the FDDI standard has alleviated much of the communications bottleneck resulting from limited transmission speeds, many commercial, educational, and scientific environments are looking for performance beyond the 100-Mbps FDDI transmission rate. The emerging *Fibre Channel Standard*, being developed by the ANSI X3T9.3 committee is a high-performance standard designed to transfer data at speeds up to 1 Gbps over distances of 10 km using fiber-optic transmission media. In addition, Hewlett-Packard Co., IBM, and Sun Microsystems Computer Corp. have announced the *Fibre Channel Systems Initiative,* which is a joint effort to advance the Fibre Channel as an affordable, high-speed interconnection standard for workstations and peripherals. The primary goals of the initiative are to advance high-speed interconnections for workstations and systems, promote open systems for distributed computing, and propose selected sets of Fibre Channel options for manufacturers to build conforming products.

The Fibre Channel defines a matrix of switches, called a *fabric*, that performs network switching functions similar to that of a telephone system. Each computer system attaches to the fabric with dedicated send and receive lines designed for point-to-point, bidirectional serial communications. The switch can route any incoming signal to any output port.

The Fibre Channel is one of the key technologies of the 1990s that will provide high-speed performance at an affordable cost. It is expected that a significant share of the workstations and servers will be using the Fibre Channel interface by the end of the decade.

10.7.6 Frame Relay

A technology that has gained acceptance in the LAN and WAN community is *Frame Relay*, a *fast-packet* switching protocol that supports a variable-length frame structure of up to 4096 bytes. Frame Relay began as a concept of a simple packet mode access to ISDN. It is defined by CCITT Recommendation I.122 as *Framework for Additional Packet Mode Bearer Services.* In contrast to the conventional packet switching discussed earlier, fast-packet protocols such as Frame Relay are used in operating environments that include digital broadband systems. Frame Relay's high throughput of 64 kbps to 2 Mbps makes it suitable for LAN and WAN data and imaging network traffic but less satisfactory for voice and real-time video. It is far more efficient than X.25 packet technology, due to its simplistic architectural framework. Roughly two-thirds of the protocol complexity of X.25 is eliminated.

The frame format for Frame Relay is shown in Figure 10-31. Note that the structure of the frame has been designed similar to that of HDLC and SDLC. This permits data encapsulation and decapsulation of these frames. The opening flag is followed by a two-byte control field used for network addressing and control. The information field of the frame is variable length up to 4096 octets (bytes). A 16-bit frame check sequence (FCS) is used for error control followed by an 8-bit closing flag.

10.7.7 SMDS

As its name implies, *Switched Multimegabit Data Service* (SMDS) is a *service* offered by several major telephone companies or local exchange carriers (LECs). SMDS is designed to offer efficient and economical connectivity between LAN and WAN systems over the PSTN. Any organization that needs to transmit large amounts of data, via the PSTN, to several locations at different times should consider SMDS as a viable service.

SMDS is currently designed to operate over the standard North American PCM Multiplex Hierarchy at DS1 and DS3 (Digital Signal 1 and 3). In Europe,

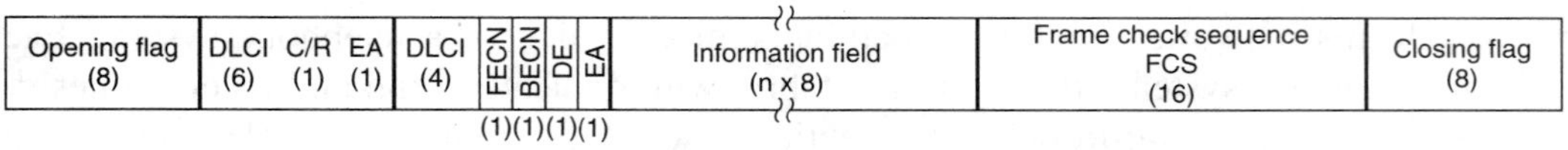

Opening flag (8)	DLCI (6)	C/R (1)	EA (1)	DLCI (4)	FECN BECN DE EA (1)(1)(1)(1)	Information field (n x 8)	Frame check sequence FCS (16)	Closing flag (8)

```
DLCI = data link connection identifier
 C/R = command/response
  EA = address extension
FECN = forward explicit congestion notification
BECN = backward explicit congestion notification
  DE = discard eligibility indicator
   n = number of octets up to 4096
```

Figure 10-31   Frame format for frame relay.

E1 and E3 are used. These transmission rates for telephone trunk lines are 1.544 Mbps and 44.736 Mbps, respectively. Because these rates are standardized between central offices, SMDS lends itself to ubiquitous public connections.

Using SMDS is very much like using a telephone. Once the user establishes the called party's "SMDS telephone number," that number (in SMDS language) need only precede packets addressed to the called party.[†] The packet reaches its destination without further ado. This type of service makes it unnecessary to add remote ends in a private network, making it ideal for subscribers. A large bandwidth at a bargain price, with the ease of connection of a telephone, makes SMDS successful and ideally suited for users looking for a wide-bandwidth digital connection worldwide.

PROBLEMS

1. Define *protocol*.
2. Which standards organization is a member of the United Nations?
3. What are the seven layers of the ISO/OSI model? Briefly describe each layer's function.
4. Which of the seven ISO/OSI layers defines the type of error control method that is used for a given protocol?
5. Which of the seven ISO/OSI layers defines the maximum voltage level that can be used on a conductor?
6. Explain the difference between a byte-oriented protocol and a bit-oriented protocol.
7. What data link control character(s) is (are) used in BISYNC:
   (a) To enter the transparent text mode?
   (b) To exit the transparent text mode?
   (c) To tell the receiver to disregard the bit pattern of data in the text field that is equivalent to DLE?
8. Explain *ACK alternating*.
9. When a time-out error occurs in BISYNC, what procedures do the transmitter and receiver go through?
10. What data link control character(s) is (are) used in BISYNC to enter the *transparent text mode*?
11. What data link control character(s) is (are) used in BISYNC to:
   (a) Contend for the use of the line?
   (b) Handshake blocks of data between a host and tributary?
   (c) Indicate that a block has been received in error?
12. What is a *time-out error*?
13. What is a *tributary*?
14. Explain the difference between device polling and device selecting.

[†]Alan J. Spiegleman, "What Is SMSD?" Networking Management, Oct. 1992, p. 44.

15. What are the names of the three switching technologies traditionally used with the PSTN?
16. What advantages does circuit switching have over other switching methods?
17. To what length is a circuit-switched connection limited?
18. What does *store and forward* mean?
19. The Telex network is an example of what type of switching technology?
20. What does *PAD* stand for?
21. What is the function of a PAD?
22. What international protocol is used for packet switching?
23. Is HDLC a byte- or a bit-oriented protocol?
24. What unique framing pattern is used in HDLC, and what is its name?
25. Refer to Figure 10-15. If an HDLC frame has a two-byte address field, how is the receiver informed that the address is a multiple-byte field?
26. What are the names of the three HDLC frame types?
27. What is the name of the HDLC process used to achieve data transparency?
28. Explain what an HDLC transmitter does when it must transmit more than five consecutive 1's after sending the opening flag.
29. Explain what an HDLC receiver does when it encounters five consecutive 1's.
30. What is the difference between an HDLC abort pattern and an idle pattern?
31. Which HDLC frame types allow use of the information field?
32. Refer to Figure 10-22. Continuing from step 12, show how a fourth terminal, A4, would be polled and sent four information transfer frames from the host computer. Use the same frame format shown. Be sure to show the acknowledgment from terminal A4 to the host computer upon receiving all four frames.
33. Repeat Problem 32, but reverse the direction of the data. That is, show how four information transfer frames would be sent from the terminal to the host computer.
34. What does *ATM* stand for?
35. Define *VP* and *VC*.
36. What is the length of an ATM cell in bytes?
37. Draw the structure of an ATM cell and briefly define its two fields.
38. Draw the structure of the ATM Header Field.
39. Briefly define each field within the Header Field of an ATM cell.
40. What does *ISDN* stand for?
41. What prevents ISDN services from being provided to residential customers?
42. Define *BRI* and *PRI*.
43. Explain the difference between ISDN B-channel and D-channel.
44. Name the five ISDN functional groupings for customer premises.
45. What are the four ISDN reference points for customer premises?
46. What does *SONET* stand for?
47. Derive the 51.84-Mbps line rate for SONET's STS-1 signal.

48. What is a *virtual tributary* in SONET?

49. How many STS-1 SONET frames must be multiplexed together to form an STS-12 frame?

50. What does *FDDI* stand for?

51. What is the transmission rate for FDDI?

52. What does *SMDS* stand for?

53. What is the *Fibre Channel Systems Initiative*?

54. Draw the frame structure for Frame Relay.

11

Local Area Networks

In recent years there has been an outgrowth in the field of communications called *networking*. Networking involves the sharing of computers, peripheral hardware, software, and switching facilities all interconnected with communications channels used to establish a connection between network users. The end result is the shared use of information and resources. The concept of a network is not new and can be understood by considering examples of those services that have been provided to us by the network for many years:

— Television
— Radio broadcast
— Telephone service
— Airline passenger and flight information
— Computer time-sharing systems
— Banking services

In each of these examples, the intention of the network is to distribute information to users requiring the network services. The structure of the network is not of importance to the user. It is the services rendered by the network that are of value. In most cases today the network is controlled by computers.

11.1 THE LOCAL AREA NETWORK (LAN)

Computers of the past were basically large, expensive, and complex mainframes that required a tremendous amount of space and specialized maintenance. A computer expert was required to run the machine for the various batch jobs that were submitted for execution. The results of programs were not immediately available. The programmer or scientist submitting the jobs were at the mercy of the computer specialist running the machine.

With advances in computer technology and supporting hardware, it has become economically feasible for industry, educational institutes, and the private sector to own and interactively operate computers directly from office desktops. A new form of network has evolved from this breakthrough called the *local area network* (LAN). A LAN is typically a privately owned network of interconnecting data communicating devices that *share* recourses (software included) within a limited physical area: a *local area*. This can be a single room within a building to several floors within a building. A LAN can also encompass a building or cluster of buildings. Most LANs link equipment within less than a few miles of each other. Why a local area? Extensive studies of the working environment indicate that 80% of the communications that take place is *within* the local environment, whereas the remaining 20% occurs *outside* the local geographical area. A LAN offers the most effective means of handling local communicating tasks. A medium to large-sized business, for example, may be flooded with intelligent computers and related products. A LAN can interconnect these resources in a manner that best suits the employee's needs for performing everyday communicating tasks. Ultimately, the company benefits from the efficient and cost-effective use of these resources. Let us consider some of these resources:

— Laser printer
— Graphics plotter
— Disk and tape storage devices
— Facsimile machines
— Personal computers
— Mainframe computers
— Modems
— Databases
— Word processing
— PSTN

The percent utilization of many of these resources is often too low and the costs are too high to justify purchasing one for each office, department, or even at times a building. Within a large corporation, for example, several buildings can share the use of a mainframe computer's speed and sophisticated database for inventory control. Each building can have a shared facsimile machine that is tied into the same LAN. Several users with personalized workstations within each building can be tied into the LAN to gain access to the shared facilities provided by the LAN. Data can be exchanged between the mainframe computer and each

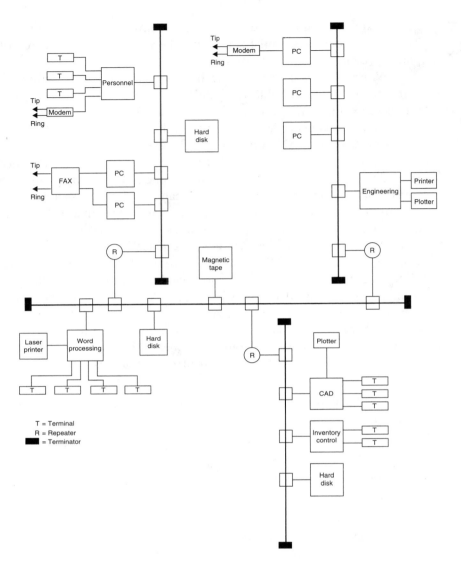

Figure 11-1 Typical LAN setup.

user, and users can send each other messages at any time via electronic mail. Figure 11-1 illustrates a LAN setup.

11.2 LAN TOPOLOGY

LAN *topology* refers to the geometric pattern or configuration of intelligent devices and how they are linked together for communications. The intelligent devices on the network are referred to as *nodes*. Nodes in a network are *address-*

able units that are *linked* together. Nodes are also referred to by some manufacturers in their specifications as *stations*. The two terms are used interchangeably. The *link* is the communications channel. In the design of a LAN, topology is considered in order to best suit a particular environment. In other words, there are advantages and disadvantages to the various topologies used. Factors such as message size, traffic volume, costs, bandwidth, reliability, and simplicity are important. Let us consider some of the most common LAN topolgies.

11.2.1 The Star

Figure 11-2 illustrates the *star* topology. The predominant feature of the star topology is that each node is radially linked to a *central node* in a point-to-point connection. The central node of the star topology is also referred to as the *central control* or *switch*. The central node, typically a computer, controls the communications within the LAN. Any traffic between the outlying nodes must flow through the central node. Central control offers a convenient base for troubleshooting and network maintenance.

The star configuration is best utilized in cases where most of the communication occurs between the central node and outlying nodes. When communications traffic is extensive between outlying nodes, a burden is placed on the central node that can cause message delays. A PBX, for example, is configured as a star. In peak demand situations, the PBX may become overburdened with calls, in which case a busy signal is issued to the person requesting the use of the PBX.

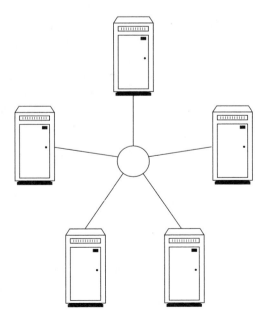

Figure 11-2   Star LAN.

The star topology can also be found in cases where there is a strong demand by several terminals for the computational power of a central computer. Time-sharing systems are configured with a star topology. Small clustered networks used for word-processing applications and database management are commonly configured as a star network.

A disadvantage to the star topology is that the central node must carry the burden of reliability for all of the nodes connected to the LAN. If the central node fails, the network fails. If an outlying node fails, the remainder of the system will continue to operate independently of that node's failure. When the failure of any portion of the network is critical to the extent that it will disable the entire network, it is referred to as the *critical resource*. Thus the central node in a star LAN is the critical resource. Communication systems with critical resources are often provided with *system redundancy*. System redundancy is a critical resource's backup protection in case of a failure. It can be hardware or software either built into the system or provided as an equivalent replacement to the device that failed. System redundancy can minimize the downtime of a network at the expense of cost and complexity.

11.2.2 The Bus

The *bus* topology is essentially a multipoint or multidrop configuration of interconnecting nodes on a *shared* channel. Figure 11-3 illustrates the bus topology. Control over the communications channel is not centralized to a particular node, as is the case with a star topology or a multipoint link with a host computer and its tributaries. The distinguishing feature of a bus LAN is that the control of the bus is *distributed* among *all* of the nodes connected to the LAN. When data are to be transmitted from one node on the bus to another node, the transmitting station *listens* to the current activity on the bus. If no other stations are transmitting data, that is, the channel is clear, then it begins its transmission of data. All nodes connected to the bus are receivers of that transmission, typically a data packet. The function of each receiver is to then determine if the received data packet corre-

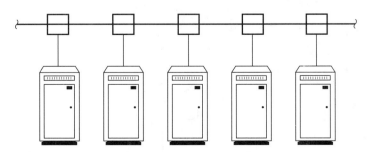

Figure 11-3   Bus LAN.

sponds to its own address. If it does, then the data are acted upon. If it does not, then the data are simply discarded. No routing or circuit switching is involved. Nor is there any transferring of messages from one node to another. The overhead and time required to perform these tasks are eliminated with the bus LAN structure, hence one of its major attractions.

With heavy volumes of traffic on the bus, the likelihood of more than one station transmitting on the channel at the same time increases. A *collision* of data occurs between stations transmitting. A priority contention scheme is used to handle this problem. Typically, the stations attempting to gain access to the channel at the same time will *back-off* for a random interval of time before attempting to access the channel again. This allows one of the stations to transmit its data while the others listen. Eventually, the traffic on the network subsides and fewer collisions occur.

Since the control of the bus LAN is not centralized, a node failure will not hinder the operation of the remaining portion of the LAN. The critical resource in this case is not a node, but rather the bus itself. A short or an open circuit along the bus can virtually cease all communications within the LAN. The degree of failure is not always fatal, however. It is possible for communications to continue between nodes up to the obstruction in the line. Once a short or break in the line occurs, the performance of the LAN is generally degraded due to the resulting reflections and standing waves.

The addition of nodes on a bus that has been previously routed within the local area necessitates gaining access to the bus cable. In some situations, this may be a cumbersome task. Typically, the bus cable is routed through walls and ceilings. A node may be tapped into the bus cable at any point, provided it meets manufacturer's specifications of minimum distance between taps. The bus may also be extended to additional areas. By branching off into other buses, a multiple bus structure called a *tree* is formed. Figure 11-4 depicts the tree structure. In Section 11.6 we look at the *Ethernet* specification and see how the bus topology can be extended with the use of *repeaters*.

11.2.3 The Ring

The *ring* topology, as its name implies, interconnects nodes point to point in a closed-loop configuration (Figure 11-5). A message is transmitted in simplex mode (one direction only) from node to node around the ring until it is received by the original source node. Each node acts as a repeater, retransmitting the message to the next node. Each node also shares the responsibility of identifying if the circulating message is addressed to itself or another node. In either case, the message is eventually received by the destination node and returned to the original source node. The source node then verifies that the message circulated around the ring is identical to the message that was originally transmitted.

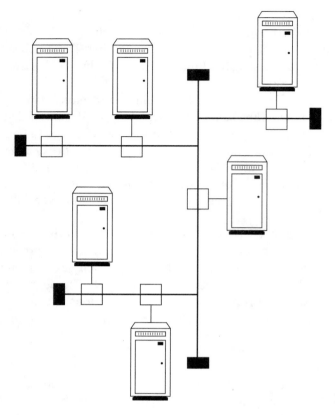

Figure 11-4   Bus tree LAN.

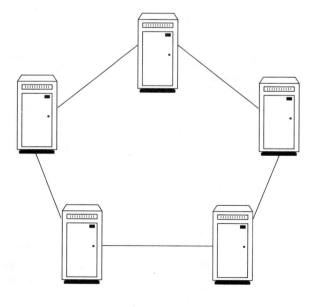

Figure 11-5   Ring LAN.

Acknowledgment bits within the message block are typically set by the destination node so that the source node has a way of verifying that the message was in fact received.

The critical resource in a ring topology is the interconnecting links and nodes. A node or link that malfunctions is typically replaced or bypassed. System redundancy is often included in the ring topology in case of a failure. This includes relays for bypassing a faulty node and additional cabling for faulty links, as shown in Figure 11-6. It can also include backup software or even backup nodes.

11.3 CHANNEL ACCESS

The data link layer of the ISO/OSI seven-layer model defines the channel access protocol used by a node to gain the use of the communications channel. The following factors govern the selection of the channel access protocol used.

— Topology
— Physical size of the LAN
— Number of nodes
— Application

11.3.1 Polling

Polling is a channel access technique whereby the host computer polls each tributary in a logical sequence for possible data. In the general sense of the word, it also includes the selection of each tributary for purposes of sending data from the host computer. The poll-select technique, discussed in Chapter 10, can also be applied to LANs. A node acting as the host can control the use of the channel by polling and selecting all other nodes within the LAN. The key to this technique is that there is only one host that controls the use of the channel at all times. When a node has been granted the use of the channel through a poll or select command from the host, its data are placed on the network and sent to their destination node either directly or indirectly through the routing of the host. Any of the topologies we have discussed thus far are suited for polling. The major disadvantage to this channel access technique is that communications is extremely time consuming due to the overhead required for controlling and acknowledging messages. Also, the critical burden of reliability is placed entirely on the host computer.

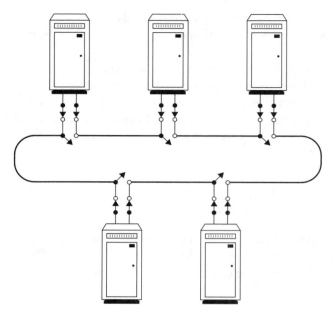

Figure 11-6   Ring LAN with bypass relays.

11.3.2 Contention

Contention is a channel access protocol whereby stations contend with each other for the use of a single communications channel. A station desiring the use of the channel asserts its message onto the channel. The message is received by a destination station on the same channel. When two or more stations inadvertently attempt to send their messages at the same time, a *collision* occurs between messages.

A common analogy used to describe contention is a *cocktail party*. At a cocktail party, when two or more people speak at the same time, their messages conflict with each other. Each person then backs off, listens for a moment, and attempts to talk again. The more crowded the party becomes, the more interference there is likely to be.

11.3.2.1 Aloha.   To share the use of a host computer located at the University of Hawaii on the island of Oahu, several surrounding islands interface their terminals to the host computer via a radio link. To minimize the costs of the network broadcasting equipment involved, all terminals share the use of the communications channel. The communications channel is a pair of frequencies. One frequency is used by all remote stations to transmit to the host, and the other frequency is used exclusively by the host for acknowledgments and transmitted messages to the remote stations. The channel access protocol used is a contention scheme called *Aloha*. In Aloha, a station wishing to transmit a message simply does so at any time, running the risk of a collision from other stations. This is referred to as *pure Aloha*: talk whenever you want. An acknowledgment from the host computer verifies that the message was received without interference from messages transmitted by other stations that share the same frequency.

In pure Aloha, when a station transmits a message, a collision from another station can occur at any time during the interval of the transmission. Even after an entire message has been transmitted, a collision can occur since the message takes time to arrive at its destination. Figure 11-7 depicts the Aloha contention scheme. When a collision occurs in Aloha, no acknowledgment is returned to any of the stations involved in the collision. All stations back off and wait a predefined period of time before attempting another transmission. Studies indicate that when station activity on the network is greater than about 20%, performance is severely degraded due to the number of collisions. In spite of its shortcomings, the Aloha contention protocol is the predecessor to more sophisticated contention schemes.

11.3.2.2 Slotted Aloha. A more sophisticated approach to pure Aloha is *slotted Aloha*. In slotted Aloha, all stations on the network are time synchronized with each other and are restricted to transmissions at predefined intervals of time called *slots*. A slot of time is based on the maximum allowable size of a message.

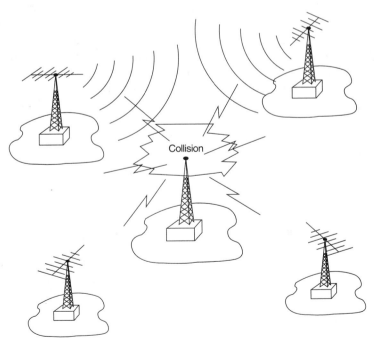

Figure 11-7 Aloha collision.

A station desiring the use of the channel must begin its transmission at the beginning of a slot and finish before the end of the slot interval. Contention of the channel still exists in slotted Aloha; however, once a station has begun its transmission without a collision occurring at the beginning of the slot, that station is guaranteed the remainder of the slot time. In slotted Aloha, any collisions that would have otherwise occurred in pure Aloha are eliminated; overlapping messages cannot occur. All other stations must begin their transmissions during another time slot. Channel utilization is effectively doubled in slotted Aloha.

11.3.2.3 Carrier sense multiple access with collision detection (CSMA/CD). One of the most widely used contention protocols for bus LANs is called *carrier sense multiple access with collision detection* (CSMA/CD). In CSMA/CD, any node can send a message, or more specifically a packet, to any other node connected to the LAN as long as the transmission media are free of signals being transmitted by other nodes.

Carrier Sense. If we consider our earlier example of the cocktail party, the meaning of *carrier sense* in CSMA/CD would be analogous to people *listening* before talking. The interruptions in a conversation could be eliminated entirely if rules at our cocktail party ensured that, in politeness, no one speaks while another one is talking. Recall that in pure Aloha, interruptions can occur at any time during the course of a transmitted packet. The highly efficient CSMA/CD eliminates much of the wasted time that would otherwise be spent having to retransmit interrupted packets. The hardware used to monitor the activity on the channel simply senses the voltage level on the channel and compares it against a reference voltage. Above or below this reference voltage determines whether or not there is activity on the channel. Figure 11-8 illustrates the concept of carrier sensing.

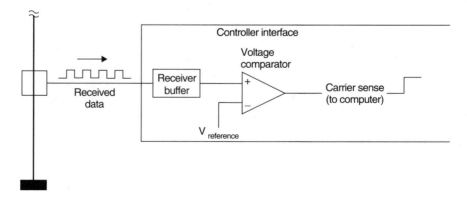

Figure 11-8 Carrier sensing in CSMA/CD.

Multiple Access. In CSMA/CD, the feature of *multiple access* allows any node on the network to access the channel at any time, provided that the channel is free. Packets are deferred if the channel is busy.

It is entirely possible for two or more nodes seeking the use of the channel to sense that the channel is free at the same time. Furthermore, there is the propagation delay time involved in the transmission of a packet from its source to its destination node. A node, for example, could have already begun transmitting a packet, and the channel, as sensed by other nodes, still appears inactive due to the propagation delay time before that packet arrives at other nodes. Technically, other nodes can begin transmitting their own packets onto the channel in this case. Collisions would then occur; hence the need for *collision detection*. Figure 11-9 illustrates a collision between packets transmitted by node B and node D.

Collision Detection. In addition to monitoring (listening) for activity on the channel prior to transmitting, each node in CSMA/CD has provisions for detecting collisions while in the *course of a transmission*. If two or more packets collide, the energy level on the channel changes. *Collision detection* involves monitoring for this change in energy level. If a collision is detected during the course of a transmission, the transmitting node continues the transmission of its current packet, consequently *jamming* its packet onto the channel. The purpose of jamming is to ensure that the nodes involved in the collision detect that a collision has occurred. The jamming signal must be asserted for a length of time necessary to propagate to all nodes connected to the LAN. Transmitting nodes then *back-off* for an interval of time before attempting to retransmit the packet again.

It is important that node back-off times not be equivalent to each other; otherwise, collisions would continue to occur between the same transmitting nodes. The back-off time is therefore random. With regard to our cocktail party exam-

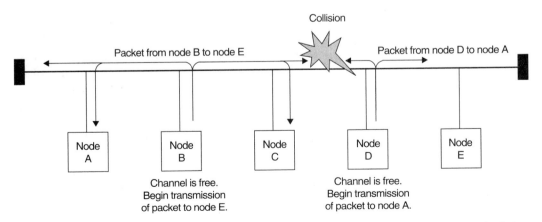

Figure 11-9 Two stations in CSMA/CD sense that the channel is free and begin their transmissions. A collision occurs between two packets.

ple, the analogy here could be two people, after noticing that no one is speaking, begin to talk at the same time. When this occurs, both are aware of the conflict and back-off a random interval of time before attempting to talk again. There is, of course, the possibility of the same two people attempting to talk again at the same time. A similar event can occur in CSMA/CD. Repeated packet collisions from the same nodes are possible in spite of random back-off times. To remedy the possibility of repeated collisions from the same nodes, the mean value of the random back-off times is increased in an exponential fashion with each repeated collision. Thus collisions involving the same nodes become even less probable.

Slot Time and Minimum Packet Size. There is a misconception among many of us in our early training that when a signal is transmitted it is at the same time being received at its destination. This is especially true in cases where signals are confined to small spaces, such as circuit cards with their respective signals between gates or signals between nodes of a network separated by a few meters. In each case, there is always a definite period of time involved for the signal to propagate from its source to its destination. That length of time is often crucial to the success or failure of the system. Several factors govern this length of time, such as distance, transmission media, and temperature. It is possible in CSMA/CD for a packet of small length to be transmitted in its entirety and a collision to occur at some time later.

In Figure 11-10a, a packet is transmitted from node A to node B, the farthest two nodes from each other on the LAN. Node A's packet size is small, and the length of time that is takes for node A to transmit its packet is *less* than the round-trip propagation delay time between the farthest two nodes in a LAN. This round-trip propagation time is defined as the *slot time*. Node B senses that the channel is free (node A's packet has not arrived yet) and begins to transmit its packet onto the channel. The collision occurs between the two packets. The collision here goes undetected by node A. Recall that collisions are monitored and reported during the *course of a transmission*. Node A thinks that its packet has been successfully transmitted, since no collision detect signal occurred during the course of its transmitting its packet. A collision, in fact, has occurred. The collision return signal arrives at node A *too late*. As far as node A is concerned, the resulting collision detect signal is caused by a collision of packets from other nodes on the bus. To avoid this problem, a minimum size requirement in CSMA/CD is placed on all packet lengths transmitted by a node. This length is slightly larger than the slot time of the LAN, as depicted in Figure 11-10b. This ensures that when a packet is transmitted by a node, it will have time not only to reach its destination, but time also to receive the change in energy level if a collision occurs. The packet can then be aborted and issued again at a later time. It follows that, once a packet has been transmitted and the slot time has elapsed, all stations connected

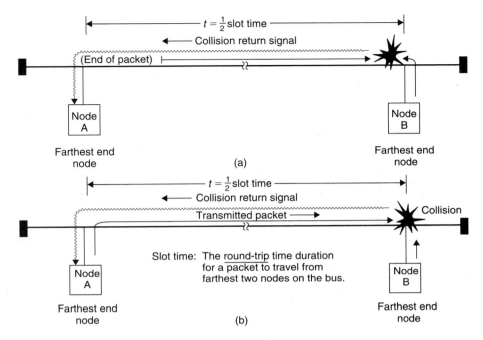

Figure 11-10 (a) A packet is transmitted from node A to node B whose length is less than the slot time. A collision occurs with node B's packet. The collision goes undetected by node A since the collision return signal arrives at node A *after* it has transmitted its packet. (b) A packet must be transmitted for a minimum duration of the slot time in order for the transmitting station to detect a collision during the course of the transmission.

to the LAN are aware of the channel being used. No collisions can occur for the remainder of the transmitted packet. Transmitted packets in CSMA/CD are typically much larger than the slot time, hence the most efficient use of the channel with the least amount of contention.

11.3.3 Token Ring Passing

In ring configured LANs, there is no contention over the use of the channel. The discipline most often used to share the channel is called *token passing*. A *token* is a special bit pattern or packet that circulates around the ring from node to node when the channel is not being used (Figure 11-11a). A node desiring the use of the channel for communications takes possession of the token and *holds* it when its turn to receive it comes up. The node then transmits its packet onto the ring. The packet circulates to its destination node on the ring and returns to the transmitting node, where it is then verified that it was received. The token is passed forward to the next node. The token cannot be used twice, and there is a limita-

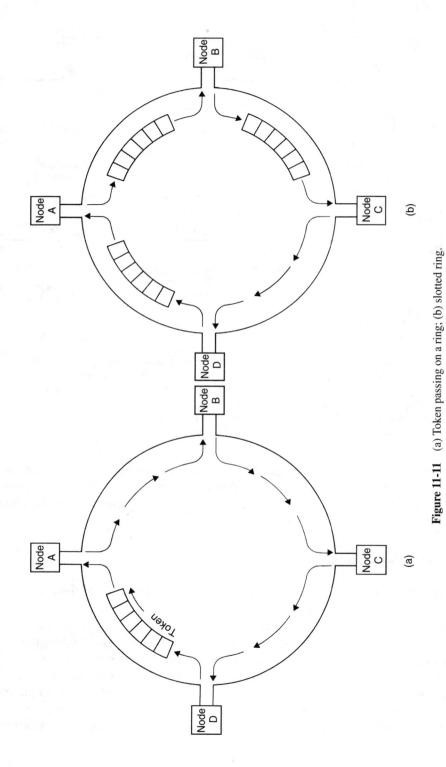

Figure 11-11 (a) Token passing on a ring; (b) slotted ring.

(a)

(b)

tion on the length of time that the token can be held before it must be passed on. This prevents any one station from "hogging" the channel. In this manner, each node has the opportunity to use the channel. Possession of the token therefore guarantees exclusive rights to the channel. Without the token, a station can only receive and transfer packets.

11.3.3.1 Slotted ring. A variation of token ring passing is *slotted ring*. In slotted ring, a fixed number of contiguous *time slots* are circulated around the ring, as shown in Figure 11-11b. Each slot is fixed in size and contains positions in the slot for packet information. This includes source and destination address, control, data, and error checking. A *busy bit* is included at the beginning of each circulating slot that indicates the availability of the slot. A slot is either full or empty depending on the state of the busy bit. When a node wishes to transmit a packet, it waits for an empty slot. The empty slot is inserted with data in the appropriate place, and the busy bit is set before it is passed on. The full slot circulates around the ring until it arrives at its destination address. The receiving node, after identifying ownership of the packet, loads the data into its buffers and resets the busy bit to indicate that the slot is empty and available for subsequent nodes to use.

11.3.3.2 Token on a bus. Token passing is also used in bus or tree topologies. Since the physical layout of a bus does not form a ring, an orderly sequence for passing the token from node to node must be established beforehand. This logical sequence does not have to be related to the physical placement of the nodes. The token, with its destination address, is simply passed in a virtual ring configuration, as shown in Figure 11-12.

Token passing is best suited for LANs that have few stations but frequent demands for communicating. Factory-oriented environments are examples where token passing is used. One such protocol that uses this concept is called *manufacturing automation protocol* (MAP), which is based on the IEEE 802.4 token-passing specification.

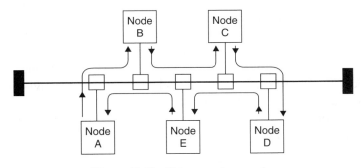

Figure 11-12 Token passing on a bus.

11.4 TRANSMISSION MEDIA

The *transmission media* of a network are the physical interconnections between nodes on a network. They provide the communications channel for the network. Some examples of transmission media are coaxial cable, twisted-pair wire, fiber optics, and air waves. There are advantages and disadvantages in using each of these media. The selection of the type of medium to use for a LAN is governed by several factors:

— Transmission rate
— Bandwidth
— Noise immunity
— Reliability
— Security
— Costs
— Service and repair
— Installation

Many of these factors will become apparent through our discussion of the various types of media.

11.4.1 Twisted-Pair Wire

Twisted-pair wire is one of the oldest and most popular transmission media used for networking. The telephone company has long used (and still does) twisted-pair wires as the main transmission medium for telephone communications.

As its name implies, twisted-pair wire, as shown in Figure 11-13, are two wires that have been twisted together to form the transmission medium. The pairs are typically bundled together with as many as several hundred pairs and placed in an insulating sheath to form a distribution cable. In multiple-pair cables, the two wires are twisted in a manner so that the pitch, or twist length, varies along the line. This helps to eliminate *crosstalk* (adjacent channel interference).

Twisting the wires together also ensures a closeness so that *both* conductors are exposed to the same noise environment. As discussed earlier, the differential amplifier will amplify the difference in signals on the pair and reject common-mode signals. RS-422-A drivers and receivers are ideally suited for twisted-pair wires and are employed in a variety of LAN applications.

Typically, 22- through 26-gauge copper wire is used for twisted pairs with polyvinyl chloride (PVC) or Teflon[†] used as the external insulating material.

[†]Teflon is a trademark of E.I. Du Pont de Nemours and Co., Inc.

Figure 11-13 Twisted-pair wire.

Twisted-pair wire is also available in single or multiple pairs with and without RFI (radio frequency interference) shielding of individual pairs. Unshielded twisted-pair wire is more formally regarded as *UTP*, whereas shielded twisted-pair wire has become known as *STP*.

Most of the growth and popularity of twisted-pair wire stems from its extensive use in the telephone system, including the private branch exchange (PBX). It has also gained widespread acceptance in RS-232-C data terminal equipment (DTE) and data communications equipment (DCE), such as terminals, computers, modems, and printers. Its main advantage over other media is that most industrial buildings, offices, and workplaces are prewired with twisted pairs through feeder networks during construction for telephone and PBX usage (Figure 11-14). Internal feeder networks are usually wired far in excess of the immediate demands for telephone usage; consequently, links are available and conveniently installed to begin with. These twisted-pair links can be used for the installation or extension of a LAN.

Recent ANSI and IEEE specifications call for the use of UTP wire in the *10BaseT standard*, a variant of the IEEE 802.3 standard. UTP wire in compliance with 10BaseT can support up to 10-Mbps transmission rates with distances up to 300 ft. Many other LAN standards are adopting UTP as an attractive low-cost transmission medium.

The primary disadvantage of twisted-pair wire is the limited bandwidth, particularly with phone lines. Data transfer rates over the phone lines are typically 300 to 9600 bps; 10-Mbps rates can be achieved on separate lines using differential line drivers and receivers.

11.4.2 Coaxial Cable

Coaxial cable is another wire medium that has long been associated with the field of communications, particularly in RF applications. Coaxial cable has several advantages to offer over twisted-pair wire. Bandwidth, data transfer rates, and noise

Figure 11-14 Twisted-pair feeder cable.

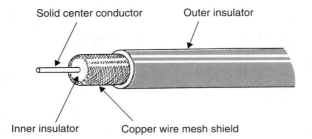

Solid center conductor   Outer insulator

Inner insulator   Copper wire mesh shield

Figure 11-15   Coaxial cable.

immunity alone are reasons for its popularity in LANs. Coaxial cable is also flexible and durable, thus permitting ease of installation.

A diagram of coaxial cable is shown in Figure 11-15. The center conductor is solid copper wire surrounded by a dielectric material made of PVC or Teflon. A braided conductor is woven over the dielectric to form a shield. The braid is usually made of copper or aluminum. The center conductor and the shield form a uniform concentric circle sharing the same center axis, hence the name *coaxial*. For the outer portion of the coax, a protective vinyl insulating sleeve forms a waterproof barrier. Coaxial cable serves in most cases as the predominant choice for transmission media used for LANs.

Coax is manufactured in a variety of different diameters and types. In addition to the flexible coax we have been discussing, there are also *rigid* and *semirigid* coax. These types of coax have solid copper material in place of the braid with no outer insulator. They are used primarily for RF applications. Coaxial cable is available in different impedances; typically, 50 or 75 Ω is used.

The frequency response of coax ranges from dc to an upper limit of approximately 500 MHz. This lends itself ideally to LAN applications.

11.4.3 Fiber-Optic Cable

With the declining costs of fiber-optic cable and equipment, we are beginning to see fiber take over applications where wire media have long been established. With regard to LANs, fiber-optic transmission media have reached an all time high. The American National Standards Institute (ANSI), with the combined efforts of other organizations, is perfecting the *fiber distributed data interface* (FDDI) standard. This standard is intended to enhance LAN technology beyond the limits of the current ISO standards. Consider some of the advantages of fiber over wire technology:

— Bandwidth: above 500 MHz for many types
— Attenuation: 0.1 to 8 dB/km
— Noise immunity
— Size

— Security

— Costs

In many industrial and government facilities, fiber-optic media have become the preferred alternative for such reasons as low cost, extended transmission length, immunity to electromagnetic and radio-frequency interference (EMI and RFI), and difficulty in tapping into classified data being transmitted. Figure 11-16 depicts various types of fiber-optic cables.

With all the advantages that fiber-optic technology has over other transmission media, one may wonder why wire technology remains the dominant media used for networking. Currently, there are still some drawbacks. Fiber-optic cables tend to be somewhat fragile and lack flexibility relative to copper wire. Coaxial cable and twisted-pair wire, as we know, can be pulled quite rigorously through conduits and walls. The same force exerted on a fiber-optic cable can result in damage and the need to splice fractured fibers. *Strength members*, as shown in Figure 11-16, are being included within the cables to overcome this problem. Splices and taps into the fiber-optic cable can amount to significant signal loss unless special equipment is used to perform this task. The substantial progress being made in fiber optics will soon resolve many of these problems. In Chapter 14 we will discuss the subject of fiber optics further.

11.4.4 Wireless

An inherent shortcoming with the transmission media discussed thus far is the burden of installation. Cables packed in conduit, walls, and ceilings leave much to be desired by most network managers. "Digital wireless" communication systems are an attractive alternative for replacing solid cabling and the extensive planning and maintenance required. Digital wireless technology for telecommunication systems has risen dramatically in the past five years. Standards organizations are scrambling to acquire portions of the available radio-frequency (RF) spectrum as the capacity of the air waves becomes increasingly congested. Wireless systems employ RF and infrared (IR) transceivers on stations, thus eliminating the need for cabling. For LAN systems, one of two wireless methods is used: *infrared* (IR) or *radio-based* systems.

IR-based systems employ one of three techniques for communicating between stations: *line-of-sight*, *scatter*, or *reflective*. Line-of-sight, as its name implies, uses optical transceivers that are aimed at each other. They offer high-speed connectivity between stations within ranges of 100 ft. Unfortunately, persons walking through the line-of-sight IR path will disrupt the signal, and therefore IR systems must be installed where the path will not be interrupted. Scatter IR systems employ transceivers at each station that transmit and receive IR that is "bounced" off walls and ceilings, similar to the manner in which light scatters,

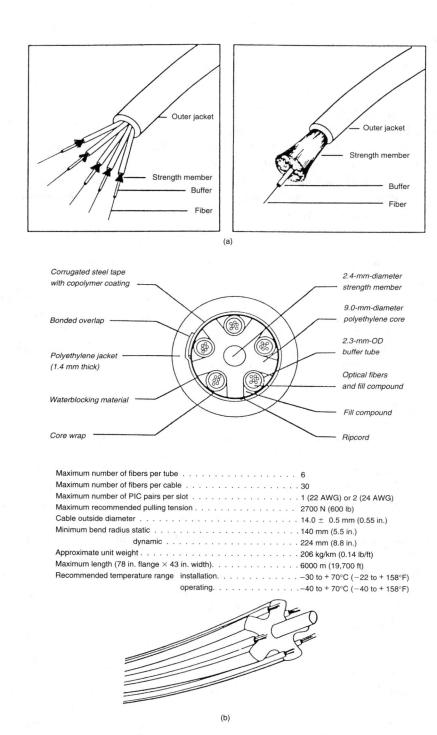

Maximum number of fibers per tube 6
Maximum number of fibers per cable 30
Maximum number of PIC pairs per slot . . . . 1 (22 AWG) or 2 (24 AWG)
Maximum recommended pulling tension . . . . 2700 N (600 lb)
Cable outside diameter . . . . 14.0 ± 0.5 mm (0.55 in.)
Minimum bend radius static . . . . 140 mm (5.5 in.)
                          dynamic . . . . 224 mm (8.8 in.)
Approximate unit weight . . . . 206 kg/km (0.14 lb/ft)
Maximum length (78 in. flange × 43 in. width). . . . . 6000 m (19,700 ft)
Recommended temperature range installation. . . . . −30 to +70°C (−22 to +158°F)
                                operating. . . . . −40 to +70°C (−40 to +158°F)

(b)

Figure 11-16 Fiber-optic cable. (Courtesy of Northern Telecom.)

except that IR is invisible to the naked eye. Scatter IR systems are best for peripheral sharing covering distances up to about 100 ft. Reflective IR systems also have optical transceivers at each node. Each workstation's transceiver is aimed at a spot on the wall off which the IR signal is reflected. Reflective systems are best suited where there are high ceilings, typically 30 to 40 ft, where persons are unlikely to disrupt the IR communications path.

Radio-based systems use microwave and UHF frequencies to transmit and receive signals between nodes. Various protocols are used for communications. In some cases a master control transceiver broadcasts and receives from a cluster of workstations. Other wireless radio-based LANs use an Ethernet-like contention scheme or token ring architecture where each node has its own wireless LAN card and plug-in antenna. The advantage of the radio-based system is that walls can be penetrated to some degree and the user does not have to worry about obstructing the path of the signal as with an IR system.

11.5 BASEBAND VERSUS BROADBAND TRANSMISSION

Data can be transmitted from one node to another within a LAN by several transmission techniques, depending on factors such as media, transmission rate, distribution area, and the volume of data. The method by which the data are transmitted in a LAN is classified into two categories: *baseband and broadband*. Let us now consider each approach.

11.5.1 Baseband

When the entire bandwidth of the transmission medium is used for a single network data channel, a LAN is classified as a *baseband LAN*. Signals in the baseband are digital and are not converted to analog signals used to modulate carrier frequencies; that is, they remain as serial binary bit streams that are transmitted onto the channel in digital form. This makes the interface to the data channel simple since most data communications devices are inherently digital. Baseband LANs are typically optimized to transfer digital data from dc to 10 MHz. The most common media used are coax cable and twisted-pair wire.

11.5.2 Broadband

In contrast to baseband LANs, *broadband* LANs use *frequency-division multiplexing* (FDM) to divide the bandwidth of a single transmission medium into multiple channels of analog signals. Each channel is a fixed band of frequencies that operates independently of each other, thus allowing various modulation techniques to be used simultaneously (Figure 11-17). These include AM, FM, and

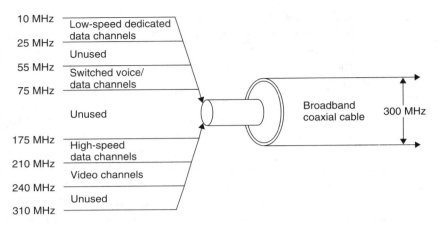

Figure 11-17 Broadband coaxial cable frequency allocation. (Courtesy of Digital Equipment Corporation.)

PM. The bands of frequencies allocated for various channels can vary in width depending on whether the intelligence that is modulated onto the channel carrier frequency is audio, video, or data.

The evolution of broadband LANs can be attributed to a technology that has been around for some time called *community antenna television* (CATV), or *cable TV*. Many years of research and development have allowed the CATV industry to evolve into the successful communications medium that it is today. Parts, servicing equipment, terminology, and technology have long been established and are easily adaptable to broadband LANs.

CATV, until recently, has been unidirectional; that is, TV signals travel in one direction only. That direction is from the CATV station antenna, where it is conditioned by a unit called the *head end* and then forwarded to community subscribers. This is referred to as *downstream transmission*. Subscribers now have the option of transmitting in the reverse direction: upstream transmission. Figure 11-18 illustrates a small portion of the frequency spectrum that has been allotted for *upstream transmission*. Since most services are forwarded to the subscriber,

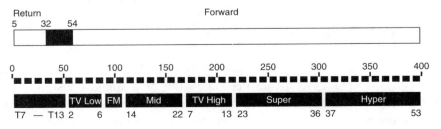

Figure 11-18 CATV frequency allocation with subsplit frequency assignment.

the frequency allocation favors downstream transmission. Essentially, broadband LANs, apart from CATV, offer bidirectional communications in much the same way, only bandwidth is more equally split.

11.6 ETHERNET

A baseband network that has dominated the LAN scene over the years and has become an accepted standard is *Ethernet*. Ethernet was developed in 1980 by the combined efforts of Xerox Corporation, Digital Equipment Corporation, and Intel Corporation. It has been accepted under the IEEE 802.3 specification. The Ethernet specification addresses the physical and data link layers of the ISO/OSI seven-layer model as illustrated in Figure 11-19. To gain further insight into the details of a LAN, the Ethernet specification, due to its widespread use, has been chosen for discussion.

Ethernet's rise in popularity stems from its simplicity. It is easy to configure, flexible, and readily extendible. Because of its popularity and acceptance as a standard, many manufacturers provide capabilities for interfacing their devices to Ethernet. This includes a broad range of devices from mainframes and PCs to mass storage devices. Figures 11-20 through 11-22 illustrate examples of small-, medium-, and large-scale Ethernet configurations.

11.6.1 Ethernet Specification

The Ethernet specification provides for a high-speed communications facility within a local area. By using CSMA/CD as access control, the need for complex routing or switching techniques is eliminated, as well as the substantial time involved in managing noncontention protocols. The management of thousands of wires is also eliminated.

Thick, double-shielded, 50 Ω coaxial cable with a solid center conductor is specified as the transmission medium for Ethernet's bus topology. The coaxial cable is typically routed through ceilings, walls, and floors, in a manner that allows easy access to the cable for the addition of nodes. A *transceiver* is used to

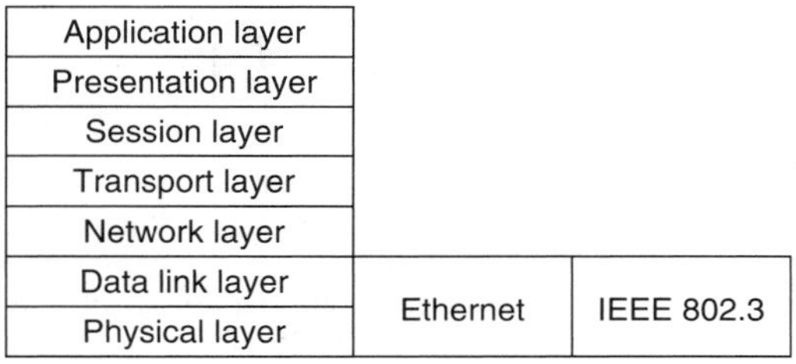

Application layer		
Presentation layer		
Session layer		
Transport layer		
Network layer		
Data link layer	Ethernet	IEEE 802.3
Physical layer		

Figure 11-19 Ethernet specification in relation to seven-layer ISO/OSI model.

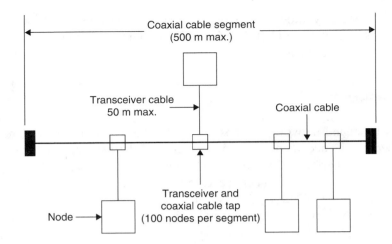

Figure 11-20 Small-scale Ethernet configuration. (Courtesy of Digital Equipment Corporation.)

interface signals to the bus cable. The transceiver is *tapped* into the Ethernet cable and connected to a node via a *transceiver cable*. A maximum of 1024 nodes can be networked together on the bus.

It is recommended that the reader refer to Figures 11-20 through 11-22 as a reference for the Ethernet specifications given in Table 11-1.

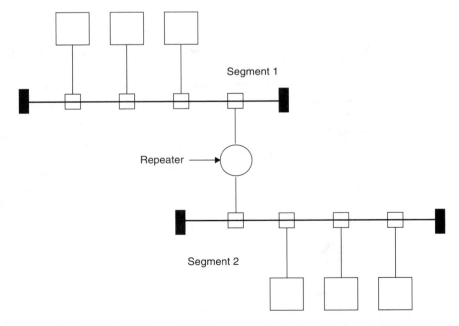

Figure 11-21 Medium-scale Ethernet configuration. (Courtesy of Digital Equipment Corporation.)

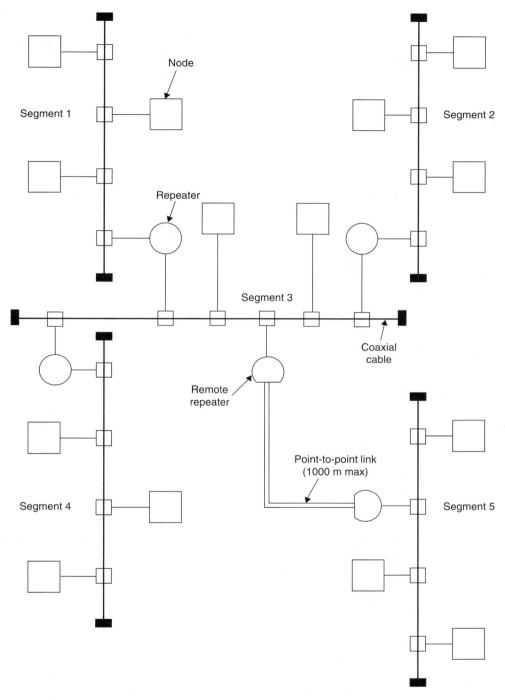

Figure 11-22 Large-scale Ethernet configuration. (Courtesy of Digital Equipment Corporation.)

TABLE 11-1   Ethernet Specifications

Topology	Bus (branching nonrooted tree)
Access control	CSMA/CD
Message protocol	Variable packet size (see data encapsulation/decapsulation)
Transmission medium	Shielded coaxial cable
Signaling type	Baseband
Signal rate	10 Mbps
Maximum number of nodes	1024
Maximum node separation	2.8 km
Slot time	51.2 μs
Maximum segment length	500 m
Maximum number of nodes per segment	100
Minimum node separation	2.5 m
Maximum length of coaxial separation between two nodes	1500 m
Maximum segment separation with point-to-point link	1000 m
Maximum number of repeaters between any two nodes	2

11.6.2 The Ethernet Segment

Each coaxial cable within a network is referred to as a *segment*. A segment (Figure 11-20) can span a maximum of 500 m and can include a maximum of 100 nodes. Nodes must be separated by a minimum of 2.5 m. When adding a node to a segment, the Ethernet cable must be tapped at multiples of 2.5 m from other nodes on the segment. This minimizes reflections and standing waves. For ease of identifying where taps may be placed for the addition of a node, the cabling manufactured for Ethernet is marked at 2.5-m intervals.

11.6.3 The Ethernet Repeater

The repeater is shown in the medium- and large-scale Ethernet configurations of Figures 11-21 and 11-22. The function of the repeater is to transmit signals from one segment to another without altering the integrity of the signal. Segments cannot be separated by more than 100 m: two 50-m transceiver cables separated by the repeater. The network can then be extended beyond 500 m by joining segments together. The repeater resides between segments and is connected to the segments via a transceiver cable. Notice in both figures that the repeater's transceiver cables are tapped into the network in the same manner as a node would be; that is, a repeater takes the position of a node and counts toward the maximum allowable 100 nodes. A repeater is not, however, a node since it is not addressable. Once the repeater is activated, it becomes *transparent* to the network.

There are two types of repeaters: the *local repeater* and the *remote repeater*. The local repeaters shown in Figures 11-21 and 11-22 are indicated by circles linking two segments together. The remote repeater shown in Figure 11-22 is indicated by the two partial circles interconnected by a full-duplex fiber-optic link. Both types of repeaters serve the function of joining segments. The remote repeater, however, allows the network to be extended to a maximum of 1000 m via a fiber-optic link. Each unit of the remote repeater has its own independent stand-alone ac power supply, thus allowing segments to be separated by the maximum 1000 m. The remote repeater is essentially identical to a local repeater with the exception of the fiber-optic interface installed in the repeater unit.

11.6.4 The Ethernet Transceiver Cable

The *transceiver cable* is used to interconnect the Ethernet transceiver to the Ethernet controller card that resides within a node. Figure 11-23 illustrates the interface. The transceiver cable consists of four twisted-pair wires that are shielded

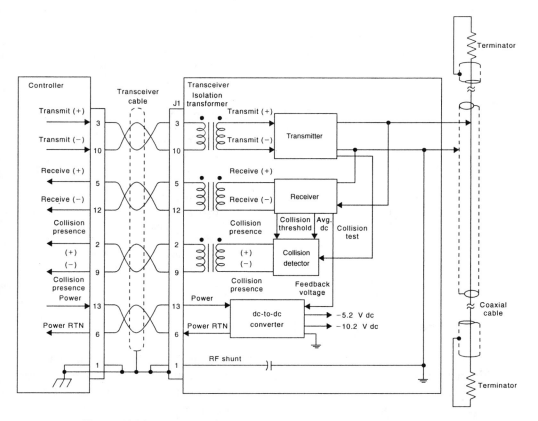

Figure 11-23 Ethernet transceiver interface. (Courtesy of Digital Equipment Corporation.)

and protected with an insulating jacket. Fifteen-pin D-type connectors are used on both ends of the cable. A plug is used on the controller end and a receptacle is used on the transceiver end.

The transceiver cable is specified for a maximum cable length of 50 m. Extending the cable length beyond this specification increases propagation delay time and results in collisions.

The transceiver cable includes the following signals:

— Power
— Transmitted data
— Received data
— Collision detection

The transceiver cable electrical connections are shown in Table 11-2.

TABLE 11-2   Transceiver Cable Electrical Connections

Connector pin assignment	Signal
Controller: 15-pin D (plug)	
Transceiver: 15-pin D (receptacle)	
Pin 1	Shield
2	Collision detect ($+$)
3	Transmitted data ($+$)
4	No connection
5	Received data ($+$)
6	Power return
7	No connection
8	No connection
9	Collision detection ($-$)
10	Transmitted data ($-$)
11	No connection
12	Received data ($-$)
13	Power
14	No connection
15	No connection
Connector shell	Shield

11.6.5 The H4000 Ethernet Transceiver

The Ethernet transceiver provides the physical and electrical interface to the Ethernet coaxial cable. Electrical circuits internal to the transceiver are powered via the transceiver cable connected to the Ethernet controller circuit card. On occasion there is a need for the test engineer to install, test, or service the transceiver. A close look at the electrical and physical construction of the transceiver is necessary. The model selected for our discussion is the widely used H4000 Ethernet transceiver manufactured by Digital Equipment Corporation. Figure 11-23 depicts a functional block diagram of the H4000 transceiver.

11.6.5.1 H4000 transceiver physical construction. The physical construction of the H4000 Ethernet transceiver is shown in Figure 11-24. A cutaway view of the transceiver and the coaxial cable shows the physical and electrical interface between the transceiver and the coaxial cable. The housing that contains the coaxial cable is removable. This allows the transceiver to be tapped into the cable if a node is added to the network. It also allows the transceiver to be easily serviced. The housing contains braid contacts and a center pin. When the Ethernet coaxial cable is secured by the housing, the braid contacts penetrate the coaxial cable outer insulating jacket and make contact with the coaxial shield. A hole is then drilled into the cable to clear the coaxial cable shield, thus allowing the center conductor pin to penetrate the dielectric material surrounding the center conductor of the cable. Contact is never actually made between the center conductor pin and the center conductor of the coaxial cable. Ethernet signals are *inductively coupled* between the transceiver and coaxial cable. The special hand tools shown in Figure 11-25 are used to tap and secure the H4000 transceiver. The drill and its bit are manufactured to penetrate the cable to the appropriate depth. Extra precaution should be taken to ensure that braided particle remains resulting from the drilling process are totally removed. This prevents the possibility of a short between the center contact pin and coaxial shield of the H4000.

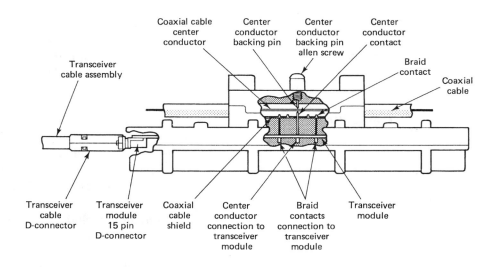

Figure 11-24 H4000 Ethernet transceiver manufactured by Digital Equipment Corporation. (Courtesy of Digital Equipment Corporation.)

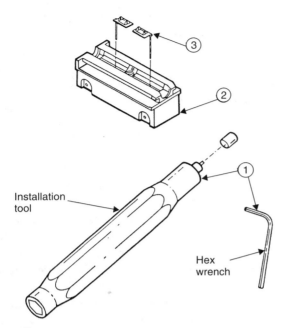

1. 1/8 in. hex wrench and combination wrench/drill (12-24664-02)
2. AMP tap III (12-24664-01)
3. Braid connector (12-24664-04)

Figure 11-25 Transceiver tool kit for tapping the Ethernet cable. (Courtesy of Digital Equipment Corporation.)

11.6.5.2 H4000 transceiver circuit. A functional block diagram of the Ethernet transceiver is shown in Figure 11-26. Also shown is the relationship between the Ethernet bus cable, transceiver cable, and controller interface. The main functions of the transceiver are discussed next.

Transmitter. The transmitter's primary function is to buffer the signals TRANSMIT(+) and TRANSMIT(−) from the transceiver cable and transmit them onto the Ethernet cable. Timing circuits in the transmitter limit the length of time that the transmitter may be on. This prevents signals from inadvertently getting stuck in the on condition beyond the maximum packet length. A circuit is also included in the transmitter to verify that the *collision detector* circuit is operational during the transmission. A protective circuit is included in the transmitter to prevent electrical damage between Ethernet and the transmitter.

Receiver. The receiver section of the H4000 couples the signals from the Ethernet coaxial cable to the receiver buffer circuit. RECEIVE(+) AND RECEIVE(−) are generated and sent to the transceiver cable pair, provided that

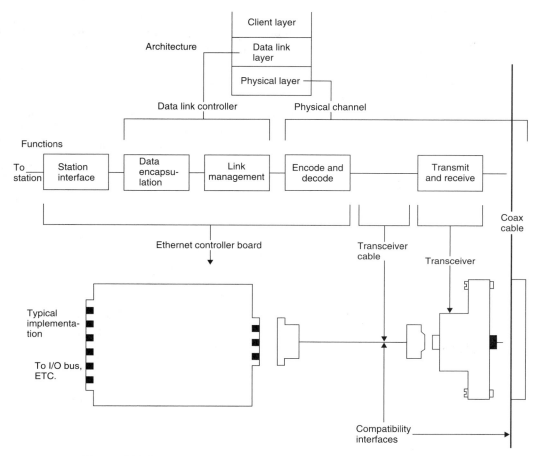

Figure 11-26 Ethernet controller layered functions. (Courtesy of Digital Equipment Corporation.)

the signals on the Ethernet cable exceed the average dc threshold for a valid signal. The average dc value of the signal and a collision threshold are sent to the collision detector circuit. A protective circuit is included in the receiver to prevent electrical damage between Ethernet and the receiver interface.

Collision Detector. The collision detector circuit monitors the average dc value from the Ethernet coaxial cable indirectly through the receiver circuit. Two stations transmitting at the same time on Ethernet increase the average dc value beyond the collision threshold, thus turning on the collision detect signals: COLLISION PRESENCE(+) and COLLISION PRESENCE(−). These signals are sent onto the transceiver cable to notify the transmitting nodes of a collision. The collision detector circuit also responds to the collision test from the transmitter.

DC-to-DC Converter. The dc-to-dc converter circuit generates the required dc voltages used by the H4000 transceiver circuits. It is sourced by the POWER and POWER RETURN signal from a node via the transceiver cable.

11.6.6 The Ethernet Controller

A final link to the Ethernet interface is the Ethernet controller. The controller (Figure 11-26) resides as a plug-in unit in an Ethernet node. Typically, the controller card utilizes a microprocessor to perform three major functions:

1. Data encapsulation/decapsulation
2. Access control
3. Manchester encoding/decoding

As illustrated in Figure 11-26, the controller functions are implemented within the data link and physical link layers of the ISO/OSI seven-layer model.

11.6.6.1 Data encapsulation/decapsulation. Data encapsulation/decapsulation refers to the task of assembling and disassembling the Ethernet frame structure depicted in Figure 11-27. Addressing and error detection are also included. The maximum frame size for an Ethernet packet is 1526 bytes (12,208 bits). The minimum packet size is 72 bytes (576 bits). For a data transfer rate of 10 MHz, this is equal to 57.6 μs (576 bits $\times \frac{1}{10}$ MHz). This time is slightly larger than the Ethernet slot time, which is specified as 51.2 μs, or 512 bit times. Recall that the slot time is the round-trip propagation time of a signal to travel between the farthest two nodes on the Ethernet bus. For Ethernet it is based on the specification for maximum separation between nodes of 2800 m. Let us now consider each field of the Ethernet frame.

Preamble/Start Byte. The preamble/start byte consists of eight bytes (64 bits) of alternating 1's and 0's used for frame synchronization.

Destination Address. This field specifies the destination of the packet. Upon receiving a packet, each controller on the bus must identify if the packet received belongs to its own physical address. If it does, the packet is taken in by the

Preamble/ start byte	Destination address	Source address	Type	Data	Frame check sequence (FCS)
8 Bytes	6 Bytes	6 Bytes	2 Bytes	46–1500 Bytes	4 Bytes

Figure 11-27 Ethernet frame structure.

controlling node for processing; otherwise, it is discarded. Ethernet also allows *multicast* addressing and *broadcast* addressing as well. In multicast addressing, a node's physical address may also correspond to a group for which the packet is intended. In broadcast addressing, all nodes receive and process the packet. Upon receiving a packet, the controller identifies the type of addressing used as determined by the following bits within the destination field:

Bit 0 = 0   If bit 0 of the destination address field is 0, the address is interpreted as a unique physical address intended for only one station on the network.

Bit 0 = 1   If bit 0 of the destination address field is 1, the address is interpreted as a multicast address. All stations that have been preassigned with this multicast address (in addition to their own physical address) are to process this packet.

Bit 0-47 = 1 If all bits of the destination field are 1's, a broadcast address has been specified. All stations are to process the packet.

Source Address.   The source address field consists of six bytes (48 bits) that correspond to the address of the station transmitting the packet.

Type.   The type field is two bytes (16 bits) that identify the protocol used in layers above the data link layer. Ethernet does not specify a protocol for these layers.

Data.   The data field is of variable length, from a minimum of 46 bytes to a maximum of 1500 bytes. The data field is transparent. Data link control characters or zero-bit stuffing is not used. Transparency is achieved by counting back from the FCS character. This is possible by knowing beforehand the number of bytes included in the data field. A total byte count of the packet is maintained by the receiving node once the packet has been taken in by the controller.

Frame Check Sequence (FCS).   The FCS is a four-byte (32-bit) CRC character used for error detection.

Interframe Spacing.   The recovery time for Ethernet controllers is 9.6 µs.

11.6.6.2 Access control.   One of the many tasks of the Ethernet controller is handling contention over the use of the bus. The access control protocol used for Ethernet is CSMA/CD. When the controller is in the process of transmitting a packet and senses that a collision has been detected by the transceiver module, the collision is enforced by continuing the transmission of the remainder of the packet. This is referred to as *jamming* the channel. Jamming the channel ensures that all *transmitting* stations involved in the collision sense the collision. Other transmitting stations follow the same procedure. As discussed in Section 11.3.2.3, the stations involved back off a random interval of time before attempting another transmission of the packet.

During the course of a collision, all receiving stations on the network continue to receive in their normal fashion. The data packets involved in the collision are fragmented as a result of the collision. Generally, they do not meet the minimal 72-byte frame size requirement for a packet and are discarded.

11.6.6.3 Manchester encoding/decoding. Since the Ethernet bus is a single coaxial cable used for half-duplex synchronous serial communications, a method of synchronizing to the serial data is necessary. No separate clock is provided on a separate line. Instead, the transmitted baseband signal is *Manchester encoded* by the Ethernet controller (Figure 11-28). Notice that each bit interval has at least one transition regardless of the data bit pattern. The center of each cell contains a state transition. A LOW-to-HIGH transition represents a logic 1 and a HIGH-to-LOW transition represents a logic 0. Manchester encoding allows the receiving device to synchronize to the serial bit stream without the use of a separate clock. The synchronous serial bit stream is often referred to as *self-clocking*. Once the receiver has achieved synchronization with the bit stream, the serial data are then decoded.

11.7 IEEE 802.3 10Base5, 10Base2, and 10BaseT

In an effort to reduce the cost of the Ethernet LAN system, the ANSI/IEEE 802.3 committee has defined alternative transmission media to the thick and expensive double-shielded coaxial cable specified in the original standard, *10Base5*.

10Base5. This is the original Ethernet physical specification that includes the "thick" 50 Ω double-shielded coaxial cable for its transmission medium. For this reason it is also referred to as *Thicknet* or *Thick Ethernet*. Recall that Thicknet supports up to 100 nodes per segment with a minimum node separation of 2.5 m. The baseband protocol supports a 10-Mbps transmission rate that uses Manchester encoding as its signaling technique.

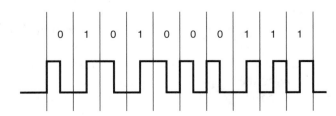

Figure 11-28 Manchester encoding.

10Base2. This standard is also known as *Cheapernet, Thin-net,* or *Thinwire Ethernet.* The 10Base2 standard supports a 10-Mbps transmission rate and also uses Manchester encoding as its signaling technique. 10Base2 specifies the use of RG-58 50 Ω characteristic impedance coaxial cable with the use of BNC-T type connectors interfaced directly to the Ethernet controller network interface card (NIC). This eliminates the expensive transceiver cable and module used in the traditional interface. It also eliminates the need to "tap" or drill into the coaxial Ethernet cable to install a network node. Cheapernet is extremely popular in the PC-based LAN environment. It uses the same 10-Mbps CSMA/CD protocol as Ethernet and offers the advantages of low cost per node, easy installation and maintenance, and ready availability of supporting hardware and software. However, Cheapernet is a scaled-down version of Ethernet. Segment lengths are limited to 185 m instead of 500 m and a maximum of 30 nodes can be placed on a segment as compared to 100 in Ethernet. Cheapernet may also be more susceptible to noise (e.g., AM radio stations). Still, there are endless applications for a Cheapernet LAN system, particularly in the office and laboratory environment.

10BaseT. Another extremely popular Ethernet transmission standard finding its way into the PC-based LAN environment is 10BaseT. The "T" stands for *twisted-pair wire* (unshielded) or simply UTP. 10BaseT also conforms to the 10-Mbps CSMA/CD Ethernet protocol standard, but uses a *concentrator* that fans out the transmission media to its end-users in a star topology. The concentrator is essentially an intelligent "hub" or multiport repeater used to expand the 10BaseT star topology network. The primary advantage of 10BaseT over 10Base5 and 10Base2 is that it uses inexpensive UTP, identical to voice-grade UTP. Modular RJ45 and RJ11 telephone wall jacks and four-pair UTP telephone wire are specified in the standard for interconnecting nodes to the LAN. The RJ45 connector plugs directly into the *network interface card* which resides in the PC. Combinations of shielded twisted-pair wire (STP) and coax can also be combined. 10BaseT is ideally suited for stand-alone PC LAN systems. Advances in high-end technology are linking systems together and making 10BaseT much more appealing for critical applications.

Figure 11-29 depicts the physical differences between the three specifications and Table 11-3 compares their specifications. The specifications for 10BaseT are a function of the degree of mixing the various transmission media.

11.8 THE LAN ENVIRONMENT

Most of our discussion of the LAN system thus far has been limited to the lower two layers of the ISO/OSI model: Layer 1, the *physical layer*, and Layer 2, the *data link layer*. Of equal importance are the upper layers of the OSI model and

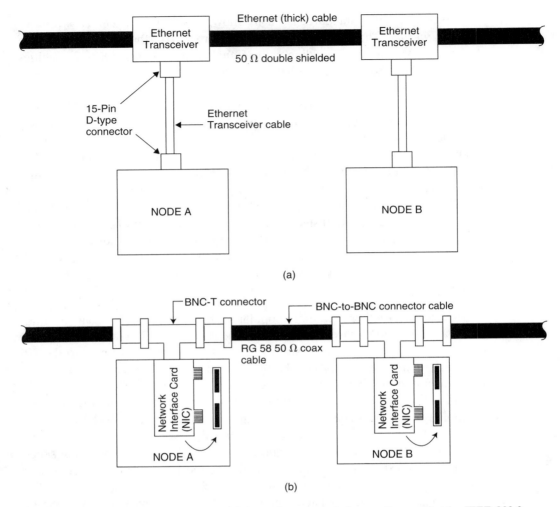

Figure 11-29 Illustration of the various transmission media specified by IEEE 802.3: (a) 10Base5 LAN setup (Ethernet); (b) 10Base2 LAN setup (Cheapernet or Thin-net); (c) 10BaseT LAN setup.

the relationship to the LAN environment. Layer 7, the *applications layer*, in particular, concerns itself with the software which provides the end-user with applications programs. Layer 7 also addresses the end-user interface between applications, its supporting software, and the underlying hardware. Examples include DOS, OS/2, NetWare, VMS, and UNIX operating systems. Although it is not the intent of this book to address layer 7 and its relationship to the LAN, there are some fundamental concepts that necessitate discussion. These include the LAN's software and hardware components in terms of network operating environments.

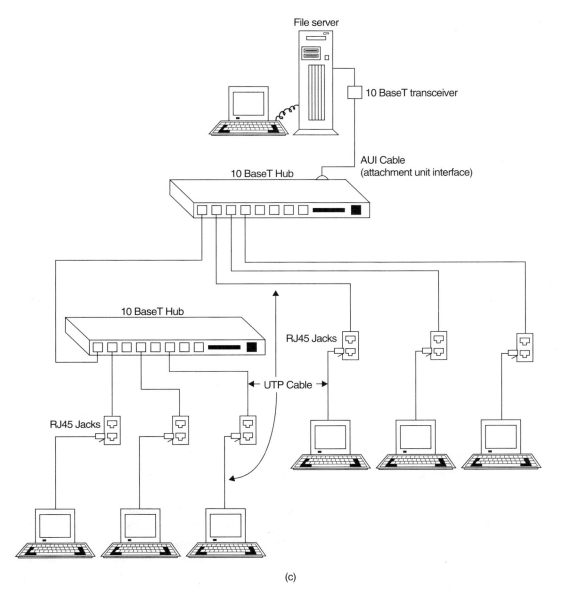

Figure 11-29 *(continued)*

11.8.1 The LAN Server

In a LAN system, the **server** is typically a high-powered computer or workstation that shares its resources as a service to the network users, also referred to as
clients. Resources can be hardware or software. The LAN relies on the server for
virtually all communications and therefore the server in most LAN systems is

TABLE 11-3 Comparison of IEEE 802.3 Physical Specifications for 10Base5, 10Base2, and 10BaseT

	IEEE 802.3		
	10Base5 (Ethernet)	10Base2 (Cheapernet)	10BaseT
Access control	CSMA/CD	CSMA/CD	CSMA/CD
Topology	Bus (branching nonrooted tree)	Bus (branching nonrooted tree)	Star
Message protocol	Variable packet size	Variable packet size	Variable packet size
Signaling rate	10 Mbps	10 Mbps	10 Mbps
Signaling type	Baseband	Baseband	Baseband
Cable type	IEEE 802.3 thick double-shielded coax	RG-58 coax	UTP Level 3 or level 4 telephone cable
Cable impedance	50 Ω	50 Ω	NA
Minimum node separation	2.5 m	0.5 m	NA
Maximum segment length	500 m	185 m	100 m (PC to hub)
Maximum nodes per segment	100	30	1024 (per network)
Maximum number of segments	5	2	NA
Maximum station separation	2500 m	925 m	NA

dedicated to "serving" the network. It cannot be used as a workstation. There are many types of servers. Figure 11-30 illustrates the LAN file server and print server.

Most network servers are **file servers**. A file server is a program that permits users to access disk drives and other mass storage devices for storing and retrieving common databases and applications programs.[†] There are many servers that can be configured on a LAN and do not necessarily have to reside within a dedicated workstation. A **print server**, for example, can be attached to the station designated as the file server, or it can be attached to any other computer on the network. Like the file server, the print server is a program that resides on the station or node that is associated with the service it renders. In this case, the service is providing the network users with shared access to the printing device. Other types of servers include terminal servers, FAX servers, high-speed modems (modem servers), compact discs or CDs (CD servers), and even *gateways* that provide services to other networks.[‡] Gateways will be discussed in the next section.

[†]John E. McNamara, *Local Area Networks* (Digital Equipment Corp., Burlington, Mass.: Digital Press, 1985), p. 93.

[‡]Emilio Ramos, Al Schroeder, Lawrence Simpson, *Data Communications and Network Fundamentals Using Novell Netware™* (New York, Macmillan, 1992), p. 354.

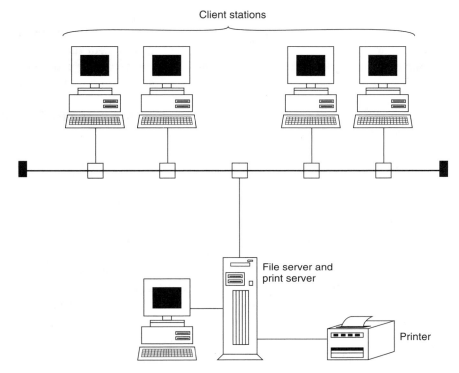

Figure 11-30 Client/server LAN system.

11.8.2 The LAN Client

A **client** is simply a station on the network that permits its user to request the shared services provided by the network server. Today's enormous number of personal computer (PC) users make up, by and large, the most common type of client on the network. Clients can also be multiple terminals connected to a terminal server.

In a typical LAN, the client station does not have (nor does it need) the high-powered capacity of the workstation or mainframe computer used as the network server. Instead, the client utilizes the network server's hardware and software. Herein lies the advantage of the network. Clients need not have an expensive laser printer, CD drive, scanner, or plotter. The client station need not have the mass storage media of the LAN server for information retrieval. Instead, of purchasing an applications program such as word processing or computer aided design (CAD) package for each client station, a single network version of the program can be purchased with a site license for a predetermined number of client stations to use. The single program is then installed on the network's file server, thus making it available on the network. The client station sends a request for a particular applications program to the file server. In turn, the file server retrieves the program from disk and sends a working copy over the network to the client requesting the service. The process is transparent to the user. It's as if the user had a stand-alone copy of the program on his or her own disk. Clearly, the services provided by the LAN offer a much more cost-effective and efficient working environment.

11.8.3 Client/Server Versus Peer-to-Peer Networking

There are basically two types of networks in terms of the network operating system: the *client/server* network and the *peer-to-peer network*. So far, our discussion has been limited to the client/server network (Figure 11-30) in which the station acting as the file server is dedicated to running the network operating system and servicing the network. The server cannot be used as a workstation, which is a drawback. Another major drawback is that if the server on a client/server network fails, the service it provides to the network is no longer available to its clients. This could be catastrophic without some degree of fault tolerance built into the system.

An alternative approach to the client/server network is to have nondedicated servers. A workstation in this environment can be configured as a client, a server, or both a client and a server. This is referred to as a peer-to-peer network. Each station is said to have "peer" status. Figure 11-31 illustrates an example of a peer-to-peer network. Peer-to-peer networking is ideally suited for the small business or working environment and has gained tremendous popularity in recent years. For example, *LANtastic* is a peer-to-peer network operating system developed and marketed by Artisoft Inc. of Tucson, Arizona. It is designed to support up to 300 DOS based PC stations on the network and operates over an IEEE 802.3 Ethernet LAN or a proprietary Artisoft system.[†] Another popular peer-to-peer network is Novell's *NetWare Lite* which is designed to network from two to 25 PCs in a small business environment. NetWare Lite is also designed to operate over an Ethernet system (10base2). Any user can set up his or her station to share applications programs, directories, and printers. NetWare Lite is easy to administer, inexpensive, and doesn't require the extensive training and software installation task of a client/server based system. However, one of the major drawbacks with peer-to-peer networks is security. Since the network can be managed from any PC on the network by virtually any user, files and directories can get corrupted as a result of poor management. Table 11-4 lists the advantages and disadvantages of the client/server network versus the peer-to-peer network.

[†]Greg Nunemacher, *LAN Primer,* 2nd ed. (New York, M&T Publishing Co., 1992), p. 97.

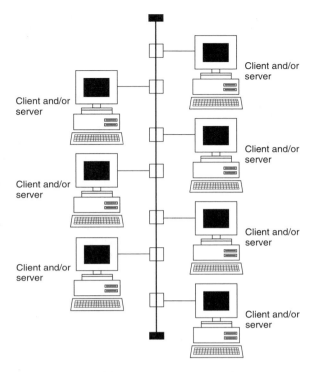

Figure 11-31   A peer-to-peer network system.

11.9 INTERNETWORKING THE LAN

In many cases, the resources shared among users of a LAN are either inadequate or can greatly be enhanced by internetworking with other LANs. And so the goal of designing or upgrading many communication systems has been to extend the network beyond the LAN so that business employees, scientists, educators, students, and other users can be tied together in a common electronics workplace, regardless of geographic location.

An example of internetworking LANs from a broad perspective is *Internet*, a collection of over 11,000 computer networks spanning the globe linking universities, businesses, libraries, museums, government agencies, and more. One does not need to be a member of one of these organizations to access the Internet and exchange files and send electronic mail to virtually anywhere in the world. Even the large consumer information bulletin board systems (BBS) such as *America Online, Prodigy, Compuserve, Delphi, GEnie,* and *eWorld* are now linked to the Internet.

TABLE 11-4   Client/Server Network versus Peer-to-Peer Network

Client/Server Network

Advantages	Disadvantages
– Security: Access rights to users and groups can be managed more effectively by a single system manager instead of several users on a peer-to-peer system. – Accounting: User and group accounts can be managed much more effectively. – Cost: For large numbers of stations on the network and internetwork, client/server LANs are cost effective. – Fault Tolerance: In the event of a power failure, client/server software is designed to interact with an uninterruptable power supply (UPS) device for back-up protection.	– Critical Resource: If a server on a client/server system fails, clients may be without service. – Cost: Software, hardware, and training can be costly, especially for small systems. – Installation: The software installation is much more extensive than a peer-to-peer system. – Training: Extensive training is often required, including certification, to administer and maintain the network.

Peer-to-Peer Network

Advantages	Disadvantages
– Installation: Easy to install with little or no training necessary. Intended for first time PC network users. – Peer Status: Any client can be a file server and any server can be a client. Servers can be used as workstations and need not be dedicated servers to the LAN. – Administration: Easy to administer without extensive training. The entire network can be managed from any PC. – Shared Resources: Directories and files within directories for each station can be networked and readily shared. Printers and other peripherals on each station can also be readily shared.	– Security: Since the network is managed by its users, users may become careless and damage other users' files and directories. Servers are usually not in a remote and secure environment. – Accounting: Several users administering the LAN may make accounting difficult to keep track of who gets rights and restrictions to the services of the LAN. – Speed: Not as fast as client/server-based systems, especially when a number of stations are requesting services from a single server.

In order to link two or more LANs together for optimum performance, several factors need to be considered:

- Bandwidth   — Throughput
- Protocol conflicts   — Buffer memory
- Transmission media   — Costs
- Distance   — Future growth

It is necessary at this time to introduce the primary devices used to link LANs together. These include the *repeater, bridge, router,* and *gateway.* The function of each of these devices can be best understood by considering its relationship with the ISO/OSI seven-layer model illustrated in Figure 11-32. Notice that the bridge, router, and gateway encompass the layers below themselves and therefore serve multiple functions.

11.9.1 The Repeater

The *repeater* is the simplest device used to extend the LAN. As illustrated in Figure 11-32, the repeater performs *physical layer* operations only. Its purpose is to prevent any ambiguity in the recognition of the transmitted bit stream by regenerating signals that have been attenuated and distorted, thus permitting the signal to be transmitted a greater distance. The repeater does not perform any data link layer functions such as error control, addressing, or flow control. It can only be used to extend the network or link portions of the network that share the same architecture and channel access protocol. Fiber optic repeaters are often used to link two LANs together that are separated by a physical distance beyond the norm for copper media.

11.9.2 The Bridge

A *bridge* is an internetworking device that is used to link two or more LANs together that use the same protocol (e.g. two IEEE 802.3 Ethernet LANs). The bridge is more sophisticated than the repeater in that it functions at the *data link*

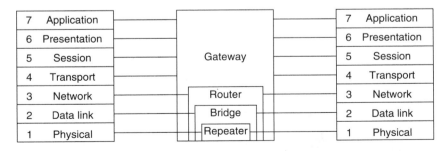

Figure 11-32   Functions of the repeater, bridge, router, and gateway in relation to the ISO/OSI seven-layer model.

layer and the *physical layer* as illustrated in Figure 11-32. Packets that are asserted by stations on one LAN that are addressed to another are recognized by the bridge (i.e., filtered) and forwarded to their destination on the interconnecting LAN. The format and content of the packet is not altered during the relay process.

11.9.3 The Router

The *router* adds another level of sophistication towards internetworking. It operates at the *network layer*, layer three, of the ISO/OSI seven-layer model. Like the bridge, the router also permits the interlinking of LANs together that have the same communications protocol. The difference, however, is that the router is not transparent to end stations on the adjoining networks like the bridge and repeater. The router is actually an addressable node on the network that is known to the end stations. It is capable of processing information embedded in the packet's control header. When a LAN is linked with several other LANs to form a WAN, the router is especially important. Packet switched networks, for example, use several routers to form intermediate nodes that route individual packets to their final destination. The router selects the optimum path for the packet to travel based on the status of the network or internetwork traffic that it monitors. A routing protocol is used to accomplish this.

11.9.4 The Gateway

The final component used for internetworking is the *gateway*. The gateway functions on all layers of the ISO/OSI seven-layer model by providing a complete hardware and software translation between networks that support entirely different architectures. It also permits the connection of a LAN to a host computer that supports a different protocol family. For example, a Novell based PC network can be interconnected by a gateway to a network or a host computer system that supports IBM's proprietary SNA (Systems Network Architecture). Gateways are extremely complex devices but offer the most diversity in internetworking.

PROBLEMS

1. What does *LAN* stand for?
2. Explain the function of a LAN.
3. Define *LAN topology*.
4. For a star LAN:
    (a) What is the critical resource?
    (b) How is this topology best utilized?

(c) What is a major disadvantage with this topology?

5. For a bus LAN:
   (a) What is the critical resource?
   (b) What is the distinguishing feature of this topology?
   (c) What is a major disadvantage with this topology?

6. For a ring LAN:
   (a) What is the critical resource?
   (b) What is the distinguishing feature of this topology?
   (c) What is a major disadvantage with this topology?

7. Explain a *collision*.

8. What is the advantage of slotted Aloha over pure Aloha?

9. What does *CSMA/CD* stand for?

10. What is the purpose of jamming?

11. To what is the round-trip propagation delay time between the farthest two nodes in a bus LAN referred?

12. When a token is passed in a ring LAN, why is a node limited in the length of time that it may possess the token?

13. In a slotted ring LAN, instead of a token, what is circulated around the ring?

14. What does *UTP* stand for?

15. What two methods of transmission are used in "wireless" LAN communications?

16. Explain the difference between a baseband LAN and a broadband LAN.

17. What does *CATV* stand for?

18. What level of the ISO/OSI seven-layer model does Ethernet fall into?

19. For the Ethernet protocol:
    (a) What is the topology used?
    (b) What is the transmission rate?
    (c) What is the maximum number of nodes that can be used?
    (d) What is the maximum length of a segment?
    (e) What is the slot time?
    (f) What is the maximum length of a transceiver cable?
    (g) How may a repeater be used to extend the length of the network?

20. Draw the Manchester encoded waveform if the bit pattern shown in Figure 11-28 were changed to 1011100101.

21. Repeat Problem 20 for a bit pattern of 1100101011.

22. What is another name for 10Base2?

23. Explain what type of connector and transmission medium is used for Cheapernet.

24. What type of transmission medium is used for 10BaseT?

25. Explain the advantage of 10BaseT over 10Base5.

12

Error Detection, Correction, and Control

A major design criterion for all telecommunication systems is to achieve error-free transmission. Errors, unfortunately, do occur. There are many types and causes originating from various sources ranging from lightning strikes to dirty switch contacts at the central office. A method of detecting, and in some cases correcting for their occurrence, is a necessity. To achieve this, two basic techniques are employed. One is to detect the error and request a retransmission of the corrupted message. The second technique is to correct the error at the receiving end without having to retransmit the message. The trade-off for either technique is the redundancy that must be built into the transmitted bit stream. This redundancy decreases system throughput.

Many of today's communication systems employ elaborate error-control protocols. Some of these protocols are software packages designed to facilitate file transfers between personal computers and mainframes. More recent error controllers are completely self-contained within a hardware module, thus relieving the CPU of the burden of error control. The entire process is transparent to the user.

In this chapter we consider some of the most common methods used for error detection and correction, including error-controlling protocols specifically designed for data-communications equipment.

12.1 PARITY

Parity is the simplest and oldest method of error detection. Although it is not very effective in data transmission, it is still widely used due to its simplicity. A single bit called the *parity bit* is added to a group of bits representing a letter, number, or symbol. ASCII characters on a keyboard, for example, are typically encoded into seven bits with an eighth bit acting as parity. The parity bit is computed by the transmitting device based on the number of 1-bits set in the character. Parity can be either *odd* or *even*. If odd parity is selected, the parity bit is set to a 1 or 0 to make the total number of 1-bits in the character, including the parity bit itself, equal to an odd value. If even parity is selected, the opposite is true; the parity bit is set to a 1 or 0 to make the total number of 1-bits, including the parity bit itself, equal to an even number. The receiving device performs the same computation on the received number of 1-bits for each character and checks the computed parity against what was received. If they do not match, an error has been detected. Table 12-1 lists examples of even and odd parity.

The selection of even or odd parity is generally arbitrary. In most cases it is a matter of custom or preference. The transmitting and receiving stations, however, must be set to same mode. Some system designers prefer odd parity over even. The advantage is that when a string of several data characters are anticipated to be all zeros the parity bit would be set to 1 for each character, thus allowing for ease of character identification and synchronization.

TABLE 12-1 Even and Odd Parity for a Seven-Bit Data Character

Data character	Odd parity bit	Data character	Even parity bit
1101000	0	1011101	1
0010111	1	1110111	0
1010110	1	0011010	1
1010001	0	1010111	1

12.2 PARITY GENERATING AND CHECKING CIRCUITS

Parity generating circuits can easily be implemented with a combination of exclusive-OR gates. Figure 12-1 illustrates an even parity generating circuit for a seven-bit data word. Odd parity can be obtained by simply adding an inverter at the output of the given circuit. Additional gates can be included in the circuit for extended word lengths. The same circuit can be used for parity checking by adding another exclusive-OR gate to accommodate the received parity bit. The received data word and parity bit are applied at the circuit's input. For even parity, the output should always be low unless an error occurs. Conversely, for odd parity checking, the output should always be high unless an error occurs.

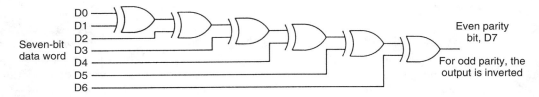

Figure 12-1 Even parity generating circuit. Odd parity generation is obtained by adding an inverter at the output.

Figure 12-2 depicts another design that can be used for parity generation and checking.

12.3 THE DISADVANTAGE WITH PARITY

A major shortcoming with parity is that it is only applicable for detecting when one bit or an *odd* number of bits have been changed in a character. Parity checking does not detect when an *even* number of bits have changed. For example, suppose that bit D2 in Example 1 were to change during the course of a transmission for an odd parity system. Example 1 shows how the bit error is detected.

Example 1

	Parity (odd)	D7	D6	D5	D4	D3	D2	D1	D0
Transmitted:	0	0	1	1	0	1	1	0	1
Received:	0	0	1	1	0	1	0	0	1

⌐Single bit error

The received parity bit, a zero, is in conflict with the computed number of 1-bits that was received: in this case, *four*, an even number of 1-bits. The parity

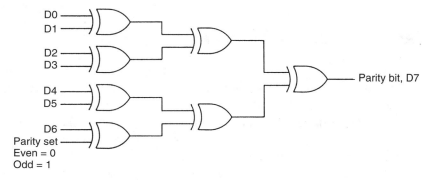

Figure 12-2 Even or odd parity generation is achieved in this circuit by setting the appropriate level at the parity set input.

bit should have been equal to a value making the total number of 1-bits odd. An error has been properly detected. If, on the other hand, bit D2 *and* bit D1 were both altered during the transmission, the computed parity bit would still be in agreement with the received parity bit. This error would go undetected, as shown in Example 2. A little thought will reveal that an even number of errors in a character, for odd or even parity, will go undetected.

Example 2

	Parity (odd)	D7	D6	D5	D4	D3	D2	D1	D0
Transmitted:	0	0	1	1	0	1	1	0	1
Received:	0	0	1	1	0	1	0	1	1

Errors: two bits or an even number of bits go undetected

Parity, being a single-bit error-detection scheme, presents another problem in accommodating today's high-speed transmission rates. Many errors are a result of impulse noise, which tends to be *bursty* in nature. Noise impulses may last several milliseconds, consequently destroying several bits. The higher the transmission rate, the greater the effect. Figure 12-3 depicts a 2-ms noise burst imposed on a 4800-bps signal. The bit time associated with this signal is 208 μs (1/4800). As many as 10 bits are affected. At least two characters are destroyed here, with the possibility of both character errors going undetected.

12.4 VERTICAL AND LONGITUDINAL REDUNDANCY CHECK (VRC AND LRC)

Thus far, the discussion of parity has been on a per-character basis. This is often referred to as a *vertical redundancy check* (VRC). Parity can also be computed and inserted at the end of a message block. In this case, the parity bit is computed based on an accumulation of the value of each character's LSB through MSB, including the VRC bit, as shown in Figure 12-4. This method of parity checking is referred to as a *longitudinal redundancy check* (LRC). The resulting word is called the *block check character* (BCC).

Additional parity bits in LRC used to produce the BCC provide extra error-detection capabilities. Single-bit errors can now be detected and corrected. For example, suppose that the LSB of the letter *y* in the message in Figure 12-4 was received as a 0 instead of a 1. The computed parity bit for the LRC would indicate that a bit was received in error. By itself, the detected LRC error does not specify which bit in the row of LSB bits received is in error. The same is true for the VRC. The computed parity bit in the column of the character in error, *y*,

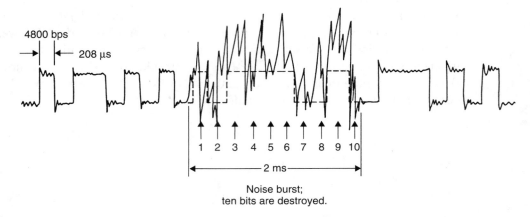

Noise burst;
ten bits are destroyed.

Figure 12-3 Effect of a 2-ms noise burst on a 4800-bps signal.

would be a 1 instead of a 0. By itself, the detected VRC error does not specify which bit in the *y* column has been received in error. A *cross-check,* however, will reveal that the intersection of the detected parity error for the VRC and LRC check identifies the exact bit that was received in error. By inverting this bit, the error can be corrected.

Unfortunately, an even number of bit errors is not detected by either the LRC or VRC check. Cross-checks cannot be performed; consequently bit errors cannot be corrected.

12.5 CYCLIC REDUNDANCY CHECKING (CRC)

Parity checking has major shortcomings. It is much more efficient to eliminate the parity bit for each character in the block entirely and utilize the redundant bits at the end of the block.

A more powerful method than the combination of LRC and VRC for error detection in blocks is *cyclic redundancy checking* (CRC). CRC is the most com-

	H	a	v	e	sp	a	sp	n	i	c	e	sp	d	a	y	!	BCC (LRC)
LSB	0	1	0	1	0	1	0	0	1	1	1	0	0	1	1	1	0
	0	0	1	0	0	0	0	1	0	1	0	0	0	0	0	0	0
	0	0	1	1	0	0	0	1	0	0	1	0	1	0	0	0	0
	1	0	0	0	0	0	0	1	1	0	0	0	0	0	1	0	1
	0	0	1	0	0	0	0	0	0	0	0	0	0	0	1	0	1
	0	1	1	1	1	1	1	1	1	1	1	1	1	1	1	1	0
MSB	1	1	1	1	0	1	0	1	1	1	1	0	1	1	1	0	1
VRC	1	0	0	1	0	0	0	0	1	1	1	0	0	0	0	1	1

Figure 12-4 Computing the block check character (BCC) for a message block with VRC and LRC odd parity checking.

monly used method for error detection in block transmission. A minimal amount of hardware is required (slightly more than LRC/VRC systems), and its effectiveness in detecting errors is greater than 99.9%.

CRC involves a division of the transmitted message block, by a constant called the *generator polynomial*. The quotient is discarded and the remainder is transmitted as the *block check character* (BCC). This is shown in Figure 12-5. Some protocols refer to the BCC as the *frame check sequence* (FCS). The receiving station performs the same computation on the received message block. The computed remainder, or BCC, is compared to the remainder received from the transmitter. If the two match, then no errors have been detected in the message block. If the two do *not* match, either a request for retransmission is made by the receiver or the errors are corrected through the use of special coding techniques.

Cyclic codes contain a specific number of bits, governed by the size of the character within the message block. Three of the most commonly used cyclic codes are *CRC-12, CRC-16,* and *CRC-CCITT.* Blocks containing characters that are six bits in length typically use CRC-12, a 12-bit CRC. Blocks formatted with eight-bit characters typically use CRC-16 or CRC-CCITT, both of which are 16-bit codes. The BCC for these three cyclic codes is derived from the following generator polynomials, $G(x)$:

CRC-12 generator polynomial: $\qquad G(x) = X^{12} + X^{11} + X^3 + X^2 + X + 1$

CRC-16 generator polynomial: $\qquad G(x) = X^{16} + X^{15} + X^2 + 1$

CRC-CCITT generator polynomial: $\quad G(x) = X^{16} + X^{12} + X^5 + 1$

A combination of multistage shift registers employing feedback through exclusive-OR gates is used to implement the mathematical function performed on the message block to obtain the BCC. Figure 12-6 depicts three CRC generating circuits for CRC-12, CRC-16, and CRC-CCITT. The BCC is accumulated by shifting the data stream into the data input of the register. When the final bit of the message block is shifted in, the register contains the BCC. The BCC is transmitted at the end of the message block, *LSB first.*

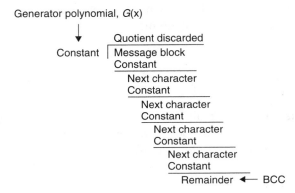

Figure 12-5 Computing the block check character (BCC) of a message block using CRC.

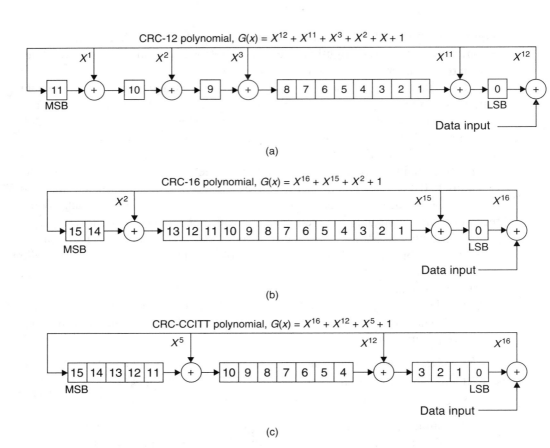

Figure 12-6 (a) CRC generating circuit for CRC-12; (b) CRC generating circuit for CRC-16; (c) CRC-CCITT.

12.5.1 Computing the Block Check Character

The generating polynomial, $G(x)$, and message polynomial, $M(x)$, used for computing the BCC include degree terms that represent positions in a group of bits that are a binary 1. For example, given the polynomial

$$X^5 + X^2 + X + 1$$

its binary representation is

$$100111$$

Missing terms are represented by a 0. The highest degree in the polynomial is one less than the number of bits in the binary code. The following discussion illustrates how the BCC can be computed using long division.

Referring to Figure 12-7, if we let n equal the total number of bits in a transmitted block and k equal the number of data bits, then $n - k$ equals the num-

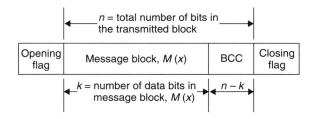

The number of bits
in the BCC is equal
to $n - k$.

Figure 12-7 Format of a message block for computing the BCC in CRC.

ber of bits in the BCC. The message polynomial, $M(x)$, is multiplied by X^{n-k} to achieve the correct number of bits for the BCC. The resulting product is then divided by the generator polynomial, $G(x)$. The quotient is discarded and the remainder, $B(x)$, the BCC, is transmitted at the end of the message block. The long-division process is *not* accomplished in the usual manner through the subtraction process. Rather, an *exclusive-OR* operation is performed. As we will see in the following examples, this will yield a BCC having a total number of bits *one* less than the number of bits in the generator polynomial, $G(x)$. This number is also equal to the highest degree of the generator polynomial. The entire transmitted block can be represented by the CRC polynomial:

$$T(x) = X^{n-k}[M(x)] + B(x)$$

where $T(x)$ = total transmitted message block
X^{n-k} = multiplication factor
$B(x)$ = BCC

Consider the following examples.

Example 3

In this example, the transmitted message block will include a total number of bits, n, equal to 14. Nine of the 14 bits are data, k. Therefore, the BCC consists of five bits ($n - k = 5$).

Given:

$$\text{Generator polynomial, } G(x) = X^5 + X^2 + X + 1$$

$$= 100111$$

$$\text{Message polynomial, } M(x) = X^8 + X^6 + X^3 + X^2 + 1$$

$$= 101001101$$

The number of bits in the BCC, $n - k = 5$ (the highest degree of the generator polynomial). Compute the value of the BCC.

Solution:

1. Multiply the message polynomial, $M(x)$, by X^{n-k}:

$$X^{n-k}[M(x)] = X^5(X^8 + X^6 + X^3 + X^2 + 1)$$
$$= X^{13} + X^{11} + X^8 + X^7 + X^5$$
$$= 10100110100000$$

2. Divide $X^{n-k}[M(x)]$ by the generator polynomial and discard the quotient. The remainder is the BCC, $B(x)$.

```
                    101111110 ← discard quotient
         100111 )10100110100000
                 100111
                  111010
                  100111
                   111011
                   100111
                    111000
                    100111
                     111110
                     100111
                      110010
                      100111
                       101010
                       100111
                        11010 ← BCC
```

3. To determine the total transmitted message block, $T(x)$, add the BCC, $B(x)$, to $X^{n-k}[M(x)]$:

$$T(x) = X^{n-k}[M(x)] + B(x)$$

$$= 10100110100000$$
$$+ \underline{11010} \qquad \text{BCC, } B(x)$$
$$ 10100110111010 \qquad \text{transmitted message block, } T(x)$$

At the receiving end, the transmitted message block, $T(x)$, is divided by the same generating polynomial, $G(x)$. If the remainder is zero, the block was received without errors.

$$
\begin{array}{r}
101111110 \leftarrow \text{discard quotient} \\
100111 \overline{)10100110111010} \\
\underline{100111} \\
111010 \\
\underline{100111} \\
111011 \\
\underline{100111} \\
111001 \\
\underline{100111} \\
111101 \\
\underline{100111} \\
110100 \\
\underline{100111} \\
100111 \\
\underline{100111} \\
0 \leftarrow \text{remainder equals zero (no errors)}
\end{array}
$$

Example 4

For simplicity, a 16-bit message ($k = 16$) using CRC-16 will be used. The total number of bits in the transmitted message block, n, is therefore 32.

Given:

Generator polynomial for CRC-16, $G(x) = X^{16} + X^{15} + X^2 + 1$

Message polynomial, $M(x) = X^{15} + X^{13} + X^{11} + X^{10} + X^7 + X^5 + X^4 + 1$

The number of bits in the BCC, $n - k = 16$ (the highest degree of the generator polynomial).

Solution:

1. Multiply the message polynomial, $M(x)$, by X^{n-k}:

$$
\begin{aligned}
X^{n-k}[M(x)] &= X^{16}(X^{15} + X^{13} + X^{11} + X^{10} + X^7 + X^5 + X^4 + 1) \\
&= X^{31} + X^{29} + X^{27} + X^{26} + X^{23} + X^{21} + X^{20} + X^{16} \\
&= 10101100101100010000000000000000
\end{aligned}
$$

2. Divide $X^{n-k}[M(x)]$ by the generator polynomial and discard the quotient. The remainder is the BCC, $B(x)$.

$$
\begin{array}{r}
1100100011011100 \leftarrow \text{discard quotient} \\
1100000000000101 \overline{)\,1010110010110001000000000000000} \\
\underline{1100000000000101} \\
11011001011001110 \\
\underline{1100000000000101} \\
11001011001011000 \\
\underline{1100000000000101} \\
10110010111010000 \\
\underline{1100000000000101} \\
11100101110101010 \\
\underline{1100000000000101} \\
10010111010111100 \\
\underline{1100000000000101} \\
10101110101110010 \\
\underline{1100000000000101} \\
11011101011101110 \\
\underline{1100000000000101} \\
1110101110101100 \leftarrow \text{BCC}
\end{array}
$$

3. To determine the total transmitted message block, $T(x)$, add the BCC, $B(x)$, to $X^{n-k}[M(x)]$:

$$
\begin{aligned}
T(x) &= X^{n-k}[M(x)] + B(x) \\
&= 1010110010110001000000000000000 \\
&\underline{ + \quad\quad\quad 1110101110101100} \leftarrow \text{BCC, } B(x) \\
& 1010110010110001110101110101100 \leftarrow \text{transmitted message block, } T(x)
\end{aligned}
$$

At the receiving end, the transmitted message block, $T(x)$, is divided by the CRC-16 generator polynomial, $G(x)$. The quotient is discarded and the remainder is checked for zero, indicating that there are no errors in the received block.

Error Detection, Correction, and Control   **Chap. 12**

```
                                        1100100011011100  ← discard quotient
1100000000000101 )1010110010110001111010101110101100
                   1100000000000101
                    1101100101100110 1
                    1100000000000101
                      1100101100100010 1
                      1100000000000101
                        10110010000000111
                        1100000000000101
                          1110010000000010 0
                          1100000000000101
                            10010000000001 10
                            1100000000000101
                              1010000000000 111
                              1100000000000101
                               1100000000000101
                               1100000000000101
                                             00  ← remainder equals zero
                                                    (no errors)
```

12.6 CHECKSUMS

Another popular method of error detection is through use of the *checksum*. The checksum is basically a summation quantity that is computed from the data and appended to the transmitted block. Like CRC and LRC, the checksum serves as the block check character (BCC). It is transmitted at the end of the message block. Data, upon arriving at their destination, undergo the same checksum computation by the receiving device. The resultant checksum is compared against the original checksum sent with the message block. If the two match, it can be assumed that the data block has been transmitted and received successfully. If the two do not match, an error has been detected (i.e., the transmitted and received data blocks are not the same and the message has been corrupted). The necessary steps are then taken to correct the error.

It should be emphasized at this point that the purpose of the checksum process is merely to detect the occurrence of an error. It by no means identifies where in the message block the error is located. This is a much more complex task that is discussed in the next section.

Checksums are commonly used in large file transfers between mass storage devices.[†] They are easily computed in hardware or software and are frequently used when blocks of data are to be transported from one location to another. There are primarily four types of checksums that we will now discuss:

[†]For a more detailed discussion on checksums, refer to Barry W. Johnson, *Designs and Analysis of Fault Tolerant Systems* (Reading, Mass.: Addison-Wesley, 1989), pp. 98–103.

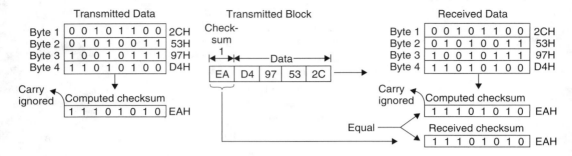

Figure 12-8 A *single-precision checksum* is generated and transmitted as a BCC at the end of a four-byte block. The receiver verifies the block by regenerating the checksum and comparing it against the original.

1. Single-precision
2. Double-precision
3. Honeywell
4. Residue

12.6.1 Single-Precision Checksum

The most fundamental checksum computation is the *single-precision checksum*. Here, the checksum is derived simply by performing a binary addition of each *n-bit* data word in the message block. Any carry or overflow during the addition process is ignored, thus the resultant checksum is also *n bits* in length. Figure 12-8 illustrates how the single-precision checksum is derived, transmitted as the BCC, and used to verify the integrity of the received data. For simplicity, a four-byte data block is used. Note that the sum of the data exceeds $2^n - 1$ and therefore a carry occurs out of the MSB. This carry is ignored and only the eight-bit (*n-bit*) checksum is sent as the BCC.

An inherent problem with the single-precision checksum is if the MSB of the *n-bit* data word becomes logically *stuck at 1* (SA1), the checksum becomes SA1 as well. A little thought will reveal that the regenerated checksum on the received data will equal the original checksum and the SA1 fault will go undetected. A more elaborate scheme may be necessary.

12.6.2 Double-Precision Checksum

As its name implies, the *double-precision checksum* extends the computed checksum to *2n bits* in length, where *n* is the size of the data word in the message block. For example, the eight-bit data words used in the single-precision checksum example above would have a 16-bit checksum. Message blocks with 16-bit data words would have a 32-bit checksum, and so forth. Summation of data words in the message block can now extend up to modulo 2^{2n}, thereby decreasing the probability of an erroneous checksum. In addition, the SA1 (stuck at 1) error discussed earlier would be detected as a checksum error at the receiver. Figure 12-9 depicts how the double-precision checksum is derived, transmitted as the BCC, and used to verify the integrity of the received data. For simplicity, a four-byte data block is used again. Hexadecimal notation is also used. Note that the carryout of the MSB position of the low-order checksum byte is not ignored. Instead, it becomes part of the 16-bit checksum result. Any carryout of the MSB of the 16-bit checksum is ignored.

12.6.3 Honeywell Checksum

The *Honeywell checksum* is an alternative form of the double-precision checksum. Its length is also *2n bits*, where *n* is again the size of the data word in the message block. The difference is that the Honeywell checksum is based on inter-leaving consecutive data words to form double-length words. The double-length words are then summed together to form a double-precision checksum. This is shown in Figure 12-10. The advantage of the Honeywell checksum is that *stuck at 1* (SA1) and *stuck at 0* (SA0) bit errors occurring in the same bit positions of all words can be detected during the error-detection process. This is true since the interleaving process places the error in the upper and lower words of the check-sum. At least two bit positions in the checksum are affected.

12.6.4 Residue Checksum

The last form of checksum in our discussion is the *residue checksum*. The residue checksum is identical to the single-precision checksum, except that any carryout of the MSB position of the checksum word is "wrapped around" and added to the LSB position. This added complexity permits the detection of SA1 errors that go undetected. This is illustrated in Figure 12-11.

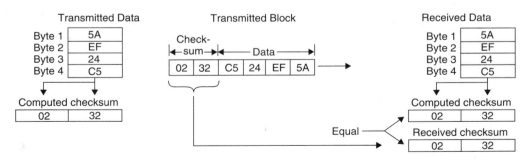

Figure 12-9 A *double-precision checksum* is generated and transmitted as a BCC at the end of a four-byte block. The receiver verifies the block by regenerating the checksum and comparing it against the original.

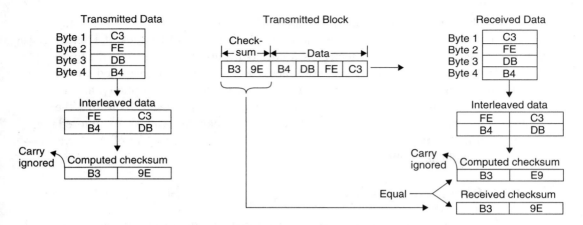

Figure 12-10 Structure of the *Honeywell checksum*. The checksum is generated and transmitted as the BCC at the end of a four-byte block. The receiver verifies the block by regenerating the checksum and comparing it against the original.

12.7 ERROR CORRECTION

Two basic techniques are used by communication systems to ensure the reliable transmission of data. They are shown in Figure 12-12. One technique is to request the retransmission of the data block received in error. This technique, the more popular of the two, is known as *automatic repeat request* (ARQ). When a data block is received without error, a positive acknowledgment is sent back to the transmitter via the reverse channel. ACK alternating in BISYNC is an example of a protocol that uses ARQ for error correction. A second technique is called

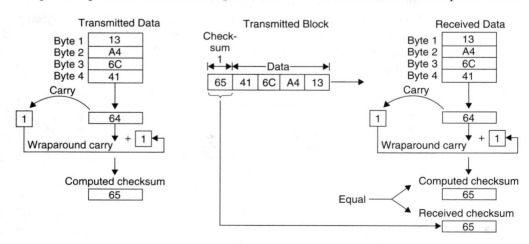

Figure 12-11 Structure of the *residue checksum*. The checksum is generated and transmitted as the BCC at the end of a four-byte block. The receiver verifies the block by regenerating the checksum and comparing it against the original.

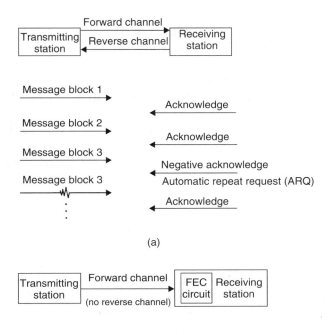

Figure 12-12 (a) Error correction using the automatic repeat request (ARQ) technique; (b) error correction using forward error correction (FEC).

forward error correction (FEC). FEC is used in simplex communications or applications where it is impractical or impossible to request a retransmission of the corrupted message block. An example might be the telemetry signals transmitted to an Earth station from a satellite on a deep space mission. A garbled message could take several minutes or even hours to travel the distance between the two stations. Redundant error-correction coding is included in the transmitted data stream. If an error is detected by the receiver, the redundant code is extracted from the message block and used to predict and possibly correct the discrepancy.

12.7.1 Hamming Code

In FEC, a return path is not used for requesting the retransmission of a message block in error, hence the name *forward error correction*. Several codes have been developed to suit applications requiring FEC.[†] Those most commonly recognized have been based on the research of mathematician *Richard W. Hamming*. These codes are referred to as *Hamming* codes. Hamming codes employ the use of redundant bits that are inserted into the message stream for error correction. The positions of these bits are established and known by the transmitter and re-

---

[†]For further details on Hamming codes, see Richard W. Hamming, *Coding and Information Theory*, 2nd ed. (Englewood Cliffs, N.J.: Prentice Hall, 1986).

ceiver beforehand. If the receiver detects an error in the message block, the Hamming bits are used to identify the position of the error. This position, known as the *syndrome*, is the underlying principle of the Hamming code.

12.7.1.1 Developing a Hamming code. We will now develop a Hamming code for single-bit FEC. For simplicity, 10 data bits will be used. The number of Hamming bits depends on the number of data bits, $m_0, m_1, \ldots, m_n$, transmitted in the message stream, including the Hamming bits. If n is equal to the total number of bits transmitted in a message stream and m is equal to the number of Hamming bits, then m is the smallest number governed by the equation

$$2^m \geq n + 1$$

For a message of 10 data bits, m is equal to 4 and n is equal to 14 bits (10 + 4):

$$2^4 \geq (10 + 4) + 1$$

If the syndrome is to indicate the position of the bit error, *check bits*, or Hamming bits, $C_0, C_1, C_2, \ldots$, serving as parity, can be inserted into the message stream to perform a parity check based on the binary representation of each bit position. How is this possible? Note in Table 12-2 that the binary representation of each bit position forms an alternating bit pattern in the vertical direction. Each column proceeding from the LSB to the MSB alternates at one-half the rate of the

TABLE 12-2 Check Bits Can Be Used as Parity on Binary Weighted Positions in a Message Stream

Bit position in message	Binary representation	Check bit	Positions set
1	0001	C_0	1, 3, 5, 7, 9, 11, 13
2	0010	C_1	2, 3, 6, 7, 10, 11, 14
3	0011	C_2	4, 5, 6, 7, 12, 13, 14
4	0100	C_3	8, 9, 10, 11, 12, 13, 14
5	0101		
6	0110		
7	0111		
8	1000		
9	1001		
10	1010		
11	1011		
12	1100		
13	1101		
14	1110		
•	•		
•	•		
•	•		

previous column. The LSB alternates with every position. The next bit alternates every two bit positions, and so forth.

To illustrate how the check bits are encoded, the 10-bit message 1101001110 is labeled m_9 through m_0, as illustrated in Figure 12-13. By inserting the check bits into the message stream as shown, the total message length n is extended to 14 bits. For simplicity, bit positions 1, 2, 4, and 8 will be used for the check bits. Even or odd parity generation can be performed on the bit positions associated with each check bit. We will use even parity here. As discussed earlier, even parity can be performed by exclusive-ORing individual bits in a group of bits. For even parity, PE0 through PE3 can serve as weighted parity checks over the bit positions listed in Table 12-2. Exclusive-ORing these bit positions together with the data corresponding to the 14 bit message stream shown in Figure 12-13, we have the following:

$$
\begin{array}{ccccccc}
13 & 11 & 9 & 7 & 5 & 3 & 1 \leftarrow \text{bit positions}
\end{array}
$$
$$\text{PE0} = 0 = m_8 \oplus m_6 \oplus m_4 \oplus m_3 \oplus m_1 \oplus m_0 \oplus C_0$$
$$
\begin{array}{ccccccc}
14 & 11 & 10 & 7 & 6 & 3 & 2 \leftarrow \text{bit positions}
\end{array}
$$
$$\text{PE1} = 0 = m_9 \oplus m_6 \oplus m_5 \oplus m_3 \oplus m_2 \oplus m_0 \oplus C_1$$
$$
\begin{array}{ccccccc}
14 & 13 & 12 & 7 & 6 & 5 & 4 \leftarrow \text{bit positions}
\end{array}
$$
$$\text{PE2} = 0 = m_9 \oplus m_8 \oplus m_7 \oplus m_3 \oplus m_2 \oplus m_1 \oplus C_2$$
$$
\begin{array}{ccccccc}
14 & 13 & 12 & 11 & 10 & 9 & 8 \leftarrow \text{bit positions}
\end{array}
$$
$$\text{PE3} = 0 = m_9 \oplus m_8 \oplus m_7 \oplus m_6 \oplus m_5 \oplus m_4 \oplus C_3$$

To determine the value of the check bits C_0 through C_3, the equations above can be rearranged as follows:

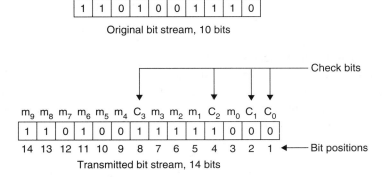

Figure 12-13 Check bits are inserted into a message stream for FEC.

$$C_0 = m_8 \oplus m_6 \oplus m_4 \oplus m_3 \oplus m_1 \oplus m_0$$
$$= 1 \oplus 1 \oplus 0 \oplus 1 \oplus 1 \oplus 0 = 0$$

$$C_1 = m_9 \oplus m_6 \oplus m_5 \oplus m_3 \oplus m_2 \oplus m_0$$
$$= 1 \oplus 1 \oplus 0 \oplus 1 \oplus 1 \oplus 0 = 0$$

$$C_2 = m_9 \oplus m_8 \oplus m_7 \oplus m_3 \oplus m_2 \oplus m_1$$
$$= 1 \oplus 1 \oplus 0 \oplus 1 \oplus 1 \oplus 1 = 1$$

$$C_3 = m_9 \oplus m_8 \oplus m_7 \oplus m_6 \oplus m_5 \oplus m_4$$
$$= 1 \oplus 1 \oplus 0 \oplus 1 \oplus 0 \oplus 0 = 1$$

even parity

Thus the check bits inserted into the message stream in positions 8, 4, 2, and 1 are

$$C_3 = 1$$
$$C_2 = 1$$
$$C_1 = 0$$
$$C_0 = 0$$

Let us now look at how a bit error can be identified and corrected by the weighted parity checks. Suppose that an error has been detected in the transmitted message stream. Bit position 7 has been lost in the transmission:

1 1 0 1 0 0 1 1 1 1 1 0 0 0 ← transmitted bit stream
↓ (lost in transmission)
1 1 0 1 0 0 1 0 1 1 1 0 0 0 ← received bit stream
↑ error in bit position 7

The receiver performs an even parity check over the same bit positions as discussed above. Even parity should result for each parity check if there are no errors. Since a bit error has occurred, however, the syndrome (location of the error) will be identified by the binary number produced by the parity checks, PE0 through PE3, as follows:

14	13	12	11	10	9	8	7	6	5	4	3	2	1	← bit position
1	1	0	1	0	0	1	0	1	1	1	0	0	0	← received bit stream

Check 0: 13 11 9 7 5 3 1 ← bit position
PE0 = $1 \oplus 1 \oplus 0 \oplus 0 \oplus 1 \oplus 0 \oplus 0 = 1$ (even parity failure) 1

Check 1: 14 11 10 7 6 3 2 ← bit position
PE1 = $1 \oplus 1 \oplus 0 \oplus 0 \oplus 1 \oplus 0 \oplus 0 = 1$ (even parity failure) 1

Check 2: 14 13 12 7 6 5 4 ← bit position
PE2 = $1 \oplus 1 \oplus 0 \oplus 0 \oplus 1 \oplus 1 \oplus 1 = 1$ (even parity failure) 1

Check 3: 14 13 12 11 10 9 8 ← bit position
PE3 = $1 \oplus 1 \oplus 0 \oplus 1 \oplus 0 \oplus 0 \oplus 1 = 0$ (correct) 0

syndrome = 0111 = 7

The resulting syndrome is 0111, or bit position 7. This bit is simply inverted and the four parity checks will result in 0000 (correct). The check bits are removed from positions 1, 2, 4, and 8, thereby resulting in the original message. One nice feature of this Hamming code is that once the message is encoded there is *no difference* between the check bits and the original message bits; that is, the syndrome can just as well identify a check bit in error.

12.7.1.2 An alternative method. Now that we have established the principle behind a Hamming code, an alternative method for correcting a single-bit error will be given here. The disadvantage with this method, however, is that it will not detect an error if it occurs in the Hamming bit. For simplicity, the same 10-bit message stream, 1101001110, will be used. Therefore, the number of Hamming bits, four, remains the same. The Hamming bits can actually be placed anywhere in the transmitted message stream as long as their positions are known by the transmitter and receiver. The procedure is outlined as follows.

1. Compute the number of Hamming bits m required for a message of n bits.

   Original message stream: 1 1 0 1 0 0 1 1 1 0 (10 bits)

   $$2^m \geq n + 1$$
   $$2^4 \geq (10 + 4) + 1, \qquad m = 4,$$
   $$n = 14$$

2. Insert the Hamming bits H into the original message stream.

   Transmitted message stream:
   14 13 12 11 10 9 8 7 6 5 4 3 2 1 ← bit position (14 bits)
    1 H 1 0 H 1 0 0 1 H 1 H 1 0

3. Express each bit position containing a 1 as a four-bit binary number and exclusive-OR each of these numbers together. Starting from the left, bit posi-

tions 14, 12, 9, 6, 4, and 2 are exclusive-ORed together. This will result in the value of the Hamming bits.

$$1110 = 14$$
$$\oplus\ \underline{1100} = 12$$
$$0010$$
$$\oplus\ \underline{1001} = 9$$
$$1011$$
$$\oplus\ \underline{0110} = 6$$
$$1101$$
$$\oplus\ \underline{0100} = 4$$
$$1001$$
$$\oplus\ \underline{0010} = 2$$
$$1011 \leftarrow \text{Hamming bits}$$

4. Place the value of the Hamming bits into the transmitted message stream shown in step 2.

14	13	12	11	10	9	8	7	6	5	4	3	2	1	← bit position
1	1	1	0	0	1	0	0	1	1	1	1	1	0	← transmitted bit stream
1	1	1	0	0	1	0	0	0	1	1	1	1	0	← received bit stream

error in bit position 6

Let us now assume that bit position 6 was received in error.

5. The Hamming bits are extracted from the received message stream and exclusive-ORed with the binary representation of the bit positions containing a 1. This will detect the bit position in error, or the syndrome.

14	13	12	11	10	9	8	7	6	5	4	3	2	1
1		1	0		1	0	0	0		1		1	0
	1			0					1		1		

← extracted Hamming bits

$$1011 \quad \text{Hamming bits}$$
$$\oplus\ \underline{1110} = 14$$
$$0101$$
$$\oplus\ \underline{1100} = 12$$
$$1001$$
$$\oplus\ \underline{1001} = 9$$
$$0000$$
$$\oplus\ \underline{0100} = 4$$
$$0100$$
$$\oplus\ \underline{0010} = 2$$
$$0110 \leftarrow \text{syndrome equals bit position 6}$$

To detect multiple bit errors, more elaborate FEC techniques are necessary. Additional redundancy must be built into the message stream. This further reduces the efficiency of the channel and lowers the transmission system's throughput. Unlike ARQ, which is extremely reliable, the best FEC techniques are not, particularly in cases where multiple bits are destroyed due to noise bursts. Generally, FEC is employed only in applications where ARQ is not feasible. The detection of multiple-bit errors is beyond the scope of this book.

PROBLEMS

1. Draw two circuit diagrams that can be used for odd parity generation for an eight-bit word.
2. Draw two circuit diagrams that can be used for even parity checking.
3. Explain why an even number of bit errors in a character will go undetected using parity for error detection. Show an example to support your explanation.
4. Using seven-bit ASCII and parity, determine the block check character using LRC for the message "Just do it!" Use even parity.
5. In Problem 4, show how a combination of LRC and VRC can detect an error if bit D2 were to change in the letter "o" during the course of transmission.
6. Given the generator polynomial $G(x) = X^5 + X^3 + X + 1$ and the message polynomial $M(x) = X^9 + X^7 + X^6 + X^2 + 1$, compute the following.
   (a) Number of bits in the BCC
   (b) Product of $X^{n-k}[M(x)]$
   (c) Discarded quotient
   (d) BCC, $B(x)$
   (e) Total transmitted message block, $T(x) = X^{n-k}[M(x)] + B(x)$
7. Repeat Problem 6 using CRC-CCITT for the message polynomial $M(x) = X^{15} + X^{14} + X^{12} + X^8 + X^6 + X^5 + X^3 + 1$.
8. For Problem 7, show how the receiver performs a CRC-16 check on the received message stream to verify that the message was received without errors.
9. Name four types of checksums.
10. What happens to any carry bit in single-precision checksums?
11. Compute the single-precision checksum for a four-byte block with hex characters: 4E, 3D, 92, and EA.
12. Compute the single-precision checksum for a five-byte block with hex characters: 7B, 2F, 37, 6A, and 4C.
13. Repeat Problem 12, but use the double-precision checksum technique.
14. What is the advantage of the double-precision checksum over the single-precision checksum?
15. Compute the Honeywell checksum for a six-byte message block with hex characters: 12, BC, F4, 8C, CA, and 68.

16. What is the advantage of the Honeywell checksum over single- or double-precision checksums?

17. Compute the residue checksum for the six-byte message block in Problem 15.

18. Explain an *SA1 error* and an *SA0 error*.

19. Explain the difference between ARQ and FEC.

20. Are Hamming codes used in ARQ or FEC?

21. Using the Hamming code developed in Section 12.7.1.1 compute the following.
    (a) Even parity check bits C_0 through C_3, for the 10-bit message 1011101011
    (b) Total message stream with the check bits included
    (c) Syndrome, through parity checks, if bit position 4 is corrupted
    (d) Syndrome, through parity checks, if bit position 13 is corrupted

22. Using the alternative method described in this chapter for Hamming codes, given the 12-bit message 101110101011, determine the following.
    (a) Number of Hamming bits, m, required
    (b) Total number of transmitted bits, n
    (c) Values of the Hamming bits if they are placed in odd *bit positions* starting from the MSB of the total transmitted bit stream and proceeding to the LSB until all Hamming bits are used
    (d) Syndrome if bit position 10 is corrupted (Be sure to show this through the exclusive-OR process outlined in the chapter.)

13

Noise

Any electrical signal transmitted from one point to another can ultimately be classified as having two parts: one that represents the original intelligence, the desirable part, and the undesirable part, which we will call *noise*. In virtually all telecommunication systems, the effects of noise superimposed on the original intelligence is of major concern to us. The magnitude of this noise is directly related to our ability to recover the intelligence without error. In this chapter we study the effects of noise, its various types, and the techniques used for evaluating system performance under the influence of noise.

13.1 EFFECTS OF NOISE

Electrical noise is inherent in all transmitted and received signals. Its effects can severely limit system performance. Today's telecommunication systems are especially vulnerable due to several factors:

— Increased volumes of traffic through central office switching centers
— Enormous channel capacities of trunk circuits
— Trend toward digital and data transmission

One unfortunate circumstance of noise is that it inevitably corrupts the signal, as shown in Figure 13-1. Great measures are taken to minimize its effects. The communication system typically employs special encoding and decoding

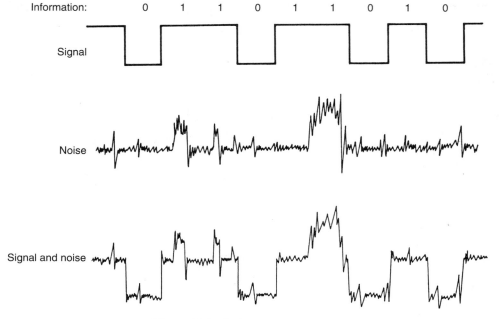

Information:　　0　1　1　0　1　1　0　1　0

Signal

Noise

Signal and noise

Figure 13-1 Effects of noise on a signal.

techniques to optimize the recovery of the signal. Extensive conditioning of the signal is performed through the use of filters and amplifiers. Careful attention is given to the transmission medium, its type, and how the signal is routed to its destination. Elaborate error detection and correction mechanisms in both hardware and software are built into the telecommunication system to identify and often correct for errors caused by noise.

13.2 NOISE MEASUREMENTS

Several mathematical tools have been developed to evaluate the effects of noise based on its magnitude relative to a given signal. The theory behind this highly complex subject is beyond the scope of this book. Several books are available on the subject and the reader is encouraged to pursue his or her needs. What we are concerned about is the practical aspects of noise measurements commonly encountered by the communications technologist.

13.2.1 Signal-to-Noise Ratio (SNR)

One of the most useful measures of noise, from a deterministic point of view, is *signal-to-noise ratio* (SNR or *S/N*). It is the ratio of signal power to noise power. SNR is of prime importance to us, since it allows us to evaluate and anticipate the extraneous effects of noise. It is typically measured at the receiving end of the communication system prior to the detection of the signal. Mathematically, SNR is expressed in decibels by the following equation:

$$\text{SNR} = 10 \log \frac{\text{signal power}}{\text{noise power}} \quad \text{dB} \quad (13\text{-}1)$$

If we assume that the composite signal (signal and noise) is measured across the same resistance, then SNR can also be expressed by the ratio of signal voltage to noise voltage. In decibels, this is computed as follows:

$$\text{SNR} = 10 \log \frac{S}{N}$$

$$= 10 \log \frac{(V_S)^2/R}{(V_N)^2/R_2}$$

$$= 10 \log \left(\frac{V_S}{V_N}\right)^2$$

and therefore

$$\text{SNR} = 20 \log \left(\frac{V_S}{V_N}\right) = 20 \log \frac{\text{signal voltage}}{\text{noise voltage}} \quad \text{dB} \quad (13\text{-}2)$$

Example 1

A 1-kHz sinusoid test tone is measured with an oscilloscope at the input of a receiver's FM detector stage. Its peak-to-peak amplitude is 3 V. With the test tone at the transmitter turned off, the noise at the same test point is measured with an rms voltmeter. Its value is 640 mV. Compute the SNR in decibels.

Solution: $V_S = 0.707 V_P = 0.707 \left(\dfrac{3 \text{ V}_{\text{p-p}}}{2}\right) = 1.06 \text{ V}$

$V_N = 640 \text{ mV}$

$\text{SNR} = 20 \log \dfrac{V_S}{V_N} = 20 \log \dfrac{1.06 \text{ V}}{640 \text{ mV}} = 4.39 \text{ dB}$

The average SNR can be measured for any composite signal with an rms voltmeter or power meter. One needs only to have control over turning the transmitted signal on and off. The resulting deflection produced on the meter's decibel scale is a direct measure of SNR.

13.2.2 Noise Factor and Noise Figure

SNR is useful in applications where the noise content of the signal is desired at a specific point in the communication system. It does not, however, characterize the amount of additional noise introduced by the various components in the overall communication system. A key parameter used for this purpose is *noise factor*. Noise factor is a measure of how noisy a device is. It is the ratio of signal-to-noise (S_i/N_i) power at the input of a device to the signal-to-noise (S_o/N_o) power at its output. Expressed in decibels, noise factor is called *noise figure*. All amplifiers, for example, contribute some degree of noise to the signal. If an amplifier generated no noise of its own, its input and output SNR would be equal. Noise factor would be 1 in this ideal case, which is equivalent to a noise figure of 0 dB. Noise factor and noise figure are governed by the following equations:

$$\text{noise factor} = F = \frac{S_i/N_i}{S_o/N_o} \tag{13-3}$$

Noise figure (NF) is related to noise factor by the expression

$$\text{NF} = 10 \log F = 10 \log \frac{S_i/N_i}{S_o/N_o} \text{ dB} \tag{13-4}$$

If the input and output of the device under consideration share the same impedance, NF can also be expressed in terms of voltage:

$$\text{NF} = 20 \log \frac{V_{S_i}/V_{N_i}}{V_{S_o}/V_{N_o}} \text{ dB} \tag{13-5}$$

Example 2

The input signal to a telecommunications receiver consists of 100 μW of signal power and 1 μW of noise power. The receiver contributes an additional 80 μW of noise, N_R, and has a power gain of 20 dB. Compute the input SNR, the output SNR, and the receiver's noise figure.

Solution: $\dfrac{S_i}{N_i} = \dfrac{100\ \mu\text{W}}{1\ \mu\text{W}} = 100$

20 dB equals a power gain, A_p, of 100. Therefore, the signal out is

$$S_o = S_i A_p = 100\ \mu\text{W} \times 100 = 10\ \text{mW}$$

The output noise is

$$N_o = N_i A_p + N_R = 1\ \mu\text{W} \times 100 + 80\ \mu\text{W} = 180\ \mu\text{W}$$

The output SNR is

$$\frac{S_o}{N_o} = \frac{10\ \text{mW}}{180\ \mu W} = 55.6$$

The noise factor is

$$F = \frac{S_i/N_i}{S_o/N_o} = \frac{100}{55.6} = 1.80$$

The noise figure is

$$\text{NF} = 10 \log F$$
$$= 10 \log 1.80$$
$$= 2.55\ \text{dB}$$

Example 3

A log amplifier is specified as having a noise figure of 3.9 dB. For an input signal of 65 μW, the amplifier produces an output power of 20 mW. If the noise out of the amplifier is measured to be 1 mW, compute the following.

(a) The power gain, A_p
(b) The power gain in decibels, A_p (dB)
(c) The noise factor, F
(d) The input noise to the log amplifier, N_i
(e) The noise contributed by the log amplifier, N_A

Solution: (a) $A_p = \dfrac{S_o}{S_i} = \dfrac{20 \text{ mW}}{65 \ \mu\text{W}} = 307.69$

(b) A_p (dB) $= 10 \log 307.69 = 24.88$ dB

(c) Since NF $= 3.9$ dB $= 10 \log$ F,

then $F = 10^{\frac{3.9}{10}} = 2.4547$

(d) Since $F = \dfrac{S_i/N_i}{S_o/N_o}$,

then $N_i = \dfrac{S_i}{(S_o/N_o)F} = \dfrac{65 \ \mu\text{W}}{(20 \text{ mW}/1 \text{ mW})(2.4547)} = 1.324 \ \mu\text{W}$

(e) The noise output, $N_o = N_i \cdot A_p + N_A$
therefore $N_A = N_o - N_i A_p$

$= 1 \text{ mW} - (1.324 \ \mu\text{W})(307.69)$

$= 592.6 \ \mu\text{W}$

13.2.3 Bit Error Rate (BER)

Another significant measure of system performance in terms of noise is *bit error rate* (BER). BER specifies the number of bits that are corrupted or destroyed as data are transmitted from their source to their destination. A BER of 10^{-6}, for example, means that one bit out of every million is destroyed during transmission. Several factors contribute to BER:

— Bandwidth

— SNR

— Transmission speed

— Transmission medium

— Environment

— Transmission distance

— Transmitter and receiver performance

A BER of 10^{-5} over switched voice-grade lines is typical. For a good digital communication system, BERs of less than a few bits per million are not uncommon.

13.2.4 Channel Capacity

If the signal-to-noise ratio for a digital communication system is large enough, it is possible to overcome any extraneous effects of noise entirely. In practice, however, this is seldom the case. Mathematical guidelines have been established to determine the maximum theoretical data transfer rate over a channel based on the channel's bandwidth and SNR. This is known as the *channel capacity*. One of the most fundamental laws used in telecommunications is *Shannon's law*. Shannon's law, governed by equation (13-6), allows us to compute channel capacity (in bps). Shannon's equation is based on a signal under the influence of *Gaussian type noise*, which will be explained later in this chapter.

$$C = \text{BW} \log_2 \left(1 + \frac{S}{N}\right) \quad \text{bps} \qquad (13\text{-}6)$$

where BW = bandwidth (Hz)
 S/N = signal power to Gaussian noise power within the given BW
 C = channel capacity (bps)

Example 4

The standard 3002 voice-grade lines have a nominal SNR of 25 dB and a bandwidth ranging from 300 to 3400 Hz. Compute the channel capacity using equation (13-6).

Solution: $C = \text{BW} \log_2 \left(1 + \frac{S}{N}\right)$

$= 3100 \log_2 \left(1 + 316\right)$ (a power gain of 25 dB = 316)
$= 3100 \log_2 317$
$= 25,755$ bps

13.3 NOISE TYPES

We have all experienced the undesirable effects of noise on our television sets, radios, and telephones. This crackling and hissing, humming and fading severely limits the performance of any communication system. It is necessary in our studies to consider some of its sources and types.

13.3.1 Atmospheric and Extraterrestrial Noise

A significant contributor to noise in a communication system stems from natural occurrences in our atmosphere.

13.3.1.1 Lightning. Lightning is a major source of noise and is caused by the static discharge of thunderclouds. Several millions of volts, with currents in excess of 20,000 A, are not uncommon for lightning discharges. A wide band of frequencies is generated during a discharge. The amplitudes of these frequency components are inversely proportional to their frequency, thus affecting mostly the low- and high-frequency bands up to 30 MHz.

13.3.1.2 Solar noise. Ionized gases of the sun produce a broad range of frequencies that penetrate the Earth's atmosphere at frequencies used by communication systems. The upper portion of our atmosphere, known as the ionosphere, is directly influenced by ultraviolet radiation of the sun. Random molecular activity in the ionosphere causes severe electrical disturbances in the high-frequency region. These disturbances are especially intense when sunspot activity peaks approximately every 11 years.

13.3.1.3 Cosmic noise. Distant stars in our universe, like the sun, also radiate intense levels of noise at frequencies that penetrate Earth's atmosphere. These frequencies affecting the communication system range from 8 MHz to as high as 2 GHz.

13.3.2 Gaussian Noise

The cumulative effect of all random noise generated external and internal to the telecommunication system, averaged over a period of time, is referred to as *Gaussian noise*. Gaussian noise includes all frequencies, similar to the manner in which white light includes all visible wavelengths of color. The distribution of Gaussian noise power for a given bandwidth forms a uniform bell curve if viewed on a spectrum analyzer. This is shown in Figure 13-2.

Noise power

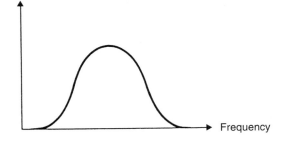

Frequency

Figure 13-2 Gaussian noise distribution producing a uniform bell curve.

13.3.2.1 Thermal noise. For electronic circuits, Gaussian noise is more specifically referred to as *white noise, Johnson noise,* or *thermal noise.* The three names are used interchangeably to represent noise generated by the *random* motion of free electrons and molecular vibrations exhibited by all electronic components, including conductors. This random motion results in frequency components that are evenly distributed over the entire radio frequency spectrum; thus the noise is said to be white. If one were to tune to a frequency in the commercial FM broadcast band, where there is no station, and turn the volume control up, the hiss heard in the background is an example of this type of noise.

All resistive components are constant noise generators due to the thermal interaction of atomic and subatomic particles that make up the resistive component; hence the name thermal noise. Since particles such as free electrons are in movement, they possess kinetic energy that is directly related to the temperature of the resistive body. This includes the resistive component of an impedance across which the thermal agitation is produced. At absolute zero (Kelvin), all motion ceases. The kinetic energy of all charged particles becomes zero, and therefore no thermal noise is generated. At temperatures above 0° Kelvin, the charged particles begin to move. This motion constitutes an electrical current. Although at any instant in time, the number of charged particles flowing in one direction may exceed those flowing in the opposite direction, the net potential created over a period of time is equal to zero. In other words, no dc current flows. An rms value does, however, exist which can be measured with an rms voltmeter. Typical values may be on the order of a few microvolts. Although this may not seem significant, for most high-gain receivers, this can be overwhelming when compared to signal amplitudes received at the antenna input.

In 1928, J. B. Johnson, a U.S. physicist, demonstrated that resistive components generate noise power in proportion to both temperature and bandwidth. Johnson's research cannot be overemphasized as it allows us to compute the *rms* noise power associated with a resistive component. Furthermore, it allows us to tailor our receiver designs accordingly. The following equation shows the mathematical relationship that Johnson developed:

$$P_n = K \cdot T \cdot \mathrm{BW} \tag{13-7}$$

where K = Boltzmann's constant of 1.38×10^{-23} J/K
   T = absolute temperature of the device (K)
   BW = circuit bandwidth
   P_n = noise power output of the resistor

An equivalent noise voltage, V_n, generated by the resistor, R, can be computed based on the maximum power transfer into an equivalent load resistor, R_L. The load resistance is assumed to be noiseless.[†]

[†]George Kennedy, *Electronic Communication Systems,* 3rd ed. (New York, McGraw-Hill, Inc., 1985), p. 13.

Rewriting equation (13-7) in terms of an equivalent noise voltage, V_n, is performed in the following manner:

$$P_n = K \cdot T \cdot \mathrm{BW} = \frac{V^2}{R} = \frac{V^2}{R_L}$$

For maximum power transfer into an equivalent load resistor, the voltage across the load, R_L, is one half the voltage across the noise-generating resistance, R, which yields the following:

$$\frac{V^2}{R} = \frac{(V_n/2)^2}{R} = \frac{V_n^{\,2}}{4R}$$

and therefore:

$$V_n = \sqrt{4K \cdot T \cdot \mathrm{BW} \cdot R} \qquad (13\text{-}8)$$

Example 5

An amplifier used to process an FDM channel group (12 voice channels) operates over a frequency range from 60 to 108 kHz. The input impedance has a resistive component of 10 kΩ. Compute the equivalent noise voltage at the input of the amplifier at an operating temperature of 24°C.

Solution: $°K = 273 + °C$
$= 273 + 24$
$= 297$
$\mathrm{BW} = 108 \text{ kHz} - 60 \text{ kHz} = 48 \text{ kHz}$
$V_n = \sqrt{4K \cdot T \cdot \mathrm{BW} \cdot R}$
$= \sqrt{4 \times 1.38 \times 10^{-23} \times 297 \times 48 \times 10^3 \times 10 \times 10^3}$
$= 2.81 \ \mu\mathrm{V}$

13.3.2.2 Shot noise. Another significant contributor to the distribution of Gaussian noise is *shot noise*. The name originates from the sound that it produces at the audio output of a receiver. The sound is similar to that of lead shot falling on top of a tin roof. This type of noise is generated as a result of the random arrival rate of discrete current carriers (holes and electrons) at the output electrodes of semiconductor and vacuum-tube devices. Although the bias currents for these devices flow at a uniform rate over a period of time, at any instant in time there is a nondeterministic number of charge carriers at their outputs. This randomness generates the noise currents associated with shot noise. For semiconductor devices, this current is equal to

$$i_n = \sqrt{2qIf} \qquad (13\text{-}9)$$

where i_n = shot noise current in rms
$\quad q$ = charge of an electron, 1.6×10^{-19} coulomb
$\quad I$ = dc current flowing through the device (A)
$\quad f$ = system bandwidth (Hz)

13.3.3 Crosstalk

Many of us have experienced listening to other telephone conversations taking place in the background of our own. These signals are not only annoying, but they can also interfere with the transmission of data. This type of noise, called *crosstalk,* occurs as a result of inductive and capacitive coupling from adjacent channels that are in proximity to each other. Subscriber loops and trunk circuits commonly multiplexed together to form bundled cables often have severe crosstalk, particularly in long lengths of cable.

Figure 13-3 illustrates how crosstalk occurs between two adjacent channels. At high frequencies, the capacitance between conductors has a low enough reactance to cause serious coupling between channels. Digital signals are especially prone to crosstalk due to the high-frequency components in the signal.

Inductive coupling between wires is based on the same principle as a transformer. If conductors are close enough to each other, the electromagnetic field generated by signals transmitted in one line will induce crosstalk currents into the others.

In telecommunication systems, crosstalk is often classified as *near end* or *far end.* As shown in Figure 13-4, near-end crosstalk occurs at the transmitting station or repeater. Strong signals radiated from the transmitter pairs are coupled into the relatively weak signals traveling in the opposite direction of the receiver pairs. Far-end crosstalk occurs at the far end receiver as a result of adjacent channel signals traveling in the same direction.

Several measures are taken to minimize crosstalk in the telecommunication system. Consider the following:

— Adjacent wire pairs are twisted at different pitches to reduce the amount of crosstalk.

— Shielding is used to prevent signals from radiating into adjacent conductors.

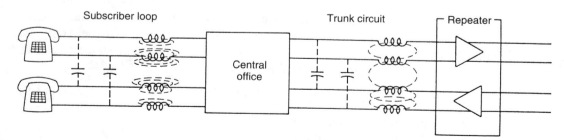

Figure 13-3 Crosstalk can result from capacitive and inductive coupling between adjacent channels in proximity to each other.

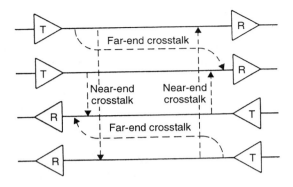

Figure 13-4   Near-end and far-end crosstalk.

— Transmitted and received signals over long distances are physically separated and shielded to prevent near- and far-end crosstalk.

— Differential amplifiers and receivers are used to reject common-mode signals.

— Balanced transformers are often used with twisted-pair media to cancel crosstalk signals coupled equally into both lines.

— Limits are set on the maximum number of channels that can be used within a cable.

— Excessive signal amplitudes and frequencies due to overmodulation are major contributors to crosstalk. Stringent regulation is placed on these parameters for digital and analog carriers.

13.3.4 Impulse Noise

Impulse noise, as its name implies, consists of sudden bursts of irregularly shaped pulses. They may last for a few microseconds to several hundreds of milliseconds, depending on their amplitude and origin. Common sources of impulse noise include transients induced from electromechanical switching relays at the central office, electric motors and appliances, ignition systems, poor solder joints, and lightning. Impulse noise is the familiar cracking and popping noise. It can be heard at a telephone set's receiver or the speaker output of a communications receiver. Its effect on voice is not nearly as damaging as it is on data. If a burst of noise, for example, were large enough, data can be completely blotted out for the entire duration of the impulse. For example, if a 9600-bps modem were transmitting data over the telephone lines and a 10-ms noise burst were to saturate the channel, nearly 100 bits of data would be corrupted. Suppose this data corresponded to a banking transaction. Without error detection and correction, this could be catastrophic!

PROBLEMS

1. The noise power at the output of a receiver's IF stage is measured at 45 μW. With the receiver tuned to a test signal, the output power increases to 3.58 mW. Compute the SNR.

2. An amplifier with a power gain of 40 dB produces a noise output power of 180 μW with an input noise power of 7 nW. The input signal power = 1 μW. Compute the noise factor, F, the noise figure, NF, and the noise power contributed by the amplifier.

3. The input signal to a repeater is made up of 150 μW of signal power and 1.2 μW of noise power. The repeater contributes an additional 48 μW of noise and has a power gain of 20 dB. Determine the following.
   (a) Input SNR
   (b) Output SNR
   (c) Noise factor, F
   (d) Noise figure, NF

4. What does *BER* stand for?

5. Compute the BER for a system that has a history of 25 bit errors out of every 2 million bits that are transmitted.

6. The telephone lines for a PBX system have a bandwidth ranging from 280 Hz to 5.1 kHz. SNR is nominally 22 dB. Compute the *channel capacity* of the line using Shannon's law.

7. A telephone channel has a bandwidth of 3 kHz and an SNR of 1023. Compute the channel capacity.

8. Explain whether or not the noise produced by lightning can be considered Gaussian.

9. An amplifier operates over a bandwidth of 750 kHz and has an input resistance of 20 kΩ. Compute the thermal noise power, P_n, and noise voltage, V_n, generated at its input at an ambient temperature of 25°C.

10. Given a bandwidth of 10 kHz and an operating temperature of 27°C, compute the thermal noise voltage, V_n, generated by the following resistors.
    (a) 1 kΩ
    (b) 20 kΩ
    (c) 100 kΩ
    (d) 1 MΩ

11. A semiconductor diode operates in a circuit having a bandwidth of 2 MHz. Its bias current is equal to 7.5 mA. Compute the rms shot noise current.

12. Explain the difference between near-end and far-end crosstalk.

13. A 4800-bps synchronous modem, transmitting data over the switched lines, is hit with 2.5 ms of impulse noise from a lightning strike. The noise saturates the channel. Compute the number of bits that would be destroyed in this case.

14

Fiber Optics

Fiber optics is one of the most explosive technologies in today's informational age. In the past few decades alone, technical advances in this field have grown from mere laboratory experiments to major industries involved with the development and production of advanced optical communication systems. Fiber optics is finding itself useful in virtually every application involving the transmission of information. Computers can now be linked together with fiber-optic cables capable of transferring data several orders of magnitude faster than copper circuits. In the medical industry, fiber-optic technology is being used to monitor and perform complex surgical operations. Throughout the world, telephone companies are laying thousands of miles of fiber underground, below oceanic floors and rivers, through manholes and existing conduit facilities. Thousands of simultaneous voice conversations are now being transmitted over these tiny strands of fiber less than the diameter of a human hair.

Fiber optics is defined as that branch of optics that deals with the transmission of light through ultrapure fibers of glass, plastic, or some other form of transparent media. From a decorative standpoint of view, most of us are familiar with the fiber-optic lamp, which uses bundles of thin optical fibers illuminated from the base end of the lamp by a light source. The light source is made to vary in color, which can be seen at the opposite ends of the fiber as a tree of illuminating points radiating various colors of the transmitted light. Although the lamp is used for decorative purposes only, it serves as an excellent model of how light can be transmitted through fiber.

14.1 HISTORY OF FIBER OPTICS

The use of light as a mechanized means of communicating has long been a part of our history. Paul Revere, as we know, used light back in 1775 to give warning of the approach of British troops from Boston. In 1880, Alexander Graham Bell invented a device called the *photophone.* The photophone used sunlight reflected off a moving diaphragm to communicate voice information to a receiver. Although the device worked, it was slightly ahead of its time.

One of the first noted experiments that demonstrated the transmission of light through a dielectric medium has been credited to a British natural philosopher and scientist named John Tyndall. In 1854, Tyndall demonstrated before the British Royal Society that light could be guided through a stream of water based on the principle of *total internal reflection:* rays of light can propagate through a transparent medium by reflecting off its boundaries. This vital principle will be discussed in detail later in this chapter. Tyndall's apparatus is shown in Figure 14-1. It included a bucket of water illuminated at its surface by a bright light source. A hole was punched and corked off near its bottom. When Tyndall released the cork, gravitational force caused the water to jet out of the hole and arc its way down to a lower glass container. Light could be seen arcing its way down to the lower container, which in turn was illuminated by the dispersion of the propagating medium. Much to the dismay of his audience, his experiment was successful and dispelled earlier theory that light travels in a straight line.

14.1.1 Milestones in Fiber Optics

Tyndall's experiment marked one of the first major milestones in the development of fiber-optic technology as it stands today. The progressive sequence of milestones to follow were few and generally regarded as interesting concepts only to the scientific elite. Not until the last few decades has fiber optics become a revolutionary trend in industry. Consider the major milestones to this date:

1854   John Tyndall demonstrated before the British Royal Society that light could be guided by its boundaries through a transparent medium based on the principle of *total internal reflection.*

1880   Alexander Graham Bell invented the *photophone,* a device that transmits voice signals over a beam of light.

1950s   Brian O'Brien, Sr., Harry Hopkins, and Naringer Kapany developed the two-layer fiber consisting of an inner *core* in which light propagates and an outer layer surrounding the core called the *cladding,* which is used to confine the light. The fiber was later used by the same scientists to develop the flexible *fiberscope,* a device capable of

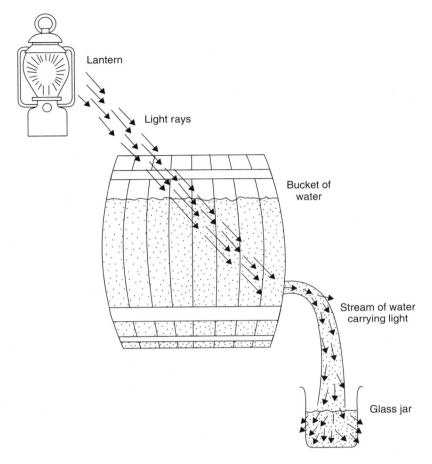

Figure 14-1 John Tyndall's demonstration before the British Royal Society proved that light could be guided by its boundaries through a transparent medium based on the principle of total internal reflection.

transmitting an image from one end of the fiber to its opposite end. Its flexibility allows peering into areas that are normally not accessible. The fiberscope, to this day, is still widely used, particularly in the medical profession to peer into the human body.

1958 Invention of the *laser* (light amplification by stimulated emissions of radiation) by Charles H. Townes allowed intense and concentrated light sources to be coupled into fiber.

1966 Charles K. Kao and George Hockham of Standard Telecommunications Laboratories of England performed several experiments to prove that, if glass could be made more transparent by reducing its impurities, light loss could be minimized. Their research led to a publication in which they predicted that optical fiber could be made pure enough to transmit light several kilometers. The global race to produce the optimum fiber began.

1967   Losses in optical fiber were reported at 1000 dB/km.

1970   Losses in optical fiber were reported at 20 dB/km.

1976   Losses in optical fiber were reported at 0.5 dB/km.

1979   Losses in optical fiber were reported at 0.2 dB/km.

1987   Losses in optical fiber were reported at 0.16 dB/km.

1988   NEC Corporation sets a new long-haul record of 10 Gbits/s over 80.1 km of dispersion-shifted fiber using a distributed-feedback laser. (*Lasers and Optronics,* Apr. 1988, vol. 7, no. 4, p. 46.)

1988   The *Synchronous Optical Network* (SONET) was published by the *American National Standards Institute* (ANSI).

1995   Multimedia applications for business have become the major impetus for increased use of optical fiber within the LAN, MAN, and WAN environment. These applications include employee training, desktop conferencing, and desktop news services.

14.2 ADVANTAGES OF FIBER-OPTIC SYSTEMS

The advantages of fiber-optic systems warrant considerable attention. This new technology has clearly affected the telecommunications industry and will continue to thrive due to the numerous advantages it has over its copper counterpart. The major advantages are discussed next.

Bandwidth.   One of the most significant advantages that fiber has over copper or other transmission media is bandwidth. Bandwidth is directly related to the amount of information that can be transmitted per unit time. Today's advanced fiber-optic systems are capable of transmitting several *gigabits* per second over hundreds of kilometers. Thousands of voice channels can now be multiplexed together and sent over a single fiber strand.

Less Loss.   Currently, fiber is being manufactured to exhibit less than a few tenths of a decibel of loss per kilometer. Imagine glass so pure that you could see through a window over 75 miles (120 km) thick! Repeaters can now be spaced 50 to 75 miles apart from each other.

Noise Immunity and Safety.   Since fiber is constructed of dielectric material, it is immune to inductive coupling or crosstalk from adjacent copper or fiber channels. In other words, it is not affected by electromagnetic interference

(EMI) or electrostatic interference. This includes environments where there are electric motors, relays, and even lightning. Likewise, since fiber-optic cables transmit light instead of current, they do not emit electrical noise, nor do they arc when there is an intermittent in the link. Thus they are useful in areas where EMI must be kept to a minimum. With the telecommunications highways of the air as congested as they are, the use of fiber is an attractive alternative.

Less Weight and Volume. Fiber-optic cables are substantially lighter in weight and occupy much less volume than copper cables with the same information capacity. Fiber-optic cables are being used to relieve congested underground ducts in metropolitan and suburban areas. For example, a 3-in. diameter telephone cable consisting of 900 twisted-pair wires can be replaced with a single fiber strand 0.005 in. in diameter (approximately the diameter of a hair strand) and retain the same information-carrying capacity. Even with a rugged protective jacket surrounding the fiber, it occupies enormously less space and weighs considerably less.

Security. Since light does not radiate from a fiber-optic cable, it is nearly impossible to secretly tap into it without detection. For this reason, several applications requiring communications security employ fiber-optic systems. Military information, for example, can be transmitted over fiber to prevent eavesdropping. In addition, fiber-optic cables cannot be detected by metal detectors unless they are manufactured with steel reinforcement for strength.

Flexibility. We normally think of glass as being extremely brittle. If one were to attempt to bend a $\frac{1}{8}$-in. thick glass window, it would certainly break at some point. One reason for this is that the glass we are familiar with has a considerable amount of surface flaws, although it appears polished. The outer surface of the glass would bend considerably more than the inside. The surface flaws would initiate the crack, in the same manner in which long scratches are intentionally used to crack and form glass windows. The surface of glass fiber is much more refined than ordinary glass. This, coupled with its small diameter, allows it to be flexible enough to wrap around a pencil. In terms of strength, a 0.005-in. strand of fiber is strong enough to cut one's finger before it breaks, if enough pressure is applied against it.

Economics. Presently, the cost of fiber is comparable to copper at approximately $0.20 to $0.50 per yard and is expected to drop as it becomes more widely used. Since transmission losses are considerably less than for coaxial cable, expensive repeaters can be spaced farther apart. Fewer repeaters means a reduction in overall system costs and enhanced reliability.

Reliability. Once installed, a longer life span is expected with fiber over its metallic counterparts since it is more resistant to corrosion caused by environmental extremes such as temperature, corrosive gases, and liquids. Many coating systems made of lacquer, silicons, and ultraviolet-cured acrylates have been developed to maintain fiber strength and longevity.

14.3 DISADVANTAGES OF FIBER-OPTIC SYSTEMS

In spite of the numerous advantages that fiber-optic systems have over conventional methods of transmission, there are some disadvantages, particularly because of its newness. Many of these disadvantages are being overcome with new and competitive technology.

Interfacing Costs. Electronic facilities must be converted to optics in order to interface to fiber. Often these costs are initially overlooked. Fiber-optic transmitters, receivers, couplers, and connectors, for example, must be employed as part of the communication system. Test and repair equipment is costly. If the fiber-optic cable breaks, splicing can be a costly and tedious task. Manufacturers, however, are continuously introducing new and improved field repair kits.

Strength. Fiber, by itself, has a tensile strength of approximately 1 lb, as compared to coaxial cable at 180 lb (RG59U). Surrounding the fiber with stranded Kevlar† and a protective PVC jacket can increase the pulling strength up to 500 lb. Installations requiring greater tensile strengths can be achieved with steel reinforcement.

Remote Powering of Devices. Occasionally it is necessary to provide electrical power to a remote device. Since this cannot be achieved through the fiber, metallic conductors are often included in the cable assembly. Several manufacturers now offer a complete line of cable types, including cables manufactured with both copper wire and fiber.

14.4 A TYPICAL FIBER-OPTIC TELECOMMUNICATIONS SYSTEM

Before we begin a detailed discussion of fiber optics, a simplified optical telecommunications system will now be considered. Figure 14-2 illustrates such a system. The analog signal generated by the telephone set is input to the coder

---

†Kevlar is a registered trademark of E.I. du Pont de Nemours Corporation. It is a nonmetallic material that is difficult to stretch.

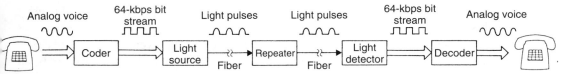

Figure 14-2 Simplified optical telecommunications system.

section. Here it is digitized and encoded into a 64-kbps binary serial bit stream. The bit stream is used to modulate the light source, which in turn transmits the series of light pulses into the optical fiber. At the receiving end, the impulses of light are converted back to an electrical signal by the light detector. The decoder section of the system converts the binary serial bit stream back to the original analog signal. This signal can now be used by the telephone set to reproduce the original sound energy.

There is nothing new about the analog-to-digital conversion process. Telephone companies have been digitizing voice signals since the early 1960s. It is the conversion from electrical impulses to light impulses, and vice versa, that the more modern technology employs.

The light source, acting as the transmitting element, must be turned on and off in accordance with the binary serial bit stream. Infrared *light-emitting diodes* (LEDs) and *injection laser diodes* (ILDs) are used for this purpose. These devices and specially designed to turn on and off several millions, and in some cases, billions of times per second.

The light detector, acting as the receiving element, converts the received light pulses back to pulses of electrical current. Two devices commonly used for this purpose are the *photodiode* and *phototransistor*. They both behave in their normal manner, except that they are activated by light instead of bias current.

If two stations are separated far enough from each other, at some point in the fiber-optic link the light pulses will become so weak and distorted that the underlying signal is not recoverable. Fiber-optic repeaters are necessary in this case. From a scientific standpoint, optical amplifiers could be used to amplify the weak light pulses. However, from a practical standpoint, this is not what is done. Instead, an optical receiving element is used to convert the signal back to electrical form. The reproduced weak and distorted electrical signal is then processed in the same manner as a repeater would in a metallic cable or microwave radio link. The reconstructed signal is then converted back to light by a transmitting element and sent back into the fiber-optic cable.

14.5 THEORY OF LIGHT

The subject of *light* remains an extraordinary phenomenon. Since the earliest known existence of people, philosophers, mathematicians, and physicists have theorized and refuted each other in an attempt to explain the nature of light. Few subjects in the scientific realm have been studied to such an extreme as light; yet to this date, there is no simple explanation.

In the seventeenth and eighteenth centuries, there were two schools of thought regarding the nature of light. Sir Isaac Newton and his followers believed that light consisted of rapidly moving *particles* (or *corpuscles*), whereas Dutch physicist Christian Huygens regarded light as being a series of *waves*.

The wave theory was strongly supported by an English doctor named Thomas Young. As a student at Cambridge University, Young's interests included the study of sound waves and their effects on the human ear. Noticing the similarities between the interaction of sound waves and waves produced by water, Young was inspired to investigate the hypothesis that light was also a series of waves. This lead to Young's famous *double-slit experiment*, shown in Figure 14-3. His experiment clearly proved that the particle theory was seemingly absurd. In addition, if two beams of light were viewed crossing each other, the particle theory would support the occurrence of collisions at the intersection. Crossing two beams, however, does not seem to influence each other.

By the twentieth century, light was fundamentally accepted as having many of the properties established by James Clerk Maxwell. In 1860, Maxwell theo-

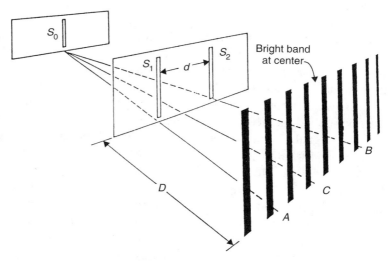

Figure 14-3 A light from a distant line source, S_0, is incident (near normal) upon a screen having two closely spaced parallel slits, S_1 and S_2. The light from the two parallel slits is observed to form alternating bright and dark bands (fringes) on a distant screen, with a bright fringe at the center, C, of the screen. According to the particle theory, there should be bright light *only* at positions A and B, not at C. (From H.Y. Carr and R. Weidner, *Physics from the Ground Up*, McGraw-Hill Book Company, New York, © 1971.)

rized that *electromagnetic radiation* consists of a series of oscillating waves made up of an electric field and a magnetic field propagating at right angles to each other. By 1905, however, quantum theory, introduced by Albert Einstein and Max Planck, showed that when light is emitted or absorbed it behaves not only as a wave, but also as an electromagnetic *particle* called a *photon.* A photon is said to possess energy that is proportional to its frequency (or inversely proportional to its wavelength). This is known as *Planck's law,* which states that

$$E = h \times v \qquad (14\text{-}1)$$

where E = photon's energy (J)
 h = Planck's constant, 6.63×10^{-34} J-s
 v = frequency of the photon (Hz)

Example 1

Compute the energy of a single photon radiating from a standard light bulb. Assume that the average frequency of the emitted light is 10^{14} Hz (which has a wavelength equal to 300 nm).

Solution: $E = h \times v$
$= 6.63 \times 10^{-34}$ J-s $\times 10^{14}$ Hz
$= 6.63 \times 10^{-20}$ J

Using the particle theory, Einstein and Planck were able to explain the *photoelectric effect*: when visible light or electromagnetic radiation of a higher frequency shines on a metallic surface, electrons are emitted, which in turn produce an electric current. The photoelectric effect is used in burglar alarms, door openers, and the like. And so it stands; the nature of light can be described as a particle or as a wave, depending on the circumstances.

14.5.1 Electromagnetic Spectrum

Maxwell's extensive research on electromagnetic wave theory has enabled other scientists to establish frequencies and wavelengths associated with various forms of light. Fundamentally, light has been accepted as a form of electromagnetic radiation that can be categorized into a portion of the entire *electromagnetic spectrum*, as shown in Figure 14-4. The spectrum, as it stands today, covers dc to cosmic rays. Each frequency has its own characteristic in terms of behavior and physical response to a given transmission medium. In addition, each frequency can be specified in terms of its equivalent *wavelength*. Frequency and wavelength are directly related to the speed of light. For most practical purposes, they are governed by the equation

$$c = f \times \lambda \qquad (14\text{-}2)$$

where c = speed of light, 3×10^8 m/s
f = frequency (Hz)
λ = wavelength (m)

Example 2

Compute the wavelength for a frequency of 7.20 MHz.

Solution: $c = f \times \lambda$

Therefore,

$$\lambda = \frac{c}{f}$$

$$= \frac{3 \times 10^8 \text{m/s}}{7.2 \times 10^6 \text{ Hz}} = 41.7 \text{ m}$$

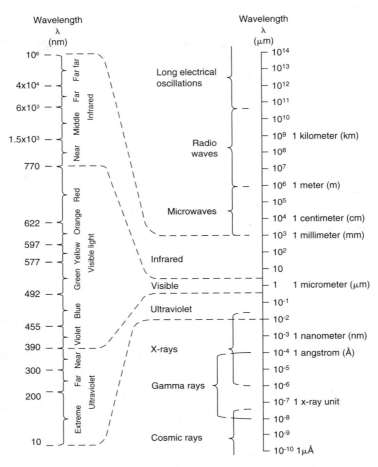

Figure 14-4 Electromagnetic spectrum showing the approximate boundaries between spectral regions. (Courtesy of RCA.)

TABLE 14-1   Units Typically Used to
Designate the Wavelength of Light

Unit	Meters	Inches
Nanometer	10^{-9}	39.4×10^{-9}
Micrometer	10^{-6}	39.4×10^{-6}
Angstrom	10^{-10}	3.94×10^{-9}

That portion of the electromagnetic spectrum regarded as light has been expanded in Figure 14-4 to illustrate the three basic categories of light:

1. *Infrared:* that portion of the electromagnetic spectrum having a wavelength ranging from 770 to 10^6 nanometers (nm). Fiber-optic systems operate in this range.

2. *Visible:* that portion of the electromagnetic spectrum having a wavelength ranging from 390 to 770 nm. The human eye, responding to these wavelengths, allows us to see the colors ranging from violet to red, respectively.

3. *Ultraviolet:* that portion of the electromagnetic spectrum having a wavelength ranging from 10 to 390 nm.

Light waves are commonly specified in terms of wavelength instead of frequency. Units typically used are the *nanometer, micrometer,* and *angstrom.* Table 14-1 lists their relationship to the meter and inch.

The light that we use for most fiber-optic systems occupies a wavelength range from 800 to 1600 nm for reasons that will be explained later. This is slightly larger than visible red light in terms of wavelength and falls within the infrared portion of the spectrum.

14.5.2 Speed of Light

The speed of light is at its maximum velocity when traveling in a vacuum or free space. It is equal to

$$(2.997925 \pm 0.000001) \times 10^8 \text{ m/s} \approx 3 \times 10^8 \text{ m/s}$$

(300 million meters per second or 186,000 miles per second). This figure has been universally established as a fact of nature despite the numerous attempts to exceed it by changing parameters such as wavelength, intensity, transmission medium, and point of reference. In mediums other than free space, the speed of

light is noticeably *reduced* and is no longer independent of wavelength. Its direction and reduced speed, however, will remain the same, just as in a vacuum, as long as the composition of the medium in which it propagates is uniform throughout. If the medium changes, both speed and direction will change.

14.5.3 Snell's Law

For light to propagate in any medium, the medium must be *transparent* to some degree. The degree of transparency determines how far light will propagate. Transparent materials can be in the form of a liquid, gas, or a solid. Some examples are glass, plastic, air, and water.

One of the most fundamental principles of light is that when it strikes the interface between two transparent mediums, such as air and water, a portion of the light energy is reflected back into the first medium and a portion is transmitted into the second medium. The path in which light travels from one point to another is commonly referred to as the *ray*. Figure 14-5 illustrates the classic example of a ray of light incident upon the surface of water. Notice that part of the light is *reflected* off the surface of water and part of it penetrates the water. The ray penetrating the water is said to be *refracted* or *bent* toward the *normal*. The normal is a line drawn perpendicular to the surface. The angle of incidence is equal to the angle of reflection, whereas the angle of refraction is bent toward the normal. This bending occurs at the surface of water as a result of a change in the speed of light. Light actually travels slower in a more dense medium, which in this case is the water. If the direction of the light ray were reversed, the opposite effect would occur: the speed of the light ray would increase as it exits the more dense medium of water into the less dense medium of air. The light ray

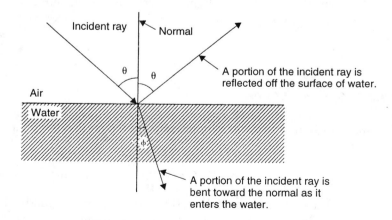

Figure 14-5   A light ray incident upon the surface of water is bent toward the normal.

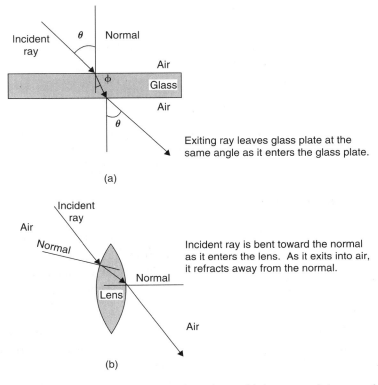

Figure 14-6 (a) A light ray enters a glass plate and is bent toward the normal. Its speed is reduced. As it emerges from the glass plate, it is bent away from the normal and returns to its original speed. (b) A light ray is bent toward the normal as it enters a lens. As it exits the lens, it is bent away from the normal.

would bend *away* from the normal. Figure 14-6 illustrates how light interacts through a glass plate and through a lens.

The amount of bending that light undergoes when entering a different medium is determined by the medium's *index of refraction,* generally denoted by the letter n. Index of refraction is the ratio of the speed of light in a vacuum, c, to the speed of light in the given medium, v. This relationship is given by the equation

$$n = \frac{c}{v} \tag{14-3}$$

Since the speed of light is lower in mediums other than a vacuum, the index of refraction in such mediums is always *greater* than 1. Table 14-2 lists the index of refraction for several common materials.

In 1621, the Dutch mathematician Willebrord Snell established that rays of light can be traced as they propagate from one medium to another based on their indices of refraction. *Snell's law* is stated by the equation

TABLE 14-2 Index of Refraction for Various Mediums

Medium	Index of refraction
Vacuum	1.0
Air	1.0003
Water	1.33
Ethyl alcohol	1.36
Fused quartz	1.46
Optical fiber	1.6 (nominal)
Diamond	2.2 (nominal)

Snell's law: $\quad n_1 \sin \theta_1 = n_2 \sin \theta_2$ $\hfill$ (14-4)

where n_1 = refractive index of material 1
θ_1 = angle of incidence
n_2 = refractive index of material 2
θ_2 = angle of refraction

Example 3

Refer to Figure 14-7. Using Snell's law, equation (14-4), compute the angle of refraction for a light ray traveling in air and incident on the surface of water at an angle of 52° with respect to the normal.

Solution: Since water is more dense than air, the light ray will bend toward the normal. From Table 14-2, the index of refraction for air is approximately 1 and for water it is 1.33.

$$n_1 \sin \theta_1 = n_2 \sin \theta_2$$
$$\sin \theta_2 = \frac{n_1 \sin \theta_1}{n_2}$$
$$\theta_2 = \sin^{-1}\left(\frac{n_1 \sin \theta_1}{n_2}\right) = \sin^{-1}\left(\frac{1.00 \sin 52°}{1.33}\right)$$
$$= 36.3°$$

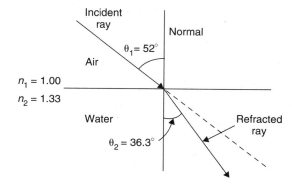

Figure 14-7 A light ray is incident upon the surface of water at an angle of 52° with respect to the normal. The light ray refracts toward the normal as it enters the more dense medium of water.

14.5.4 Total Internal Reflection

The importance of Snell's law with regard to fiber optics cannot be overstated. We can now consider how light is guided and contained within an optical fiber strand. In Example 3, if we reverse the direction of the light ray from water to air, as shown in Figure 14-8, according to Snell's law, we would expect the light ray to bend *away* from the normal since it travels from a more dense medium to a less dense medium.

In Example 4, Snell's law is used to compute the refracted angle of the light ray as it exits the water and enters the air. In our solution, we can see that it is impossible to have an angle whose sine is greater than 1. When this happens, a phenomenon known as *total internal reflection* occurs. Instead of the light ray refracting away from the normal as it enters the air, the ray is *reflected off the interface and bounces back into the water.*

Example 4

Refer to Figure 14-8. The direction of the light ray in Example 3 is reversed so that the light ray emerges from the water into air. Compute the angle of refraction in air for a ray with the same incident angle of 52°.

Solution: Since air is less dense than water, we would expect the emerging ray to refract away from the normal as shown. From Snell's law, however, the sine of the

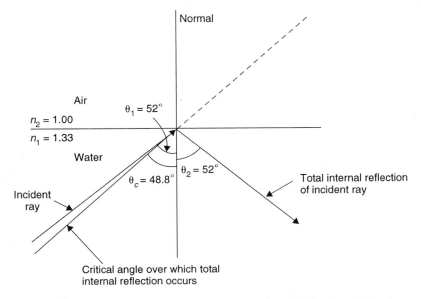

Figure 14-8 Light entering a medium whose index of refraction is less than the medium from which it exits will refract away from the normal unless the angle of incidence exceeds the critical angle. When this occurs, as in Example 4, total internal reflection occurs.

refracted angle is greater than 1; therefore, there is no mathematical solution to this problem using Snell's law.

$$n_1 \sin \theta_1 = n_2 \sin \theta_2$$

$$\sin \theta_2 = \frac{n_1 \sin \theta_1}{n_2}$$

$$\theta_2 = \sin^{-1}\left(\frac{n_1 \sin \theta_1}{n_2}\right) = \sin^{-1}\left(\frac{1.33 \sin 52°}{1.00}\right)$$

$$= \sin^{-1}\left(\frac{1.05}{1}\right) \qquad \text{no such angle!}$$

When the angle of incidence, θ_1, becomes large enough to cause the sine of the refracted angle, θ_2, to exceed the value of 1, total internal reflection occurs. This angle is called the *critical angle, θ_c*. The critical angle, θ_c, can be derived from Snell's law as follows:

$$n_1 \sin \theta_1 = n_2 \sin \theta_2$$

$$\sin \theta_1 = \frac{n_2 \sin \theta_2}{n_1}$$

When $\sin \theta_2 = 1$, then $\sin \theta_1 = n_2/n_1$. Therefore,

$$\textit{\textbf{Critical Angle:}} \quad \theta_c = \sin^{-1}\left(\frac{n_2}{n_1}\right) \tag{14-5}$$

Example 5

Refer to Figure 14-8. Using equation (14-5), compute the critical angle above which total internal reflection occurs.

Solution: $\theta_c = \sin^{-1}\left(\frac{n_2}{n_1}\right) = \sin^{-1}\left(\frac{1.00}{1.33}\right) = 48.8°$

Example 6

Refer to Figure 14-9. Using equation (14-5), compute the critical angle above which total internal reflection will occur for light traveling in a glass slab surrounded by air. The glass slab has an index of refraction equal to 1.5 and for air, it is equal to 1.

Solution: $\theta_c = \sin^{-1}\left(\frac{n_2}{n_1}\right) = \sin^{-1}\left(\frac{1.00}{1.5}\right) = 41.8°$

By surrounding glass with material whose refractive index is *less* than that of the glass, total internal reflection can be achieved. This is illustrated in Figure 14-9. Ray A penetrates the glass–air interface at an angle exceeding the critical

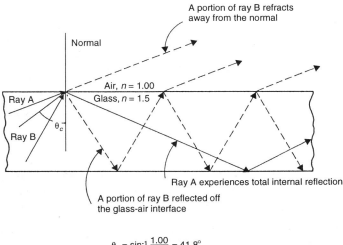

$$\theta_c = \sin^{-1} \frac{1.00}{1.5} = 41.8°$$

Figure 14-9 Light rays are transmitted into a glass slab surrounded by air. Ray A exceeds the critical angle and experiences total internal reflection. Ray B is less than the critical angle. Total internal reflection does not occur. A portion is reflected back into the glass and a portion exits the glass into air and refracts away from the normal.

angle, θ_c, and therefore experiences total internal reflection. On the other hand, ray B penetrates the glass–air interface at an angle less than the critical angle. Total internal reflection does not occur. Instead, a portion of ray B escapes the glass and is refracted away from the normal as it enters the less dense medium of air. A portion is also reflected back into the glass. Ray B diminishes in magnitude as it bounces back and forth between the glass–air interface. The foregoing principle is the basis for guiding light through optical fibers.

Two key elements that permit light guiding through optical fibers are its *core* and its *cladding*. The fiber's core is manufactured of ultrapure glass (silicon dioxide) or plastic. Surrounding the core is a material called cladding. A fiber's cladding is also made of glass or plastic. Its index of refraction, however, is typically 1% less than that of its core. This permits total internal reflection of rays entering the fiber and striking the core–cladding interface above the critical angle of approximately 82° [$\sin^{-1}(1/1.01)$]. The core of the fiber therefore *guides* the light and the cladding *contains* the light. The cladding material is much less transparent than the glass making up the core of the fiber. This causes light rays to be absorbed if they strike the core–cladding interface at an angle less than the critical angle.

In Figure 14-10, a light ray is transmitted into the core of an optical fiber. Total internal reflection occurs as it strikes the lower index cladding material.

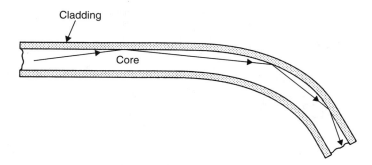

Cladding

Core

Figure 14-10 A light ray is transmitted into the core of an optical fiber strand. The critical angle between the glass core and its cladding is exceeded; therefore, total internal reflection occurs.

Notice that the ray is totally contained within the core and behaves in accordance with the property: the angle of incidence is equal to the angle of reflection. The ray continues to propagate down the fiber by bouncing back and forth against the core–cladding interface.

14.5.5 Propagation Modes and Classifications of Fiber

In fiber optics, it is important to consider the term *mode*. A fiber's mode describes the propagation characteristics of an electromagnetic wave as it travels through a particular type of fiber. There are basically two modes of transmission in a fiber. When a ray of light is made to propagate in one direction only, that is, along the center axis of the fiber, the fiber is classified as *single-mode* fiber. If there are a number of paths in which the light ray may travel, the fiber is classified as *multimode* fiber.

Several mathematical equations are used to represent the various modes of propagation in multimode fiber. From equation (14-6), it can be seen that the number of modes is a function of core diameter, index of refraction, and the wavelength of the light.

$$f = \frac{\left[\frac{\pi d}{\lambda} \sqrt{(n_1)^2 - (n_2)^2}\right]^2}{2} \tag{14-6}$$

where n_1 = index of refraction of the core
n_2 = index of refraction of the cladding
d = core diameter (m)
λ = wavelength (m)
f = number of propagating modes

Example 7

Using equation (14-6), compute the number of transmission modes for a light ray transmitted into a multimode step-index fiber having a core diameter of 50 μm, a core index of 1.60, and a cladding index of 1.584. The wavelength of the light ray is 1300 nm.

Solution: $f = \dfrac{\left[\dfrac{\pi d}{\lambda}\sqrt{(n_1)^2 - (n_2)^2}\right]^2}{2} = \dfrac{\left[\dfrac{\pi(50 \times 10^{-6})}{1300 \times 10^{-9}}\sqrt{(1.6)^2 - (1.584)^2}\right]^2}{2} = 372$

Let us now consider the light source shown in Figure 14-11. Rays of light corresponding to electromagnetic waves are incident on the surface end of a multimode step-index fiber. There are several modes in which the rays travel through the fiber. Ray A does not penetrate the core or the cladding and is lost in the surrounding air. Ray B is incident on the cladding. Notice the refraction of the ray is toward the normal due to the higher index of refraction of the cladding material over air. These rays are lost in the cladding through absorption, which will be discussed later. Ray C penetrates the core of the fiber and immediately bends toward the normal. However, as it strikes the core–cladding interface, its angle of incidence does not exceed the critical angle. Consequently, ray C is lost in the cladding. Ray D strikes the interface at an angle exceeding the critical angle and therefore experiences total internal reflection. Ray E propagates directly through the center of the core.

Fibers are further classified by the refractive index profile of their core, which can be either *step index* or *graded index*. Figure 14-12 illustrates these profiles for the three main types of fibers: *multimode step-index fiber, single-mode step-index fiber,* and *multimode graded-index fiber.* These are described next.

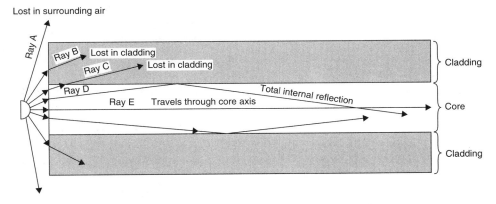

Figure 14-11 Transmission modes shown for various rays of light.

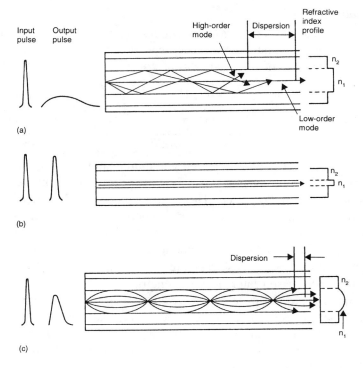

Figure 14-12 Three major classifications of optical fibers and their index profiles: (a) multimode step index; (b) single-mode step index; (c) multimode graded index. (Courtesy of AMP Incorporated.)

14.5.5.1 Multimode step-index fiber. Multimode step-index fiber is useful in local applications that do not require enormous transmission speed. Core diameters range from 50 to 1000 μm. This relatively large core size, as shown in Figure 14-12a, supports many propagation modes and permits the use of simple and inexpensive LED transmitters and PIN diode receivers.

The index of refraction is manufactured uniformly throughout the core of multimode step-index fiber, as shown by its index profile in Figure 14-12a. Light rays entering the fiber and exceeding the critical angle will bounce back and forth to the end of the fiber. This abrupt change in index of refraction between the core and the cladding has resulted in the name *step-index* or *multimode step-index fiber.* Each ray will obey the law stating that the angle of incidence is equal to the angle of reflection. Notice that light rays exhibiting the steepest angles relative to the central axis of the core will have a longer distance to travel as they propagate to the end of the fiber, consequently taking more time to reach the receiving element. These rays may be reflected or bounced back and forth against the cladding several thousands of times before they reach the end of the fiber. Rays that have a shallower angle have less distance to travel and therefore arrive at the receiver much sooner. The effect of this on a light pulse is distortion. Transmitted light pulses will broaden as they reach their destination. This is referred to as *pulse spreading* or *modal dispersion* due to the various propagation modes that the light rays take. The greater the pulse spreading, the farther the transmitted pulses must be separated from each other, and therefore the less information that can be transmitted per unit time.

Good attenuation and bandwidth performance are obtained in the 820-nm region and even better performance in the 1300-nm range for multimode step-index fiber.

14.5.5.2 Single-mode step-index fiber.

For single-mode step-index fiber, the core of the fiber is manufactured substantially smaller relative to multimode fiber. In addition, the index of refraction of the core is further reduced, thus increasing the critical angle or decreasing the angle at which the ray must penetrate the core with respect to its central axis. The intent is to permit light rays to propagate in one mode only: through the central axis of the fiber, as illustrated in Figure 14-12b. The advantage to this is that all light rays travel the same path and therefore take the same length of time to propagate to the end of the fiber. Modal dispersion is minimized and higher transmission speeds are attained.

Today's single-mode fiber dominates the telecommunications industry. Although it is costly, it offers the best performance in terms of information capacity (bandwidth) and transmission distance. The disadvantage, however, is that the small core diameter tightens the requirements for coupling light energy into the fiber. A laser is typically used as the light source. Another disadvantage with single-mode fiber is the task of splicing and terminating the fiber. Alignment of the fiber requires precise control, as we will see.

14.5.5.3 Multimode graded-index fiber.

To reduce the amount of pulse spreading or modal dispersion arising from various propagation delay times associated with multimode step-index fiber, *multimode graded-index* fiber is used. This type of fiber serves as an intermediary between single-mode and multimode step-index fiber in terms of cost and performance. Multimode graded-index fiber is characterized by its core having an index of refraction that is *graded* from its center out to the cladding interface; that is, the index of refraction is highest at the center and gradually tapers off toward the perimeter of the core. A bending effect is produced on light rays as they deviate from the central axis of the fiber, as illustrated in Figure 14-12c.

For simplicity, consider the two rays shown in Figure 14-13. Ray 1 propagates directly through the central axis of the core, whereas ray 2 travels in a sinu-

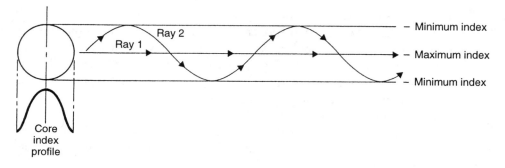

Figure 14-13 Multimode graded-index fiber. Ray 1 travels much less distance than ray 2. However, ray 2 increases in velocity as it propagates farther away from the core. This is caused by the graded index of the fiber. Ray 2 makes up in distance by its increase in velocity. Pulse spreading is minimized.

soidal manner. Clearly, ray 2 has a greater distance to travel to the end of the core than ray 1. However, since the index of refraction is constantly reduced (roughly parabolically) from the center of the core out, ray 2 makes up for distance by its increase in velocity as it deviates from the central axis. The net effect is that light rays take the same amount of time to propagate to the end of the fiber regardless of the path they take. Modal dispersion is considerably reduced.

Unfortunately, graded-index fiber is not without its trade-offs. Optical sources of light, such as lasers and especially LEDs, emit a range of frequencies and consequently a range of wavelengths. Different wavelengths travel at different velocities. The differences in refractive indices at different wavelengths cause the spreading of a light pulse as it travels to its destination. This is called *chromatic dispersion*. Chromatic dispersion can be reduced by using lasers with very narrow spectral widths or by using step-index fiber.

14.6 FIBER-OPTIC CABLE CONSTRUCTION

Figure 14-14 illustrates the basic construction of a typical fiber-optic cable. The fiber itself is generally regarded as the core and its cladding. The material composition of these two layers can be any of the following:

— Glass cladding and glass core
— Plastic cladding and glass core
— Plastic cladding and plastic core

Surrounding the fiber's cladding is a *coating* that is typically applied to seal and preserve the fiber's strength and attenuation characteristics. Coating materi-

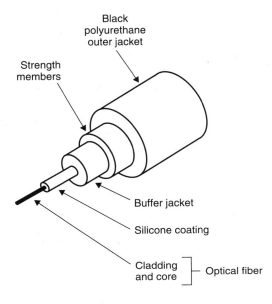

Black
polyurethane
outer jacket

Strength
members

Buffer jacket

Silicone coating

Cladding
and core — Optical fiber

Figure 14-14 Basic construction of a fiber-optic cable. (Courtesy of Hewlett-Packard Optical Communication Division.)

als include lacquer, silicones, and acrylates. Some fibers have more than one protective coating system to ensure that there are no changes in characteristics if the fiber is exposed to extreme temperature variations.

Sealing the fiber with a coating system also protects the fiber from moisture. A phenomenon called *stress corrosion* or *static fatigue* may occur if the glass fiber is exposed to humidity. Silicon dioxide crystals will interact with the moisture and cause bonds to break down. Spontaneous fractures can occur over a prolonged period of time.

The coating is further surrounded by a *buffer jacket*. The buffer provides additional protection from abrasion and shock. Some fibers have *tightly buffered jackets* and some have *loosely buffered jackets*.

Surrounding the buffer jacket is the *strength member*. As its name implies, it gives the fiber strength in terms of pulling. Various methods used to strengthen a fiber-optic cable are considered next.

Finally, the *outer jacket* is used to contain the fiber and its surrounding layers. The outer jacket is usually made of polyurethane material.

14.6.1 Strength and Protection

One disadvantage of fiber is its lack of pulling strength, which may be on the order of 1 lb. (This is much larger than copper and comparable to steel having the same diameter.) Although this may seem relatively large for a strand of material having a diameter the thickness of a human hair strand, it is not sufficient for most installation requirements. The fiber must therefore be reinforced with strengthening material so that it can withstand the mechanical stresses from being pulled through ducts, hung on telephone poles, and buried underground.

Several materials are used by manufacturers to strengthen and protect the fiber from abrasion and environmental stress. The extent to which the fiber is protected depends on the application. Materials commonly used are listed next.

— Steel
— Fiberglass
— Plastic
— FR-PVC (flame-retardant polyvinyl chloride)
— Kevlar yarn
— Paper

Figure 14-15 depicts how some of these materials are used in relationship to the fiber. Notice that tensile strengths in excess of several hundreds of pounds can be achieved by reinforcing the fiber with strengthening materials.

14.7 ATTENUATION LOSSES IN OPTICAL FIBERS

One of the most important performance specifications for an optical fiber is its attenuation rating. *Attenuation* is the term used by fiber manufacturers to denote the decrease in optical power from one point to another. This loss per unit length is expressed logarithmically in decibels per kilometer at a given wavelength and is given by the standard power equation:

$$\text{loss} = 10 \log \frac{P_{\text{out}}}{P_{\text{in}}} \quad \text{dB} \tag{14-7}$$

A 10-dB/km loss, for example, corresponds to 10% of the light making it to the end of a 1-km length of fiber. If the same fiber were doubled in length, 1% of the transmitted light (20 dB) would emerge from the end of the fiber. By today's standards, this amount of attenuation is enormous. It is not uncommon for optical fibers used for telecommunication systems to have an attenuation specification of less than 0.5 dB/km.

Attenuation losses in a fiber are wavelength dependent, as shown in Figure 14-16. Notice that there are peaks and troughs in the curve, meaning that there are optimum wavelengths that transmitters and receivers must be tailored to in order to minimize attenuation losses. For example, there is significantly less at-

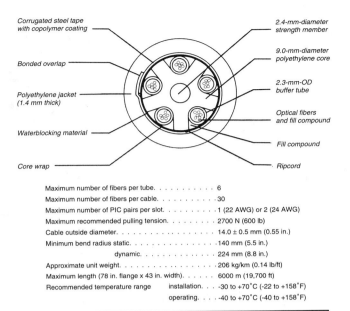

- Corrugated steel tape with copolymer coating
- Bonded overlap
- Polyethylene jacket (1.4 mm thick)
- Waterblocking material
- Core wrap
- 2.4-mm-diameter strength member
- 9.0-mm-diameter polyethylene core
- 2.3-mm-OD buffer tube
- Optical fibers and fill compound
- Fill compound
- Ripcord

Maximum number of fibers per tube. 6
Maximum number of fibers per cable. 30
Maximum number of PIC pairs per slot. . . . . 1 (22 AWG) or 2 (24 AWG)
Maximum recommended pulling tension. . . . . 2700 N (600 lb)
Cable outside diameter. . . . . 14.0 ± 0.5 mm (0.55 in.)
Minimum bend radius static. . . . . 140 mm (5.5 in.)
                    dynamic. . . . . 224 mm (8.8 in.)
Approximate unit weight. . . . . 206 kg/km (0.14 lb/ft)
Maximum length (78 in. flange x 43 in. width). . . . . 6000 m (19,700 ft)
Recommended temperature range   installation. . . -30 to +70°C (-22 to +158°F)
                                 operating. . . . -40 to +70°C (-40 to +158°F)

TubeStar Design

Northern Telecom TubeStar Optical Fiber Cable combines the strength, durability and ruggedness of Northern Telecom's proven slotted core design with the convenience of a buffer tube cable. It features a proprietary core design with a central strength member surrounded by an oscillating slotted polyethylene core. The central strength member of steel or fiberglass rod provides for maximum pulling tensions, equal or superior to industry standards. The oscillating slots surrounding the central axis eliminate bending effects on fibers, and minimize the effects of temperature variations and mechanical loads. Attenuation remains constant regardless of temperature.

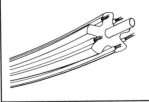

Features

- Enhanced tube cable design with improved fiber protection provided by a star core.
- Fibers are isolated from cable bending, twisting and temperature variations resulting in reliable consistent performance.
- Superior crush and impact resistant cable suitable for all applications.
- High grade flexible buffer tubes prevent crimping and fiber damage during cable preparation and splicing.
- Decisive PIC color coding for both tubes and fibers provide for easy identification and segregation of working and protection fibers.
- Popular steelpeth and polysteelpeth cable jackets of composite heat bonded steel and polyethylene are both rugged and flexible.
- Medium density outer jacket provides maximum resistance to external forces encountered in harsh environmental terrains and underground duct structures.
- Designed to meet BellCore TR-TSY-000020, EIA mechanical tests and REA requirements.

(a)

Figure 14-15 Various optical fiber cable configurations: (a) Northern Telecom's TubeStar design can be used for underground ducts, aerial, direct burial, and submarine applications (courtesy of Northern Telecom Inc.); (b) Remfo's series 10 light-duty construction indoor simplex cable (courtesy of Remfo Fiber Optic Division); (c) Remfo's series 12 indoor/outdoor multifiber cable. (Courtesy of Remfo Fiber Optic Division.)

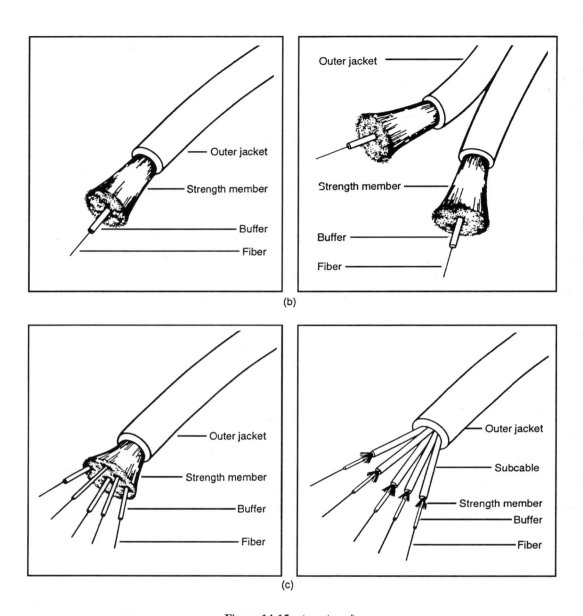

(b)

(c)

Figure 14-15 (*continued*)

tenuation for glass fibers at 820 nm than there is at 1000 nm, and even less at 1300 and 1550 nm. Most fiber-optic components are therefore designed to operate in these regions. Table 14-3 lists the attenuation characteristics for various classifications of fiber.

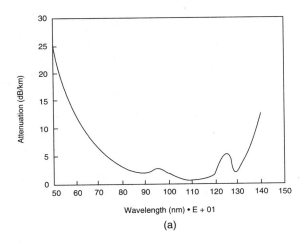

(a)

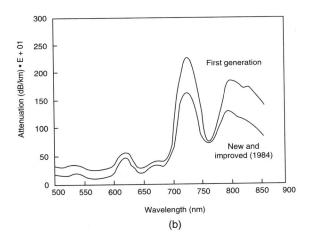

(b)

Figure 14-16 Attenuation loss as a function of wavelength for (a) glass fibers and (b) plastic fibers. (Courtesy of Hewlett-Packard Optical Communication Division.)

14.7.1 Absorption Loss

The intrinsic impurities of fiber only amount to a few parts per billion. These impurities, however, *absorb* photons of light at specific wavelengths, depending on the type of impurity. Molecular agitation resulting from absorption produces heat that is lost in its surroundings.

Figure 14-17 depicts how absorption losses in fiber are wavelength dependent. At about 1.4 μm, a very minute residual of water (OH) in the fiber absorbs a significant portion of the light. Above 1.6 μm, glass is no longer transparent to infrared light. Instead, the light is absorbed by the glass and converted to heat.

TABLE 14-3 Attenuation Characteristics for Various Classifications of Fiber

Type	Core diameter (μm)	Cladding diameter (μm)	Buffer diameter (μm)	NA	Bandwidth (MHz-km)	Attenuation (dB/km)
Single mode	8	125	250		6 ps/km[a]	.5 @ 1300 nm
	5	125	250		4 ps/km[a]	.4 @ 1300 nm
Graded index	50	125	250	.20	400	4 @ 850 nm
	63	125	130	.29	250	7 @ 850 nm
	85	125	250	.26	200	6 @ 850 nm
	100	140	250	.30	20	5 @ 850 nm
Step index	200	380	600	.27	25	6 @ 850 nm
	300	440	650	.27	20	6 @ 850 nm
PCS	200	350	–	.30	20	10 @ 790 nm
	400	550	–	.30	15	10 @ 790 nm
	600	900	–	.40	20	6 @ 790 nm
Plastic		750	–	.50	20	400 @ 650 nm
		1000	–	.50	20	400 @ 650 nm

[a]Dispersion per nanometer of source width.

14.7.2 Rayleigh Scattering

By 1970, the transparency of optical fibers was so pure (99.9999%) that the presence of impurities no longer set the lower limits of attenuation. At that time, this loss was about 1.5 dB/km for light transmitted in the 800-nm region. A mechanism called *Rayleigh scattering* prevents any further improvement in attenuation loss. Rayleigh scattering is caused by micro irregularities in the random molecular structure of glass. These irregularities are formed as the fiber cools from a molten state. Normally, electrons in glass molecules interact with transmitted light by absorbing and reradiating light at the same wavelength. A portion of the light, however, strikes these micro irregularities and becomes scattered in all directions of the fiber, some of which is lost in the cladding. Consequently, the intensity of the beam is diminished.

Rayleigh scattering is wavelength dependent and decreases as the fourth power of increasing wavelength. This is illustrated in Figure 14-17. At about 800 nm, Rayleigh scattering sets the lower limits of attenuation to approximately 1.6 dB/km. By doubling the wavelength of light from 800 to 1600 nm, one-sixteenth ($1/2^4$) the amount of Rayleigh scattering loss occurs. Attenuation loss can be reduced to less than a few tenths of a decibel per kilometer.

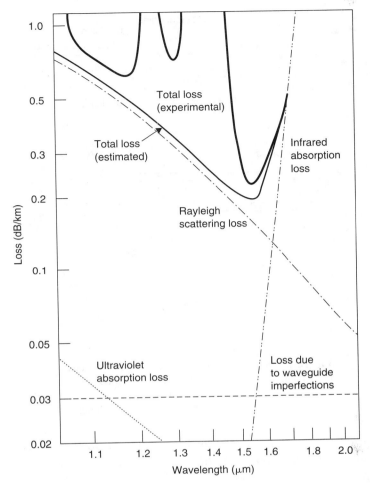

Figure 14-17 Factors contributing to fiber attenuation in the 1500-nm region. (Courtesy of Corning Glass Works.)

14.7.3 Radiation Losses

A phenomenon called *microbending* can cause radiation losses in optical fibers in excess of its intrinsic losses. Microbends are miniature bends and geometric imperfections along the axis of the fiber that occur during the manufacturing or installation of the fiber. Mechanical stress such as pressure, tension, and twist can cause microbending (Figure 14-18a). This geometric imperfection causes light to get coupled to various unguided electromagnetic modes that radiate and escape the fiber. With a good coating system and jacketing around the fiber, microbend-

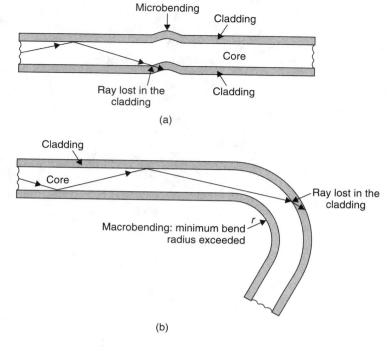

Figure 14-18 Radiation losses occur in microbends or macrobends: (a) microbending occurs when there are miniature bends and geometric imperfections along the axis of the fiber; (b) macrobending occurs when a fiber is bent to a radius that is less than the fiber's minimum bend radius (typically 10 to 20 cm).

ing losses normally contribute less than 20% to the overall attenuation. For a fiber-optic cable having an attenuation rating of 0.5 dB/km, microbending losses typically contribute less than 0.1 dB/km.

Fiber curvatures on a larger scale can also cause radiation losses. This is referred to as *macrobending*. Macrobending occurs when a fiber is bent to a radius less than the fiber's *minimum bend radius* specification, as shown in Figure 14-18b. This may be on the order of 10 to 20 cm.

14.8 NUMERICAL APERTURE

An important characteristic of a fiber is its *numerical aperture* (NA). NA characterizes a fiber's light-gathering capability. Mathematically, it is defined as the sine of half the angle of a fiber's light *acceptance cone,* which is shown in Figure 14-19. Equations (14-8) and (14-9) are used to compute NA.

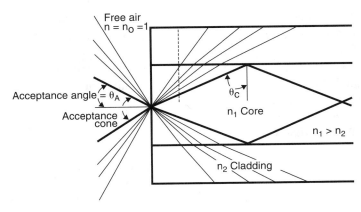

Numerical aperture = sin θ$_A$
NA is always less than 1

Free air
$n = n_O = 1$

Acceptance angle = θ$_A$

Acceptance cone

θ$_C$

n_1 Core

$n_1 > n_2$

n_2 Cladding

Figure 14-19 Acceptance cone for measuring numerical aperture (NA). (Courtesy of Hewlett-Packard Optical Communication Division.)

$$NA = \sin \theta_A \qquad (14\text{-}8)$$

Also,

$$NA = \sqrt{n_1^2 - n_2^2} \qquad (14\text{-}9)$$

where n_1 is the refractive index of the core and n_2 is the refractive index of the cladding.

Equations (14-8) and (14-9) are straightforward for multimode step-index fiber. For graded-index fiber, however, n_1 depends on the core profile. The largest acceptance angle is measured at the core center. Therefore, the index of refraction at the core center must be used to compute NA. Typical values for NA are 0.25 to 0.4 for multimode step-index fiber and 0.2 to 0.3 for multimode graded-index fiber (see Table 14-3).

Although a fiber having a large NA gathers more light, a greater amount of modal dispersion occurs due to the large number of propagation modes. Conversely, a fiber having a low NA makes it more difficult to couple light into the fiber, but offers greater propagation efficiency since there are fewer modes.

14.9 FIBER-OPTIC CONNECTIONS

Once a fiber-optic cable has been manufactured, a connection must be made. This connection may involve a splice to lengthen or repair the cable or to mount a connector for mating purposes. A splice is used in cases where the connection is permanent, whereas a connector is used for temporary connections. This may involve another fiber's connector, a transmitter, or a receiver. Whether the connection is made by a splice or a connector, the two ends must be physically aligned with enough precision that an appreciable amount of light energy is coupled from one fiber end to the other. Unlike copper, fiber is so fine in diameter that several complexities arise in making the connection. Special tools, training, manual dexterity, and practice are essential requirements for the technician to make good connections.

14.9.1 Splice Connection

Two methods are used for splicing fiber ends together: the *mechanical splice* and the *fusion splice*. In a mechanical splice, two ends of fibers are brought together, aligned with a mechanical fixture, and glued or crimped together. In a fusion splice, alignment is performed under a microscope. An electric arc is drawn that melts the two glass ends together and forms a strong bond. In general, the fusion splice offers better performance specifications in terms of splice loss. Both methods are commonly used in the field, and one method may not necessarily be better than the other; that is, some technicians feel more comfortable with one method over the other. The more sophisticated mechanical and fusion splices often include prealignment under a microscope and final precision alignment performed automatically.

14.9.1.1 Fiber preparation. Cable preparation is necessary prior to splicing. Cable manufacturer's procedures should be carefully followed for each cable design. Fiber ends are typically stripped of their plastic jacket and strength member and *cleaved* to a 90° angle. It may be necessary to remove any coating material. A number of methods are used to do this, such as mechanical stripping tools, thermal stripping equipment, and chemical strippers.

The goal of cleaving is to produce a flat, smooth, perpendicular fiber end face.[†] The scribe-and-break method is generally used to cleave fibers. Both manual and automated tools are available. The quality of the cleave is one of the most important factors in producing high-quality, low-loss fusion or mechanical splices. Whatever the method used, the cleave must be a clean break with no burrs or chips. The fiber end angle should be less than 1°. The ends are then polished and cleaned with a cleaning agent such as isopropyl alcohol or freon.

14.9.1.2 Mechanical splice. Dozens of mechanical splicing techniques are used to join both single-mode and multimode fiber ends. Each has its advantages and disadvantages. To achieve performances comparable to fusion splic-

---

[†]Corning Glass Works, *Application Note 113.*

ing, large, bulky, and expensive equipment is often used. This may include fixtures with built-in microscopes, micrometers, digital alignment test circuitry, and ultraviolet curing lamps. A considerable amount of operator training and judgment is required. The current trend by many manufacturers is to produce inexpensive mechanical splicing kits that make splicing efficient and easy to perform with little training. Two of the more popular types of mechanical splices are the *V-groove splice* and the *tube splice*.

14.9.1.3 V-groove splice. The V-groove splice, shown in Figure 14-20, uses a block or substrate with a precisely cut groove through its center. For multiple splices, in the case of ribbon cables, several V-grooves are manufactured into the substrate for alignment. The fibers are placed in the grooves and butted

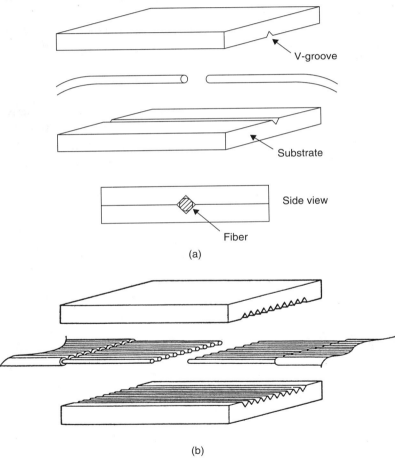

Figure 14-20 (a) V-groove splice; (b) multiple V-groove splice for ribbon cable.

up against each other. Index matching glue is often used to bond the ends together and reduce reflection losses. Assuming the core and cladding of the fibers are the same diameter, the groove will cause them to align with each other on the same axis. A top matching plate is placed over the spliced fiber and secured. The entire splice is placed into a box called a *splice enclosure* which is used for environmental protection and strain relief.

14.9.1.4 Tube splice. Fibers can also be aligned and spliced together through the use of capillary tubes. The tubes are machined to the diameter of the fiber and funneled at each end to allow the fibers to be easily inserted. For single-mode fiber, the capillary is typically round-holed, whereas for multimode fiber the capillary is triangular-holed. Figure 14-21 illustrates various types of tube splices.

Fiber ends are initially prepared in the manner described above. Index matching fluid is then inserted into the tube and the fibers are pushed to the center of the tube against each other. The splice is exposed to ultraviolet light, which is used to polymerize or cure the epoxy or index matching fluid, thus forming a strong bond.

Some of the latest tube splice designs have become extremely popular due to their simplicity. Many do not require epoxy, thus eliminating the need for ultraviolet curing equipment. Instead, the fibers are held in place by a mechanical fixture or a crimp, as shown in Figure 14-22. Losses tend to be slightly higher for these types of designs (0.1 to 0.2 dB); however, they are extremely reliable, economical, and easy to install.

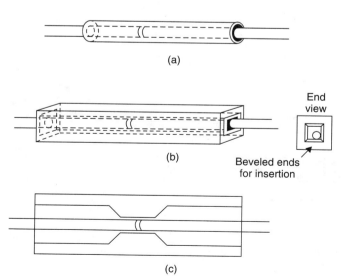

(a)

End view

(b)

Beveled ends for insertion

(c)

Figure 14-21 Various tube splices: (a) snug tube splice; (b) loose tube splice; (c) transparent capillary tube.

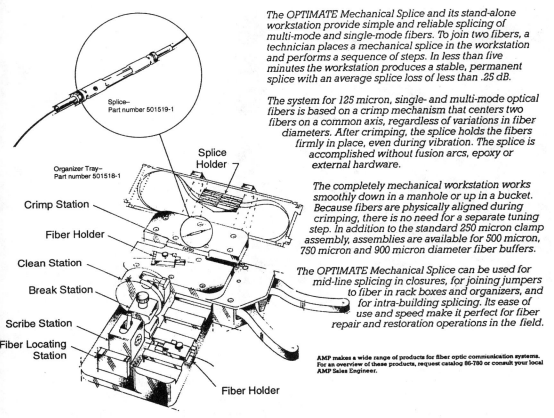

The OPTIMATE Mechanical Splice and its stand-alone workstation provide simple and reliable splicing of multi-mode and single-mode fibers. To join two fibers, a technician places a mechanical splice in the workstation and performs a sequence of steps. In less than five minutes the workstation produces a stable, permanent splice with an average splice loss of less than .25 dB.

The system for 125 micron, single- and multi-mode optical fibers is based on a crimp mechanism that centers two fibers on a common axis, regardless of variations in fiber diameters. After crimping, the splice holds the fibers firmly in place, even during vibration. The splice is accomplished without fusion arcs, epoxy or external hardware.

The completely mechanical workstation works smoothly down in a manhole or up in a bucket. Because fibers are physically aligned during crimping, there is no need for a separate tuning step. In addition to the standard 250 micron clamp assembly, assemblies are available for 500 micron, 750 micron and 900 micron diameter fiber buffers.

The OPTIMATE Mechanical Splice can be used for mid-line splicing in closures, for joining jumpers to fiber in rack boxes and organizers, and for intra-building splicing. Its ease of use and speed make it perfect for fiber repair and restoration operations in the field.

AMP makes a wide range of products for fiber optic communication systems. For an overview of these products, request catalog 86-780 or consult your local AMP Sales Engineer.

Splice–
Part number 501519-1

Organizer Tray–
Part number 501518-1

Splice Holder

Crimp Station

Fiber Holder

Clean Station

Break Station

Scribe Station

Fiber Locating Station

Fiber Holder

Figure 14-22 Tube splice held together by a crimp. (Courtesy of AMP Incorporated.)

14.9.2 Fusion Splice

The fusion splice uses an electric arc to fuse or weld two fiber ends together. The resulting splice is usually stronger than that of unspliced fiber. Losses under 0.01 dB can be achieved with a fusion splice. The trade-off, however, is that operator training and judgement are relatively high compared to that required for the mechanical splice. In addition, equipment is bulky and often very expensive. Sophisticated microprocessor controllers are used to eliminate operator judgment by automating most of the process. Figure 14-23 illustrates a fusion splicer.

14.9.3 Connectors

Connectors are necessary in fiber optics to interface the fiber to transmitting and receiving components. They are also used for interfacing to other optical fibers. There are a myriad of connectors on the market making it difficult to discuss any

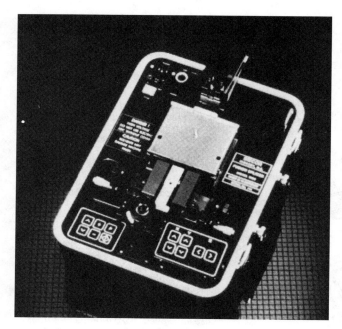

Figure 14-23 Fusion splicer manufactured by Siecor Corporation, model M67. (Courtesy of Siecor Corporation.)

one type. Unfortunately, connector standards for the fiber-optic community have been slow in the making. For the most part, manufacturers have tailored their designs toward existing mechanical standards. Among the most popular connector types[†] are

— USA: BICONIC, ST, SMA
— Japan: FC, D4, PC
— Europe: DIN/IEC, RADIALL, STRATOS

Today, most connectors are based on the physical contact of two well-cleaved fibers in order to allow the direct transition of optical power between each other. Figure 14-24 illustrates a popular SMA-type connector assembly. The following set of procedures are typically used to install a connector assembly to a fiber end.

1. Strip the jacket and coating until only core and cladding are left.
2. Center and fix the uncoated fiber in the ferrule.

[†]*Fiber Optics Handbook,* Hewlett-Packard, © 1988.

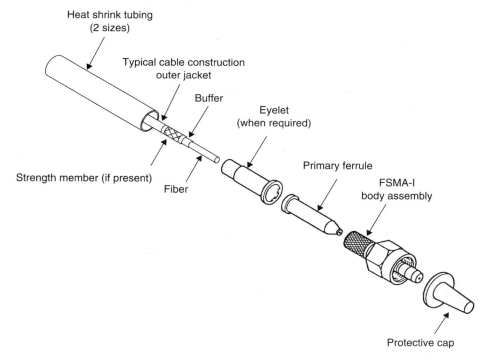

Figure 14-24 SMA-style connector. (Courtesy of AMP Incorporated.)

3. Grind and polish the ends of the fiber and ferrule.
4. Clean the fiber ends with a cleaning agent such as freon or isopropyl alcohol.
5. Assemble the connector housing.

14.9.4 Insertion Losses for Connectors and Splices

Connecting losses fall into two categories: *intrinsic* and *extrinsic*. The two combined make up the total insertion loss of a connection. They are described next.

Intrinsic Losses. These types of losses are due to factors that are beyond the control of the user. They include variations in:

— Core diameter
— Numerical aperture
— Index profile

— Core/cladding eccentricity

— Core concentricity

These factors can be expected to vary at random, even between fibers with the same specifications or from the same cable spool. They are influenced by the manufacturing process and quality control rather than the nature of the connection.

Extrinsic Losses.   Extrinsic losses are those resulting from the splicing and connector assembly process. They include the following errors:

— Mechanical offsets between fiber ends

— Contaminants between fiber ends

— Improper fusion, bonding, and crimping methods

— End finishes

Most connection losses can be attributed to extrinsic losses caused by the user and the equipment being used to perform the splice. They can be reduced or eliminated by ensuring that equipment is maintained and manufacturing procedures are followed.

14.9.4.1 Mechanical offsets.   Regardless of the method used to splice or connect fibers, in most cases, major extrinsic losses can be attributed to mechanical offsets, which are classified as follows:

— Angular misalignment

— Lateral misalignment

— End separation

— End face roughness

Figure 14-25 illustrates these types of losses. In each case, a portion of the light escapes from its normal path and contributes to the total insertion loss.

14.9.4.2 Losses due to reflections.   Whenever light enters glass from air, it can be shown mathematically that about 4% of the light is reflected back into the air. The same effect occurs in the reverse direction: when light exits glass into air, 4% of it is reflected back into the glass. We will not dwell on the mathematics behind this, but rather on the importance of correcting for some of this loss. Since light must exit and reenter the surface of the fiber ends in a connection, there is an 8% overall reduction, which amounts to approximately 0.4-dB

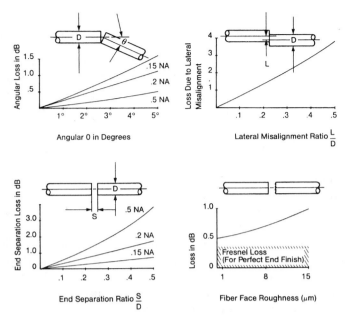

Figure 14-25 Four major extrinsic losses that can occur in a fiber connection. (Courtesy of AMP Incorporated.)

loss. For splices, this reflective loss can be reduced by applying an index matching material between the fiber ends. This material is usually a transparent glue or epoxy that has an index of refraction of approximately 1.5, which is equivalent to that of the fiber. The glue also serves the purpose of bonding the fiber ends together. We do not do this between connectors since periodic disconnection is required.

14.10 LIGHT-EMITTING DEVICES

Several devices are emitters of light, both natural and artificial. Few of these devices, however, are suitable for fiber-optic transmitters. What we are interested in is a light source that meets the following requirements:

— The light source must be able to turn on and off several tens of millions and even billions of times per second.

— The light source must be able to emit a wavelength that is transparent to the fiber.

— The light source must be efficient in terms of coupling light energy into the fiber.

— The optical power emitted must be sufficient enough to transmit through optical fibers.

— Temperature variations should not affect the perfomance of the light source.

— The cost of manufacturing the light source must be relatively inexpensive.

Two commonly used devices that satisfy the above requirements are: the LED (light-emitting diode) and the ILD (injection laser diode). Both are semiconductor devices. Each has advantages over the other depending on the application.

14.10.1 LED versus ILD

The major difference between the LED and the ILD is the manner in which light is emitted from each source. The LED is an *incoherent* light source that emits light in a *disorderly* way, as compared to the ILD, which is characterized as a *coherent* light source that emits light in a very *orderly way*. The ILD can therefore launch a much greater percentage of its light into a fiber than an LED. Figure 14-26 illustrates the differences in radiation patterns. Both devices are extremely rugged, reliable, and small in size.

In terms of spectral purity, the LED's half-power spectral width is approximately 50 nm, whereas the ILD's spectral width is only a few nanometers. This is shown in Figure 14-27. Ideally, a single spectral line is desirable. As the spectral width of the emitter increases, attenuation and pulse dispersion increase. The spectral purity for the ILD and its ability to couple much more power into a fiber make it better suited for long-distance telecommunications links. In addition, the injection laser can be turned on and off at much higher rates than an LED. The drawback, however, is its cost, which may approach several hundreds of dollars as compared to a few dollars for LEDs in large quantities. Table 14-4 lists the differences in operating characteristics between the LED and the ILD.

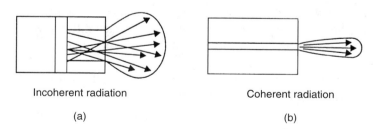

Incoherent radiation

Coherent radiation

(a)

(b)

Figure 14-26   Radiation patterns for the (a) LED and (b) ILD.

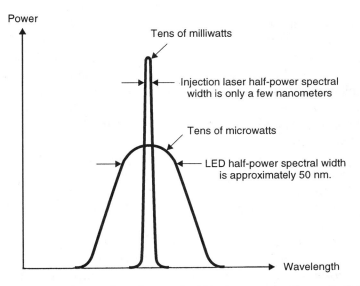

Figure 14-27 Comparison of spectral widths between the fiber-optic light-emitting diode (LED) and injection laser diode (ILD).

14.10.2 Light-Emitting Diode

Because of its simplicity and cost, the LED is, by far, the most widely used light-emitting device in the fiber-optic industry. Most of us are familiar with LEDs capable of emitting visible light. They are used in calculators, watches, and a multitude of other visible displays. The LEDs used in fiber optics operate on the same principle. By passing current through the LED's PN junction, recombination occurs between holes and electrons. This causes particles of light energy

TABLE 14-4 Typical Source Characteristics for LEDs and ILDs

Type	Output power (μW)	Peak wavelength (nm)	Spectral width (nm)	Rise time (ns)
LED	250	820	35	12
	700	820	35	6
	1500	820	35	6
Laser	4000	820	4	1
	6000	1300	2	1

called photons to be released or emitted. The wavelength of these photons is a function of the crystal structure and composition of the material.

Extensive research in material science, physics, and chemistry has made it possible to grow highly reliable crystals used in the manufacturing of fiber-optic LEDs. These LEDs are designed to emit light in the infrared region for reasons explained earlier. Various semiconductor materials are used to achieve this. Pure *gallium-arsenide* (GaAs) emits light at a wavelength of about 900 nm. By adding a mixture of 10% aluminum (Al) to 90% GaAs, *gallium-aluminum-arsenide* (GaAlAs) is formed, which emits light at a wavelength of 820 nm. Recall that this is one of the optimum wavelengths for fiber-optic transmission. By tailoring the amount of aluminum mixed with GaAs, wavelengths ranging from 800 to 900 nm can be obtained.

To take advantage of the reduced attenuation losses at longer wavelengths, it is necessary to include even more exotic materials. For wavelengths in the range 1000 to 1550 nm, a combination of four elements is typically used: *indium, gallium, arsenic,* and *phosphorus.* These devices are commonly referred to as *quaternary devices.* Combining these four elements produces the compound *indium-gallium-arsenide-phosphide* (InGaAsP). By tailoring the mixture of these elements, 1300- and 1550-nm emissions are possible. Figure 14-28 illustrates the construction of various fiber-optic LEDs. Table 14-5 lists their differences.

TABLE 14-5   LED as a Fiber-Optic Source

	Material			
	GaAsP	GaAlAs		InGaAsP
Device	Surface LED	Surface LED	Etched-well LED	Transparent substrate LED
1. Wavelength (nm)	665	820	820	1300
2. Spectral line width (nm)	30	40	40	120
3. Ext. quantum efficiency, $\eta(\%)$	0.1–0.2	1.5–2.0	4	4
4. Response time (nsec)	<40	5–15	<15	<10
5. Relative cost	Low	Medium	Higher	Highest

Source: Hewlett-Packard Optical Communication Division.

14.10.3 Injection Laser Diode (ILD)

The term *laser* is an acronym for *light amplification by stimulated emissions of radiation.* There are many types of lasers on the market. They are constructed of gases, liquids, and solids. For many of us, when we think of a laser, what comes

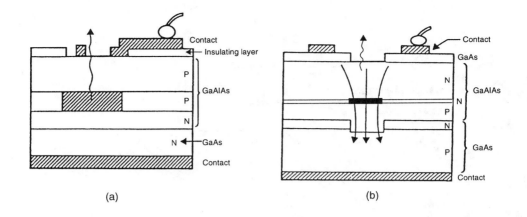

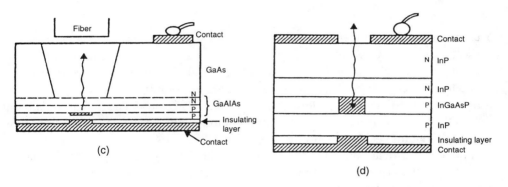

Figure 14-28 Construction of various LEDs: (a) S-PUP (Silox P side up) emitter, 820 nm (surface LED); (b) enhanced emitter design, 820 nm; (c) etched-well emitter, 820 nm; (d) 1300-nm emitter. (Courtesy of Hewlett-Packard Optical Comunication Division.)

to mind is a relatively large and sophisticated device that outputs a highly intense beam of visible light. Although this is in part true, the laser industry is currently devoting a great deal of effort toward the manufacture of miniature semiconductor laser diodes. Laser diodes are also called *injection laser diodes* (ILDs), because when current is injected across the PN junction, light is emitted. Figure 14-29 illustrates the construction of an ILD. ILDs are ideally suited for use within the fiber-optic industry due to their small size, reliability, and ruggedness. Various ILD package designs are shown in Figure 14-30.

The ILD is basically an oscillator. A cavity built into its layered semiconductor structure serves as a feedback mechanism to sustain oscillation. The amplification necessary for oscillation is produced by creating a condition in the cavity called a *population inversion*. A large density of holes and electrons in the

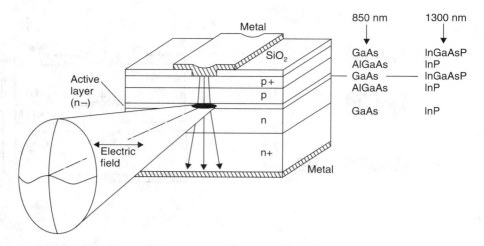

Figure 14-29 Construction of the injection laser diode. (Reprinted with permission from Hewlett-Packard.)

cavity of the ILD is waiting to recombine. By injecting a large density of current (holes and electrons) into the cavity, some holes and electrons in the cavity recombine and release photons of light. A population inversion occurs when a greater percentage of these holes and electrons combine to release photons of light instead of creating additional holes and electrons. The released photons stimulate other holes and electrons to recombine and release even more photons, thus producing gain or amplification. To achieve oscillation, the ends of the crystal structure are polished to a mirror finish. This provides a feedback mechanism which causes light to reflect back and forth in the cavity, stimulating other electrons and holes to recombine and release more photons. *Lasing* is said to occur.

14.11 LIGHT-RECEIVING DEVICES

The function of the fiber-optic receiving device is to convert light energy into electrical energy so that it may be amplified and processed back to its original state. A device commonly used for this purpose is the *photo diode* or *photodetector*. Photodetectors are made of semiconductor materials. Here again the composition of the structure determines the wavelengths that it is sensitive to.

Photodetectors use the reverse mechanism of transmitting devices. Instead of stimulated emissions of radiation, *absorption* of photons occur. When photons are absorbed, holes and electrons are created, thus producing current. This is known as the *photoelectric effect.*

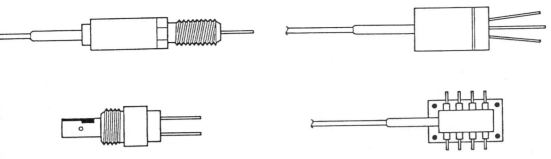

Figure 14-30   Typical injection laser diodes.

14.11.1 Receiver Sensitivity

Receiver sensitivity can be characterized by one of two parameters: *quantum efficiency* or *responsivity*. Both parameters are essentially the same. They are both a measure of an optical receiver's sensitivity to a particular wavelength.

   If each particle of light (photon) illuminating the surface of a photodetector were converted to a useful electron-hole pair, the quantum efficiency would be equal to 1, or unity. If 10% of the light were reflected off of its surface, the quantum efficiency would be equal to 90%, and so forth.

   From a system designer's point of view, a more practical way of interpreting a receiver's sensitivity is to consider the amount of optical power that is directly converted to current at a specific wavelength. Most manufacturers specify their receivers in this manner, which is called *responsivity*. Responsivity has the units of amperes per watt (A/W) or microamperes per microwatt ($\mu A/\mu W$). Typical values may range from 0.2 A/W to as high as 100 A/W.

14.11.2 PIN Photodiode

The most common photodetector used in fiber optics is the *PIN photodiode*. PIN is an acronym for P-type, intrinsic (I), N-type semiconductor material. The P and N regions shown in Figure 14-31 are heavily doped. The I region (shown as the depletion region) is a lightly doped (near intrinsic) N-type material. Its purpose is to increase the depletion and absorption regions when a reverse-biased potential is applied to the diode. In comparison to the LED, which emits light in the forward-biased condition, the PIN photodiode is used in the reverse-biased condition. Current does not flow unless light (photons) penetrates the depletion region. When this occurs, electrons are raised from their valence band to the conduction band, thus leaving excess holes. These current carriers will begin to flow due to the attractive force of the applied potential. The magnitude of the current is proportional to the intensity of the light. Typical values for responsivity are in the order of 0.5 A/W. Bias voltages may range from 5 to 20 V.

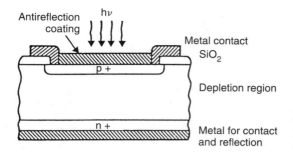

Figure 14-31   Structure of the PIN photodiode. (Reprinted with permission from Hewlett-Packard.)

The degree of absorption in a PIN photodiode depends on the wavelength and material of which the device is made. For example, if the diode is made of silicon, wavelengths of light in the range 800 to 900 nm will penetrate. PIN diodes made of indium-gallium-arsenide absorb light in the 1300 nm range. Sensitivity to longer wavelengths is achieved with indium-gallium-arsenide-phosphide. Figure 14-32 gives the dimensions and a cross-sectional view of a typical silicon PIN photodiode sensitive to light in the 800 nm range.

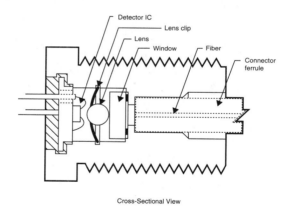

Cross-Sectional View

HFBR-2208 SMA Style Compatible

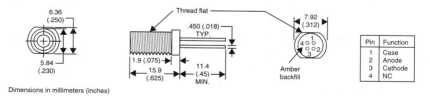

Figure 14-32   Hewlett-Packard's HFBR-2208 PIN photodiode receiver. (From *Hewlett-Packard Optoelectronics Designer's Catalog*, 1988–1989, p. 8-98.)

14.11.3 Avalanche Photodiode (APD)

Although the PIN photodiode is extremely well suited for most fiber-optic applications, its sensitivity to light (responsivity) is not as great as the *avalanche photodiode* (APD). Due to their inherent gain, typical values of responsivity for APDs may range from 5 A/W to as high as 100 A/W. This is considerably higher than the PIN photodiode, which makes it extremely attractive for fiber-optic communications receivers.

The APD functions in the same manner as the PIN photodiode, except that a larger reverse-biased potential is necessary. This is usually on the order of over 100 V. The APD is constructed in a manner that causes an avalanche condition to occur if a sufficiently large reverse-biased potential is applied to it. As the reverse bias increases, electron-hole pairs gain sufficient energy to create additional electron-hole pairs, thus creating additional ions (positive and negative charged particles). A multiplication or avalanche of carriers occurs. This effect is called *impact ionization,* which is considered to be an internal gain advantage over the PIN photodiode.

Although the APD is extremely responsive to light, it is not without its drawbacks. Unfortunately, temperature and bias stabilization are necessary with the APD, as both of these parameters influence its performance. Also, costs are considerably higher than for PIN photodiodes. In terms of size, the APD is similar to the PIN photodiode.

14.12 A LOW-COST FIBER-OPTIC LINK

There are several applications that do not require high-speed transmission over long-distances. For many of these applications, a low-cost fiber-optic link would be a solution. Figure 14-33 depicts Hewlett-Packard's low-cost *Versatile Link.*

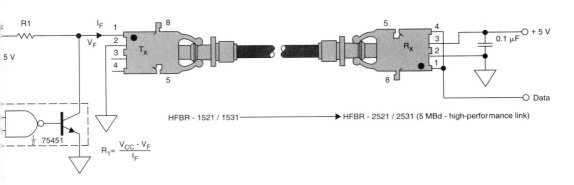

Figure 14-33 Hewlett-Packard's Versatile Link. (Courtesy of Hewlett-Packard Optical Communication Division.)

This link[†] will function at speeds ranging from dc to 5 Mbps. Extended distance links can be up to 82 m. This low-cost fiber-optic link is TTL and CMOS compatible.

PROBLEMS

1. Name at least six advantages of using fiber.
2. Name at least two disadvantages of using fiber.
3. What are the two classic theories of light?
4. Using equation (14-1), compute the energy of a single photon having a frequency of 3.65×10^{14} Hz.
5. Refer to Figure 14-4. In Problem 4, what portion of the electromagnetic spectrum does the light emitted by this photon fall under?
6. Using equation (14-2), compute the wavelength of the photon in Problem 4.
7. Compute the wavelength for the following frequencies.
   (a) 1.6 MHz
   (b) 88 MHz
   (c) 4 GHz
   (d) 6 GHz
   (e) 1.94×10^{14} Hz
8. Refer to Figure 14-4. What are the wavelengths associated with:
   (a) Infrared light?
   (b) Visible light?
   (c) Ultraviolet light?
9. Convert 1300 nanometers to inches.
10. What is the velocity of light in free space?
11. The speed of light in a given medium is measured at 2.00×10^8 m/s. Compute the index of refraction using equation (14-3).
12. Refer to Table 14-2. Compute the speed of light:
    (a) In water.
    (b) In an optical fiber.
    (c) In fused quartz.
13. Refer to Figure 14-5. Using Snell's law, compute the angle of refraction in water, for a light ray traveling in air and incident upon the surface of water at an angle of 60°.
14. Refer to Figure 14-5. A ray of light is refracted in water at an angle of 45°. Compute the angle of the incident ray in air, θ.
15. In Problem 14, the direction of the ray of light is reversed so that the light ray emerges from water into air. Compute the angle of refraction in air, θ, for a light ray striking the interface at 45°.
16. Explain *total internal reflection*.

[†]*Hewlett-Packard Optoelectronics Designer's Catalog, 1988–1989, p. 8-13.*

17. Explain the *critical angle.*

18. A glass slab, having an index of refraction of 1.55, is surrounded by water whose index of refraction is 1.33. Compute the *critical angle*, θ_c, above which total internal reflection occurs in the glass slab.

19. A glass fiber has an index of refraction of 1.62. It is surrounded by cladding material having an index of refracion of 1.604. Compute the critical angle, θ_c.

20. Explain the difference between single-mode fiber and multimode fiber.

21. Using equation (14-6), compute the number of transmission modes for a multimode step-index fiber having a core diameter of 62 μm, a core index of 1.60, and a cladding index of 1.584. Light having a wavelength of 1300 nm is used.

22. Repeat Problem 21 for light having a wavelength of 820 nm.

23. Explain *graded-index fiber.*

24. Name two strengthening materials used with fiber-optic cables.

25. An LED launches 180 μW of power into the end of a fiber.
    (a) How much power would emerge at the end of the fiber at a distance of 500 m if the fiber were specified as having a loss of 0.2 dB/km?
    (b) What would the loss of the fiber be in decibels per kilometer if 172 μW were measured at the output of the fiber 500 m away?

26. Explain what the following losses are caused by.
    (a) Absorption loss
    (b) Rayleigh scattering
    (c) Radiation loss

27. Compute the numerical aperture (NA) for the fiber in Problem 19.

28. Refer to Figure 14-19. A multimode fiber has a NA of 0.25. Compute the angle of the fiber's acceptance cone.

29. What are the two comonly used methods for splicing fiber ends?

30. Explain the difference between *intrinsic* and *extrinsic* losses in connectors and splices.

31. What semiconductor materials are used in LEDs to emit light at 820 nm?

32. To take advantage of the reduced attenuation losses at wavelengths above 1000 nm, what four elements are typically used?

33. What does *laser* stand for?

34. Explain a *population inversion.*

35. What specification is used to characterize the sensitivity of photodiodes?

36. Explain the difference between a PIN diode and an APD.

Glossary

Absorption Loss The loss of signal energy caused by a medium's impurities.

Acoustically Coupled Modem A low-speed modem (1200 baud or less) that connects indirectly to the telephone lines through the use of an *acoustic coupler*. Two rubber cup-size sockets on the modem are used to mount a telephone set's handpiece to the coupler.

Acoustic Coupler A transducer that converts sound energy to electrical energy, and vice versa. Used with *indirect connect modems*.

A/D Converter *Analog-to-Digital Converter* Used to convert an analog signal into digital format.

Adaptive Equalizer A line conditioning circuit used in a receiver to provide post-equalization of a signal. It is implemented in the receiver section of a high-speed modem.

Aloha A *LAN* protocol that uses a contention scheme to gain access to the channel.

Analog Any continuous and variable quantity of information.

Analog Loop Back Test (ALB) A test used by a transmitting and receiving device (e.g., a modem) that verifies the operation of its transmitter and receiver sections by internally looping back the transmitter's analog signal back to its own receiver section.

ANSI *American National Standards Institute* The official U.S. agency and voting representative for ISO. ANSI develops information exchange standards above 50 Mbps.

Answer Modem The modem that has been designated to answer a call initiated by an *originate modem.*

APD *Avalanche Photodiode* A photodetector used as a fiber-optic receiver. The APD exhibits high *responsivity* in comparison to the PIN photodiode. When reverse-biased, current carriers flow when light penetrates the depletion region.

ARQ *Automatic Repeat Request* Any error correction technique used whereby a receiver requests the retransmission of a block received in error.

ASCII *American Standard Code for Informational Interchange* A seven-bit alphanumeric code used extensively in data communications. A parity bit is often added to the seven-bit code for error detection.

ASK *Amplitude Shift Keying* A form of amplitude modulation whereby a carrier's amplitude is shifted in accordance with a serial binary bit stream.

ASL *Adaptive Speed Leveling* A U.S. Robotics' proprietary technology that boosts modem speeds from 19,200 bps to 21,600 bps.

Asynchronous Transmission The transmission of characters separated by time intervals that vary in length, usually in accordance with the key entries of a terminal operator. Start and stop bits are used to identify the beginning and end of the asynchronously transmitted character.

ATM *Asynchronous Transfer Mode* A high-speed form of packet switching developed as part of *BISDN*. ATM uses short fixed-length packets called "cells" that find their way through the network.

Attenuation Distortion A parameter used by the PSTN to specify the limits of amplitude variations in the passband of the telephone lines. Ideally, a flat frequency response is desired.

Balanced Electrical Circuit A circuit that has two signal paths on which the transmitted signal propagates. Each signal path has the same impedance with respect to signal ground.

Bar Code A series of consistently sized white and black bars that are pasted, painted, or burned onto items for purposes of identification. The wide and narrow bars, representing 1's and 0's, are read by an optical scanner and sent to a computer for processing.

Baseband An encoded analog or digital signal transmitted directly onto a channel without modulating a carrier frequency beforehand.

Baudot A five-bit code used in the field of telegraphy and Radio Teletype (RTTY).

Baud Rate The rate at which a signal is changed or modulated. Baud rate is directly related to the number of bits transmitted per second.

BCC *Block Check Character* A redundant character sent at the end of a message block for error detection. Its value is computed based on the bit pattern of the message stream.

BCD *Binary Coded Decimal* A code used to express a decimal number with four binary bits. It is also referred to as 8421 code, due to the binary weighting of each bit.

BER *Bit Error Rate* Specifies the number of bits that are corrupted or destroyed as data are transmitted from source to destination. BER is typically expressed as a power of ten (e.g., 10^{-5} means that one bit out of 100,000 was destroyed during transmission).

bis French for "second revision."

BISDN *Broadband Integrated Digital Services Network* A broadband PSTN service intended to support transmission speeds from 150 Mbps to 2.5 Gbps.

BISYNC *Binary Synchronous Communications* (also referred to as BSC) protocol. A byte-oriented synchronous serial communications protocol developed by IBM.

Bit-Oriented Protocol Any protocol that uses special groups of uniquely defined bit patterns to control the framing, error checking, and flow between devices.

Bit Rate The number of bits transmitted per second.

Bit Stuffing The technique used in bit-oriented protocols to achieve *transparency* in the data field of a transmitted block. A 1-bit is "stuffed" by the transmitter after any succession of five consecutive 1-bits that are sent. This ensures that no pattern of the opening or closing flag is sent within the data field of the block. Also referred to as *zero bit insertion.*

Bit Time The length of time required to transmit one bit.

Blocking An event that occurs when a central office is overburdened with more calls than it can accommodate. A blocked call is identified by a fast *busy tone* that is sent to the caller.

Block Mode of Transmission For ASCII terminals, a high-speed mode of synchronously transmitting groups of contiguous characters. The characters typed at the terminal are stored in the terminal's buffer and transmitted all at once rather than asynchronously.

BOC *Bell-operating companies* Also known as RBOC (Regional BOC). BOCs are the local telephone companies spun off from AT&T as a result of divestiture. BOCs are organized into seven regions within the United States.

BRI *Basic Rate Interface* Delivers *ISDN* services to the subscriber over a standard twisted-pair telephone wire. The BRI carries two 64-kbps B-channels and one 16-kbps D-channel.

Bridge An internetworking device that is used to link two or more LANs together that use the same protocol. The bridge functions at layer two of the ISO/OSI seven-layer model.

Broadband In contrast to baseband, broadband transmissions use one or more carrier frequencies to convey the intelligence. Frequency-division multiplexing (FDM) is employed.

Bus LAN A network topology that uses a multipoint or multidrop configuration of interconnecting nodes on a shared channel.

Byte-Oriented Protocol Any protocol in which the transmission of data blocks are controlled by special characters (e.g., ASCII or EBCDIC control characters such as SYN, SOH, ETX, etc.).

CATV *Community Antenna Television or Cable TV* Broadband television transmission over 75 Ω (typ.) coaxial cable.

CCITT *Consultative Committee for International Telephony and Telegraphy* An agency of the United Nations that develops recommended worldwide standards and protocols for the telecommunications industry. CCITT is now known as *ITU-TS*.

Cell A geographical area that services cellular mobile telephones.

Cell Site A base station used in cellular telephony that is equipped to transmit and receive calls from any mobile unit within the cell.

Cellular Telephone A telephone technology that employs low-power mobile radio transmission as an alternative to a subscriber loop connection to the central office.

Central Office The local telephone switching exchange.

Channel Any communications link over which information is transmitted.

Channel Capacity The maximum theoretical data transfer rate over a channel based on the channel's bandwidth and SNR. Channel rate is expressed in bits per second (bps).

Character Mode of Transmission An asynchronous mode of transmission featured by a terminal. Characters are transmitted to a host computer only when a key is depressed by an operator.

Cheapernet A scaled-down version of *Ethernet* that utilizes RG58 50 Ω coaxial cable. Cheapernet is also known as *Thin-net* or *Thin Ethernet*.

Check Bit A redundant parity bit inserted into a message stream for the purpose of *forward error correction* (FEC).

Checksum An error-detection technique that basically computes the sum of data and appends this value to the end of a transmitted block of data.

Chromatic Dispersion The spreading of a light pulse caused by the difference in refractive indices at different wavelengths.

Circuit Switching A switching technology utilized by the PSTN that permits DTE to establish an immediate full-duplex connection to another station on a temporary basis. Exclusive use of the channel is available until the channel is relinquished by one of the stations.

Cladding The surrounding layer of a fiber-optic core which is used to confine light. Its index of refraction is less than that of the core, thus permitting *total internal reflection* of the transmitted light within the fiber's core.

Client A station on the network that permits its user to request the shared services provided by the network server.

Client/Server Network A network in which the server is dedicated to serving the network and cannot be used as a workstation.

CMRR *Common-Mode Rejection Ratio* A measure of a differential amplifier's ability to reject signals common to its input pair of lines.

Coaxial Cable Wire consisting of an inner conductor, typically copper wire, surrounded by a dielectric insulator. A wire mesh or copper tube, acting as an electrical shield, encases the dielectric. A protective insulating material, such as PVC, often surrounds the entire construction for protection.

Code 39 An alphanumeric *bar code* consisting of 36 characters and seven special characters.

Collision The event that occurs when two transmitters are transmitting at the same time over the same channel. Data are destroyed.

Companding The process of COMpressing a signal's dynamic range before it is transmitted and exPANDING it back to its original range at the receiver. SNR is greatly improved at the receiver.

Compromise Equalizer Provides pre-equalization of a signal. It is implemented in the transmitter section of a high-speed modem.

Contention The process whereby multiple stations vie for the use of the communications channel.

CPE *Customer Premises Equipment* Equipment on the customer premises, such as PBXs, LAN gateways, and host computers.

CRC *Cyclic Redundancy Check* The most commonly used method for error detection in block transmission. The transmitted message block is divided by a polynomial. The quotient is discarded and the remainder is transmitted at the end of the message block as the *block check character* (BCC).

Critical Angle In fiber optics, the angle above which *total internal reflection* occurs for light rays.

Critical Resource Any part of a LAN system that fails and consequently disables the entire system.

Crossbar Switch An electromechanical switch used in older central office switching systems. The crossbar switch consists of a lattice of horizontal and vertical crossbars with contact points that make and break contact for telephone calls.

Crosstalk Electrical noise or interference caused by inductive and capacitive coupling of signals from adjacent channels.

CSMA/CD *Carrier Sense Multiple Access with Collision Detection* One of the most widely used protocols for bus LAN systems. Used in Ethernet LAN systems.

Current Loop A serial interface standard that has evolved from the old electromechanical Teletype, which used current to activate its relays. Typically, 20 or 60 mA is turned on and off in accordance with the binary serial data.

Data Communications The transmission and reception of digital signals from one location to another.

Data Compression The mechanism used to reduce the number of data bits that is normally used to carry a given amount of information.

Data Encapsulation/Decapsulation The process of assembling and disassembling the Ethernet frame structure.

Data Set The name given to a modem by telephone companies.

DCE *Data Communications Equipment* Typically, a modem or data set used to interface a terminal or computer to the telephone lines.

Dial Tone An audible tone acknowledging connection to the central office when a caller goes off-hook.

Dial-Up Line The standard unconditioned two-wire switched voice-grade line.

Dibit Two bits encoded into one of four possible phase changes.

Digital Information that is encoded into two discrete binary levels: a "1" or a "0."

Digital Channel Bank Equipment used by the central office to digitize and time-division multiplex several voice channels together.

Digital Loop Back Test (DLB) A test used to verify the transmitter and receiver sections of a local and remote station. A command, issued by the local station, is sent to the remote station to *loop back* the detected digital data. The communications channel, modem interface, and I/O device used to issue the command are also tested.

Direct Connect Modem A modem that is capable of electrically connecting directly to telephone lines.

DLE *Data Link Escape* An ASCII and EBCDIC control character used to control the flow of data between two stations.

Double Buffering The process of using two registers to store data. One register is used to hold the data and the other is used to process it.

Double-Ended Amplifier An amplifier used to transmit or receive signals on a balanced electrical circuit. Differential amplifiers and receivers fall into this category.

DPSK *Differential Phase Shift Keying* A modulation technique in which the phase of the carrier frequency is shifted in accordance with the binary serial data. Groups of bits are encoded into a single phase change relative to the phase of the previously encoded interval.

DTE *Data Terminal Equipment* In data communications, an end user or terminating circuit, typically a terminal or computer.

DTMF *Dual-Tone Multifrequency* The technology that utilizes telephone numbers on a pushbutton keypad that are encoded into audible tones when a key is depressed. Each tone is comprised of a pair of sine waves called a DTMF tone. DTMF is also referred to as *Touch Tone.*

Dynamic Range The functional operating range over which a device operates.

EBCDIC *Extended Binary-Coded Decimal Interchange Code* An eight-bit alphanumeric code (no parity) developed by IBM and used in many of its mainframe computers and peripherals.

Echo A reflected signal. In long-distance transmission over the PSTN, impedance mismatches can cause a signal to be reflected back to the caller. This echo is especially annoying when the round-trip delay time exceeds about 50 ms. This occurs at distances beyond 1500 miles.

Echo Cancellation A technique used by high-speed modems to achieve true full-duplex operation over switched lines. An inverted replica of the transmitted signal is added to the received signal to eliminate interference.

Echo Canceller A device used in long-haul networks to electrically cancel any returned echo from a long-distance contact.

Echo Suppressor A device used in long-haul networks to suppress or attenuate any returned echo from a long-distance contact.

EIA *Electronic Industries Organization* A U.S. organization of manufacturers that establishes and recommends industrial standards.

End Office Same as *central office.*

Envelope Delay Distortion A parameter used by the PSTN to specify the limits of phase variations that can occur over the passband of the telephone lines. Ideally, a constant phase response is desired.

Equalizer A circuit used by the PSTN to equalize any variations in gain or phase across the passband of the telephone lines.

Ergonomics The study of people adjusting to their working environment. Ergonomics seeks ways to improve the working environment.

Escape Sequence Escape sequences are used to perform special command and control functions, such as cursor movement on a terminal, clearing the screen, controlling a printer, and so on. For ASCII terminals, escape sequences begin with the ASCII character ESC and end with a single character or string of characters.

ESS *Electronic Switching System* A solid-state computer-controlled switching system used by the PSTN for routing and processing telephone calls.

Ethernet A baseband LAN system standardized by the IEEE 802.3 Committee. Ethernet employs a bus topology and uses a 10-Mbps transmission rate over coaxial cable. Channel access is established through a contention protocol called *CSMA/CD.*

Excess 3 Code A BCD code similar to 8421 BCD, except that 3 is added to the decimal number before it is encoded into a four-bit word.

Extrinsic Losses In optical fibers, these types of losses are a result of the splicing and connector assembly process. They include mechanical offsets between fiber ends, contaminants, improper fusion, gluing, crimping methods, and so on.

FCS *Frame Check Sequence* A *block check character* (BCC), typically 16 bits in length, used in bit-oriented protocols for error-detection purposes.

FDDI *Fiber Distributed Data Interface* An ANSI standard designed to enhance LAN technology through the use of fiber optics.

FDM *Frequency-Division Multiplexing* This type of multiplexing utilizes the *frequency domain* to send multiple signals simultaneously over a channel's available bandwidth.

FEC *Forward Error Correction* Any error correction technique used in simplex communications whereby a redundant code, inserted into the message, is extracted and used to predict and correct an error without having to request a retransmission.

Fiber Optics The communications technology that utilizes the transmission of light over glass or plastic fibers.

Fibre Channel An ANSI standard designed to transfer data at speeds up to 1 Gbps using fiber-optic transmission media.

File Server A program that permits users to access disk drives and other mass storage devices for storing and retrieving common databases and applications programs.

Four-Wire Circuit Two physically separate pairs of lines designed for full-duplex operation: one for transmit and one for receive.

Frame In data communication systems, a group of synchronous serial binary data bits representing information of some form. A preamble and postamble mark the beginning and end of the frame, respectively. A frame is also referred to as a *packet* or *block* by some protocols.

Frame Relay A fast-packet switching protocol defined by CCITT that supports variable-length frame structures transmitted on LAN, MAN, and WAN networks.

Framing The procedure used to identify the beginning and end of a group of data bits.

Framing Error This type of error occurs when a receiver loses synchronism to the incoming data.

FSK *Frequency Shift Keying* A modulation technique used with low-speed modems (300 to 1800 bps). The carrier frequency is shifted between two discrete frequencies in accordance with the binary serial data.

Full-Duplex Simultaneous two-way transmission.

Fusion Splice A method of splicing two optical fiber ends together with an electric arc. The fusion splice is typically performed under a microscope.

Gateway An internetworking device that functions at all seven layers of the ISO/OSI seven-layer model. The gateway permits the interlinking of LANs together that have completely different architectures.

Gaussian Noise This type of noise is the accumulative effect of all random noise, averaged over a period of time, generated internal and external to the telecommunication system.

Geosynchronous Orbit Modern telecommunications satellites are positioned 22,300 miles above the equator. At this altitude (geosynchronous orbit), the satellite travels at a velocity that maintains a fixed position relative to a point above the equator.

Go Back N An ARQ (automatic repeat request) error-correction protocol that requests the retransmission of *N* blocks from the transmitter.

GPIB *General-Purpose Interface Bus* See *IEEE-488.*

Gray Code An unweighted code that provides a single bit change between successive counts. Gray code reduces switching noise and bit errors.

Half-Duplex Alternating two-way transmission.

Hamming Code A redundant set of bits inserted into a message stream for *forward error correction* (FEC) purposes. If a receiver detects an error in a message stream, the Hamming code is used to identify the location of the error. This location is referred to as the *syndrome.*

Handoff When a mobile unit employing an active cellular telephone approaches the perimeter of the cell it is traveling in, its signal strength diminishes. As it enters a new cell, the mobile telephone switching office (MTSO) automatically assigns it a new channel within the new cell. This is called a *handoff.* Handoffs are transparent to the user.

Handshaking A process that regulates and controls the flow of data between two devices.

Harmonic A multiple of the fundamental frequency.

HDLC *High-Level Data Link Control* A bit-oriented protocol proposed by *ISO.* HDLC is an internationally recognized standard. It is used to implement the X.25 packet switching network.

Head End   In broadband communication systems, that part of the network which serves as the origin and destination of all RF signals distributed to and from devices connected to the network.

Huffman Encoding   One of the oldest data compression techniques. Huffman encoding is used to reduce the number of bits representing those characters that have a high frequency of occurrence.

Hybrid   In telephony, a *hybrid* is a balancing network used to convert a two-wire circuit into a four-wire circuit. Transmitted and received signals are electrically separated by the hybrid and placed onto separate pairs of wires.

IEEE   *Institute of Electrical and Electronic Engineers*   A U.S. professional organization of engineers.

IEEE-488   An asynchronous parallel bus interface standard for interconnecting digital programmable instruments (DPI). The original standard, *HP-IB*, was developed by Hewlett-Packard. The bus is also known as the *General Purpose Interface Bus* (GPIB).

ILD   *Injection Laser Diode*   A semiconductor laser diode used as a transmitting device for fiber-optic communications.

Impulse Noise   Any sudden burst of noise induced into a circuit from electromagnetic switching relays, electric motors, lightning, and so on. Impulse noise is the familiar cracking and popping noise heard at the output of a receiver.

Index Matching Fluid   A glue or epoxy used to bond two fiber ends together. To minimize loss, the glue's *index of refraction* is made to match that of the fiber.

Index of Refraction   The ratio of the speed of light in a vacuum to the speed of light in a given medium. As light passes from one medium to another, it is bent at the surface, where the refractive indices change. The amount of bending is directly related to a medium's index of refraction.

Indirect Connect Modem   An acoustically coupled modem.

Infrared Light   That portion of the electromagnetic spectrum having a wavelength ranging from 770 to 10^6 nm. Fiber-optic systems operate in this range.

Interoffice Trunk   A circuit interconnecting two class 5 central office switching centers.

Intrinsic Losses   In optical fibers, these types of losses are beyond the control of the user. They include variations in core diameter, numerical aperture (NA), core concentricity, and so on.

ISDN   *Integrated Services Digital Network*   A CCITT standard developed to provide a standardized interface and signaling protocol for delivering integrated voice and data via the PSTN.

ISO   *International Standards Organization*   One of the largest and most widely recognized standards organizations in the world.

ISO/OSI Seven-Layer Model   A seven-layer hierarchy of data communication protocols that encourages an *Open Systems Interconnect* (OSI) between computers, terminals, and networks for the exchange of information.

ITU-TS   *International Standards Union–Telecommunication Standardization Sector*   Formerly *CCITT.*

Jamming   The process of completing the transmission of a packet onto the communications channel, even though a collision has occurred with another packet. This ensures that all system nodes detect the collision.

Johnson Noise   See *Thermal Noise.*

Kermit   A data communications protocol developed at Columbia University for file transfers between microcomputers and mainframes.

LAN   *Local Area Network*   A privately owned network of interconnecting data communicating devices that share resources (software included) within a local area. The local area can be a single room, building, or group of buildings, generally within less than a few miles of each other.

LAP B   *Link Access Procedure*   A data communications protocol used in the X.25 public data network.

LAP D *Link Access Procedure* A data communications protocol used in *ISDN*.

LAP M *Link Access Procedure for Modems* A data communications protocol recognized by CCITT's V.42 standard: Error Correction Procedures for DCEs.

Laser *Light Amplification by Stimulated Emissions of Radiation* A coherent light source used as a transmitter in fiber-optic communication systems.

LED *Light-Emitting Diode* A device that converts electrical energy to light energy.

Limpel-Ziv A data compression technique based on the principle of assigning numbers to strings of characters of varying length.

Line Conditioning Electrical compensation for attenuation and phase delay distortion exhibited by the PSTN. Conditioning is performed through the use of *equalizer* circuits.

Link The communications channel.

Local Echo When a terminal is configured to route its transmitted character internally around to its receiver section for display, a local echo is said to be generated.

Long-Haul Network This portion of the PSTN includes class 1 through 4 switching exchanges. Long-distance calls beyond the local switching exchange are routed through these higher levels of switching centers and are subject to toll charges.

LRC *Longitudinal Redundancy Check* A parity check performed on an accumulation of the value of individual bit positions in a message steam. The resulting word is used as a *block check character* (BCC) transmitted at the end of the message stream.

MAN *Metropolitan Area Network* A network of data communicating devices connected to service a metropolitan area.

Manchester Encoding A method of encoding a binary serial bit stream such that each bit interval exhibits at least one signal-level transition regardless of the data bit pattern. This allows a receiving device to synchronize to the bit stream without the use of a separate clock. The signal is said to be self-clocking.

Mark A logic 1.

M-ary Derived from the word "bi-nary," M-ary is used to denote the number of encoded bits used to modulate a carrier frequency.

Mechanical Splice A method of splicing two optical fiber ends together by aligning them with a mechanical fixture and gluing or crimping them together.

Message Switching A switching technology utilized by the PSTN that permits the transfer of messages between DTE by temporarily "storing" the message at the exchange and "forwarding" it to the next exchange. Message switching is also known as "store and forward."

Microbending Losses In optical fibers, this type of loss is caused by miniature bends and geometric imperfections along the axis of the fiber. These imperfections cause light to get coupled to various unguided electromagnetic modes that radiate and escape the fiber.

MNP *Microcom Network Protocol* A data communications protocol for modems designed by Microcom Inc. MNP is recognized by CCITT's V.42 standard: Error Correction Procedures for DCEs.

Modal Dispersion In multimode fiber, pulse stretching occurs as a result of the different transit lengths for different propagating modes throughout the fiber.

Modem A contraction of the words *modulator/demodulator*. The modem converts a computer's digital bit stream into an analog signal suitable for the telephone lines, and vice versa.

Morse Code A digital code made up of a series of dots and dashes representing the alphabet, punctuation, and decimal number system.

MTSO *Mobile Telephone Switching Office* In cellular telephony, the MTSO coordinates all mobile calls between cell sites and the central office.

Multidrop A communications link in which a single channel is shared by several stations (e.g., a computer shared by several terminals). Only one station may transmit at a time. Multidrop is also referred to as *multipoint*.

Multimode Fiber Optical fibers that permit light to travel in many paths (modes) throughout the core of the fiber.

Multimode Graded-Index Fiber Optical fibers whose cores are manufactured with an index of refraction that is graded from the center of the core, out to the *cladding* interface. This permits light rays to travel faster as they deviate from the central axis of the fiber's core, thus minimizing pulse spreading.

Multimode Step-Index Fiber Optical fibers that are manufactured with an index of refraction that is uniform throughout the core. The index of refraction is said to "step" from the core value to the cladding value. Light rays bounce back and forth to the end of the fiber.

Node An addressable device on a communications network. Nodes are also referred to as *stations*.

Noise Any extraneous and undesirable portion of a signal that does not contribute to the original intelligence.

Noise Factor A measure of how noisy a device is. It is the ratio of a device's input SNR to its output SNR.

Noise Figure The same as *noise factor* but expressed in decibels.

Noise Immunity A circuit with a high *noise margin* is said to have a high noise immunity.

Noise Margin A quantitative measure of a circuit's ability to tolerate noise transients.

Normal A perpendicular line drawn to the surface of an object or medium.

Null Modem A cable that replaces the modem interface, thus allowing two DTEs to communicate with each other without the use of DCE. Connector wires are interchanged to achieve this.

Numerical Aperture (NA) A measure of an optical fiber's light-gathering capability.

Nyquist Sampling Rate A theorem which states that a signal must be sampled at a rate that is at least twice the highest-frequency component that it contains in order for it to be fully recovered.

Octet In HDLC, a byte or eight bits.

Optocoupler A sealed infrared emitter and photodetector used to couple an electrical signal by light. The optocoupler or *optoisolator* is used to isolate high voltage and noisy circuits from their respective controlling circuit.

Originate Modem The modem that initiates the telephone call for data communications.

Overrun Error An overrun error occurs when a CPU does not read an available character from a buffer before the next one is available. The previous character is destroyed.

Packet A block of data designed for independent processing in the *X.25 packet switching network*. See also *Frame*.

Packet Switching A communications protocol used in networks such as the telephone system. Messages are divided up into discrete units called *packets* and routed independently of each other to their final destination.

PAM *Pulse Amplitude Modulation* A modulation technique that produces a series of pulses whose amplitudes are in proportion to the modulating voltage at the time of the sample.

Parity An error-detection method whereby a single bit is added to a group of bits to make the total number of 1-bits, either even or odd.

Parity Error Indicates that the total number of 1-bits in a received character does not agree with the even or odd parity bit sent with the character.

PCM *Pulse Code Modulation* The process of converting an analog signal into an encoded digital value for transmission.

"Peer-to-Peer" Network A network operating system in which any workstation can be configured as a client or a server or both.

Photodetector A device used to convert light energy to electrical energy. *Photodiodes* and *phototransistors* are commonly used for this purpose.

Photoelectric Effect The transformation of light, incident on a metallic surface, to an electric current.

Photon An electromagnetic particle that possesses energy in proportion to its frequency. Photons are regarded as particles of light.

Photophone A device invented in 1880 by Alexander Graham Bell that uses sunlight reflected off a moving diaphragm to communicate voice information.

PIN Photodiode *P-type Intrinsic N-type diode* A *photodetector* used as a fiber-optic receiver. The PIN diode is used in the reverse-bias condition. Current carriers flow when light (photons) penetrate the depletion region.

Ping-Pong A *pseudo full-duplex* mode of transmission. Full-duplex operation is simulated between two modems by buffering each modem's data and rapidly turning their modulated carriers on and off in a successive fashion using flow control procedures.

Planck's Law A law which states that the energy of a photon is directly proportional to its frequency in the electromagnetic spectrum.

PLL *Phase-Locked Loop* A circuit designed to achieve phase and frequency synchronization to an incoming signal.

Point-to-Point In contrast to *multidrop or multipoint,* a communications link connecting two stations.

Polling A communications control procedure in which a host computer systematically addresses one of several *tributaries* to enquire (ENQ) if it has any data to send. Used in conjunction with *device selecting:* the addressing of a tributary to enquire if it is ready to receive data.

Population Inversion The condition that occurs in a laser when holes and electrons recombine and release photons of energy (light). The released photons further excite other holes and electrons to recombine, thus releasing more light. Light amplification occurs.

PRI *Primary Rate Interface* An ISDN trunking technology that delivers ISDN services to digital PBXs, host computers, and LANs. The PRI carries 23 64-kbps B-channels and one 64-kbps D-channel.

Private Line A hard-wired leased line that bypasses the normal central office switching facility.

Protected Field When a terminal is configured for *block mode,* certain areas on the terminal's display cannot be written to. These areas are called protected fields.

Protocol A set of rules that govern the manner in which data are transmitted and received.

Protocol Efficiency A measure of how efficient a protocol is in terms of transmitting data from source to destination, without error, in a minimal amount of time.

Protocol Overhead A measure of how much redundancy is added to a message stream for framing, flow control, and error detection.

Pseudo Full-Duplex A transmission mode used by some modems (e.g., the Bell 202 modem) whereby a low-speed reverse channel is used in the lower portion of the available passband. This is also referred to as *statistical duplexing.* See also *Ping-Pong.*

PSK *Phase Shift Keying* A modulation technique in which the carrier frequency remains constant and its phase is shifted in accordance with the binary serial bit stream.

PSTN *Public Switched Telephone Network* The dial-up telephone network.

Pure Aloha A protocol in which stations must contend for the use of the communications channel. A station wishing to transmit a message does so at any time, running the risk of a collision from other stations.

QAM *Quadrature Amplitude Modulation* A modulation technique employed in high-speed modems. A combination of ASK and DPSK is used to encode four bits into one of 16 signaling state changes.

Quadbit Four binary bits encoded into one of 16 possible signaling state changes. The CCITT V.22bis standard uses quadbit encoding for its QAM modulation technique.

Quantization The division of a signal's dynamic range into discrete numerically encoded binary values.

Quantization Error The error resulting from quantizing an analog signal. This error can be a maximum of one-half of a quantized level.

Quantization Noise The noise resulting from a reconstructed signal that was quantized prior to transmission.

Quantum Efficiency A measure of how efficient an optical receiver is in terms of converting particles of light (photons) to current carriers. Quantum efficiency is expressed as a percentage.

Quintbit Five binary bits encoded into one of 32 possible signaling state changes. The CCITT V.32 standard uses *trellis encoding* of five bits in its 32-point signal constellation.

Rayleigh Scattering A type of loss in optical fibers caused by micro irregularities that are formed as the fiber cools from its molten state during manufacturing. As light strikes these irregularities, it becomes scattered in all directions of the fiber.

Redundancy In data transmission, any part of a message stream that is not part of the original intelligence being conveyed. This includes framing, error control, addressing, and so on.

Refraction The bending of light, either toward or away from the normal, as it exits one transmission medium and enters another.

Regenerative Repeater A repeater that uses *threshold detection* to recover, reconstruct, and retransmit a PCM signal.

Remote Echo An *echo* generated from a remote terminal as a result of receiving an asynchronous character.

Repeater A device used to extend the transmission distance of a signal by conditioning the signal through amplifiers, filters, and so on, and retransmitting the conditioned signal.

Responsivity A measure of how sensitive to light an optical receiver is in terms of converting optical power to electrical current. Responsivity has the units of amperes per watt (A/W).

Ring LAN A network *topology* in which nodes are connected *point-to-point* in a closed-loop configuration.

Roaming Operation of a cellular mobile telephone outside a registered metropolitan area. A roam LED indicator on the cellular telephone unit turns on when roaming. Roaming is possible throughout the country, provided that there is a prearranged agreement between the user and the telephone company beforehand.

Router An internetworking device that is used to link two or more LANs together that have the same communications protocol. Unlike a repeater or bridge, the router is an addressable device that is not transparent to the end user. Instead, the router acts as a node that is capable of processing information embedded in a packet's control header. A router performs functions at layer three of the ISO/OSI seven-layer model.

RS-232-C A serial interface standard that specifies the electrical, functional, and mechanical interface specifications between data communicating devices.

Screen Capacity The maximum number of characters that can be displayed on a terminal's screen, typically 1920 (24 rows by 80 columns) characters.

SDLC *Synchronous Data Link Control* A *bit-oriented protocol* developed by IBM and used in its *Systems Network Architecture* (SNA). SDLC is functionally identical to HDLC. See *HDLC*

SDM *Space-Division Multiplexing* Signals that are multiplexed together by physically combining conductors into a bundled cable. Space is required for this type of multiplexing.

Segment A coaxial cable link used in *Ethernet* that can span a maximum of 500 m and can include up to 100 *nodes*.

Server A high-powered computer or workstation that shares its resources as a service to network users.

S/H Amplifier *Sample-and-Hold Amplifier* An amplifier that samples an analog signal and holds its value so that it may be processed, typically by an A/D converter.

Shannon's Law A fundamental law used in telecommunications to compute *channel capacity.*

Shot Noise This type of noise results from the random arrival rate of discrete current carriers (holes and electrons) at the output

electrodes of semiconductor and vacuum tube devices.

Sidetone The small amount of signal fed back from a telephone set's transmitter to its receiver that allows a person speaking into the handset to hear himself or herself.

Simplex Transmission in one direction only.

Single-Ended Amplifier An amplifier with a single output referenced to a common ground and used to drive an *unbalanced electrical circuit.*

Single-Mode Fiber Optical fiber that is manufactured so that rays of light travel only along the central axis of the fiber.

Slotted Aloha A refinement of the *Aloha* contention scheme whereby all stations on the network are time synchronized to each other. Transmissions are restricted to predefined intervals of time, called "slots."

Slotted Ring A variation of token passing in a *ring LAN*. A fixed number of contiguous time slots are circulated around the ring LAN. A *node* wishing to transmit a *packet* simply fills an empty slot. The filled slot circulates around the ring to its destination.

Slot Time The round-trip propagation time for a *packet* to travel between the farthest two *nodes* in a *bus LAN*.

SMDS *Switched Multimegabit Data Service* A PSTN service designed to offer customers efficient and economical connectivity between LAN and WAN systems.

Snell's Law A fundamental law used in optics to predict the path of light rays as they travel from one medium to another based on each medium's *index of refraction.*

SNR *Signal-to-Noise Ratio* The ratio of signal power to noise power. Also referred to as *S/N.*

SONET *Synchronous Optical Network* Published by *ANSI* in 1988, SONET is a fiber-optic communications standard designed to allow the internetworking of optical communications equipment between different vendors.

Space A logic 0.

Star LAN A network *topology* in which several *nodes* are radially linked to a central node through *point-to-point* connections.

Start Bit The first bit used to frame an asynchronously transmitted character. Its logic level is a 0 (space).

Statistical Duplexing See *Pseudo Full-Duplex.*

Stop Bit The last bit used to frame an asynchronously transmitted character. Its logic level is a logic 1 (mark).

Stress Corrosion. A defect in an optical fiber caused by the exposure of the glass fiber to humidity. Also known as *static fatigue.* Spontaneous fractures can occur as a result of stress corrosion.

Strowger Switch An electromechanical step-by-step switch used in older central office switching systems.

STS *Synchronous Transport Signals* Signals designed to be transported on the SONET network.

Subscriber Loop The set of wires connecting the telephone set to the central office.

Superframe Twelve contiguous *T1 carrier frames.*

Synchronous Transmission High-speed communication whereby data characters are sent in direct succession to each other without the use of *start* and *stop bits.*

Syndrome Identifies the bit location of an error in a message stream in systems that employ *forward error correction* (FEC).

System Redundancy A *critical resource's* backup protection in case of a failure of that portion of the communication system. It can be hardware or software built into the system or provided as an equivalent replacement for the failed part.

T1 Carrier A *time division multiplexed* (TDM) *PCM* standard developed by Bell Laboratories. The T1 carrier is used in most major telephone circuits today. The transmission rate for the standard T1 carrier is 1.544 Mbps.

T1 Carrier Frame The basic T1 carrier frame consists of 24 channels, eight bits per channel, making a total of 192 bits. A 193rd framing bit is added to the *frame* for synchronization. Each voice channel is sampled at an 8-kHz rate, making a total of 1.544 Mbps (8 kHz $\times$ 193 bits).

Tandem Office A class 5 switching center used to interconnect central offices in tandem, thereby minimizing the number of trunk circuits that a local call must be routed through to reach its destination.

Tandem Trunk A circuit used to interconnect two class 5 switching centers in tandem.

TDM *Time Division Multiplexing* This type of multiplexing utilizes the *time domain* to interleave multiple signals together for transmission over a single channel.

Telecommunications Long-distance communications via a conglomeration of information-sharing networks tied together.

Telex An international Teletype exchange that uses *baudot* code.

ter French for "third revision."

Terminal An input/output device used by an operator to communicate with a host computer. It consists of a keyboard and a display to monitor alphanumeric characters entered at the keyboard or received from a remote device.

Thermal Noise Also referred to as *Johnson noise* or *white noise*. This type of noise is generated by the random motion of free electrons and molecular vibrations in resistive components. Thermal noise power is proportional to both temperature and bandwidth.

Threshold Detection The process used to compare the voltages of two signals, one of which is used as a threshold. When one exceeds the other (i.e., the threshold), a logic 1 is output; otherwise, a logic 0 is output.

Throughput A measure of transmission rate based on the time required to transmit and receive successfully a maximum amount of data per unit time. This includes the added time required for acknowledgments, framing, and error control.

Time-out Error This type of error occurs when a device fails to respond to a message within a given period of time.

Token A special bit pattern or *packet* that circulates around a *ring LAN* from *node* to node.

Token Passing The channel access scheme used in *ring LANs* for sharing the use of the channel. A *token* is circulated around the ring from *node* to node. Possession of the token gives a node exclusive rights to the channel for communications.

Toll Center A class 4 central office switching center that processes calls outside the local area.

Toll Trunk A circuit interconnecting a class 5 central office switching center to a class 4 *toll center*.

Topology The geometric pattern or configuration of intelligent devices in a LAN.

Total Internal Reflection The principle behind fiber optics that governs how rays of light propagate through a transparent medium by reflecting off its boundaries.

Transparency In synchronous serial communications, data are said to be *transparent* when any received bit patterns, equivalent to control characters, are disregarded in terms of the normal control procedures.

Transportable A *cellular telephone* that can be used outside the mobile unit.

Trellis Encoding A 32-point signal constellation specified by the CCITT V.32 recommendation for modems. Five consecutive binary digits, called a *quintbit,* are encoded into one of 32 signaling states represented in the constellation.

TRIB *Transmission Rate of Information Bits* A formula used to compute system *throughput.*

Tribit Three bits encoded into one of eight possible phase changes.

Tributary A terminal or station, other than the controlling station, that must be *polled* or *selected* by a host computer in order to transmit or receive data, respectively.

Trunk A circuit interconnecting two telephone switching centers.

TTY *Teletype* Used in the *Telex* exchange. An electromechanical terminal consisting of a keyboard, printer, paper tape reader, and punch. Teletype is a trade mark of the former Teletype Corporation.

Turnaround Time For modems, the time that it takes for transmission direction to change between two stations operating over a *half-duplex* link.

Twisted-Pair Wire Insulated pairs of wire twisted together to minimize *crosstalk* interference. Twisted-pair wire is widely used in *subscriber loop* and short-haul trunk circuits.

Twistor Memory Superpermanent memory that stores the control program for *ESS* systems.

Two-Wire Circuit *Twisted-pair* wire that connects the subscriber's telephone to the central office. Signals are transmitted and received over the same pair.

Ultraviolet Light That portion of the electromagnetic spectrum having a wavelength of 10 to 390 nm.

Unbalanced Electrical Circuit A circuit in which the transmitted signal propagates down a single transmission line accompanied by its signal return line or common.

Unprotected Field When a terminal is configured for *block mode,* certain areas of the display can be written to or edited. These areas are called unprotected fields.

Unweighted Code Any code whose bits do not have any numerical weighting.

UPC *Universal Product Code* A 12-digit numerical bar code adopted by the *National Association of Food Chains.* UPC is most commonly found on grocery items.

USART *Universal Synchronous Asynchronous Receiver Transmitter* A programmable device used to convert parallel data from a CPU to a serial data stream for transmission. The USART can also receive serial data and convert it back to parallel format for a CPU to read.

UTP *Unshielded Twisted-Pair Wire* A transmission medium that has gained popularity in LAN systems (e.g., the IEEE 802.3 10BaseT standard).

VDT *Video Display Terminal* Same as a *terminal.*

Visible Light That portion of the electromagnetic spectrum having a wavelength ranging from 390 to 770 nm.

VRC *Vertical Redundancy Check* A *parity* check performed on a per-character basis.

WAN *Wide Area Network* A long-distance network of data-communicating devices designed to service users beyond the local area.

Wavelength The distance between two successive peaks (or troughs) of a sinusoidal waveform.

Wireless Telecommunications A digital wireless technology that eliminates the use of cabling. Wireless technology in telecommunications includes digital and analog cellular telephony, cordless telephony, wireless LANs, wireless PBXs, and wireless personal communication services (PCSs).

Workstation An intelligent terminal that features a self-contained operating system, high-resolution graphics, and 32-bit microcomputer architecture.

X.25 The CCITT recommendation for *packet switching networks.*

Xmodem A modem protocol used in data communications to ensure the reliable transfer of files between PCs.

XOFF *Transmit Off* A device control character (DC3) used to control the flow of data between two devices. XOFF is used as a *handshake* with *XON.*

XON *Transmit On* A device control character (DC1) used to control the flow of data between two devices. XON is used as a *handshake* with *XOFF.*

Ymodem A modem multiple file transfer protocol developed by Chuck Fosberg of Omen Technology Inc.

Ymodem/G Same as Ymodem but supports MNP data compression and error control in the modem's hardware.

Zero-Bit Insertion See *Bit Stuffing.*

Zmodem A multiple file transfer protocol developed by Chuck Fosberg of Omen Technology Inc. Zmodem uses a 32-bit CRC for error checking.

Answers to Odd-Numbered Problems

CHAPTER 1

1. Telecommunications has affected our lifestyles and will continue to do so for the foreseeable future. Many of our jobs, educational courses, and everyday events require skill levels necessary to communicate with each other using computers, telephones, radio, and television.

3. *Telecommunications* can be defined as long-distance communications via a conglomeration of information-sharing networks tied together.

5.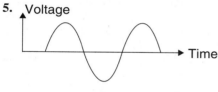

7.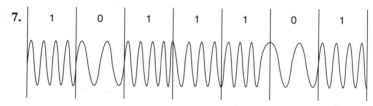

9. LAN stands for *local area network*, MAN stands for *metropolitan area network*, and WAN stands for *wide area network*.

CHAPTER 2

1. (a) 0111
   (b) 0001 0010 0011
   (c) 1001 0110

3. Decimal   Excess-3

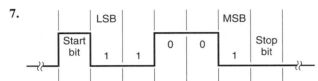

       9   1100
     + 8   + 1011
     17   1 0111

When adding two decimal numbers whose sum exceeds nine, a valid answer in BCD is produced if the two numbers are added together using excess-3 code. This is true since the addition of two excess-3 characters has an inherent correction factor of six (three plus three), which is necessary for invalid BCD results.

5. (a) 1001
   (b) 11111110
   (c) 11101011
   (d) 100010

7.

9. Upper- and lowercase letters in EBCDIC have a difference of 40H. Bit 6 is used to distinguish between the two cases. For example, b = 82H and B = C2H.

11. There are 32_{10} ASCII control characters. In hexadecimal, their codes are distinguished by having values less than 20H.

13. The ASCII character **Nul** is represented by all zeros.

15. ASCII *format effectors* control the position of a terminal's cursor.

17. Two advantages of bar code are low cost and high throughput. Two disadvantages of bar code are the extra equipment required, and its use is intended primarily for fixed repetitive data input.

19. A bar code that has intercharacter spaces or gaps between characters; consequently, each character within a bar code symbol is independent of every other.

21. 100000110

23. $ = 010101000.

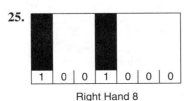

25.

Right Hand 8

CHAPTER 3

1. The distinguishing feature of the smart terminal over the dumb terminal is its ability to transmit and receive blocks of data.

3. Computer-aided design; simulation; modeling.

5. 1920 characters (80 columns by 24 rows).

7. Special function keys are software defined. They may be programmed by the manufacturer or user to perform special functions, such as communications control, editing, printing, and so on. They are tremendous time-savers since they can be programmed to perform several instructions at the touch of a key.

9. <u>Odd parity</u>

   1 1 0 1 0 1 0 <u>1</u>
   0 0 1 1 0 1 0 <u>0</u>
   1 1 1 0 0 0 0 <u>0</u>
   1 0 1 1 0 1 0 <u>1</u>

   <u>Even parity</u>

   0 0 1 1 0 1 0 <u>1</u>
   1 0 1 0 1 1 0 <u>0</u>
   1 1 1 1 0 0 0 <u>0</u>
   0 0 1 1 0 1 0 <u>1</u>

11. A local echo is produced in half-duplex mode.

13. Characters are duplicated on the screen as keys are typed. The first character is produced by the local echo and the second is the remote echo.

15. An escape sequence is a sequence of characters used to perform special command or control functions. An escape sequence begins with the ASCII character **ESC**, followed by one or more characters as defined by the escape sequence.

17. RS-232-C; IEEE 488; current loop.

19. A *listener.*

CHAPTER 4

1. A *protocol* is a set of rules that governs the manner in which data are transmitted and received.

3. 36.4%

5. The term *modem* is a contraction of the words *modulator–demodulator.*

7. A printer is considered DTE.

9. RS-232-C cable lengths should be kept under 50 ft (15 m).

11.

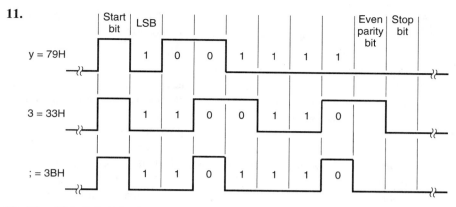

13. The EIA RS-449 nine-pin connector is designed to carry secondary interchange circuits.

15. A *differential amplifier* typically has a CMRR specification.

17. 10 Mbps

CHAPTER 5

1. *Double buffering* is the mechanism used to provide temporary storage for two data words. While one data word is being processed, the other is being held. This prevents the destruction (writing over) of one of the data words.

3. **(a)** 9600 baud
   (b) 873 characters per second (cps)

5. Figure 5-6b, 26 μs; Figure 5-6c, 13 μs; Figure 5-6d, 3.26 μs.

7. Transmit Ready (TxRdy).

9. P.E. = 0; PARITY = 0.

CHAPTER 6

1.

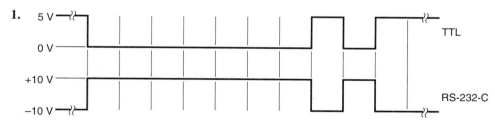

3. **(a)** When C/$\overline{\text{D}}$ is high, a control command can be issued to the 8251A (command word or a mode word) or its status can be read.
   (b) When C/$\overline{\text{D}}$ is low, data can be written or read from the 8251A.

5. A mode instruction must be issued to the 8251A to program it for asynchronous operation. The least significant two bits of the mode word determine the asynchronous baud rate factor.

7. DAH

9. Since all control words written after a mode word are interpreted as command words, the 8251A has been designed to provide an internal reset bit in the command word format. This allows the UART to be internally reset through a command word, thus allowing mode word parameters to be reprogrammed without having to perform a power-on reset.

11. *Framing error:* A framing error occurs when the logic state of the asynchronous character's stop bit is tested by the UART and is not a logic 1. This generally implies that the received character's baud rate is different than what the UART has been programmed to receive. *Overrun error:* This type of error occurs when the CPU does not read the available character before the next one is available. The previous character is destroyed. *Parity error:* A parity error occurs when the parity bit level does not correspond to the total number of 1-bits set in the character.

CHAPTER 7

1. Hybrid.

3. The diaphragm of a telephone set's receiver fluctuates in accordance with the electrical current representing sound. This is caused by the varying magnetic field produced by the receiver's coil that is wound around a permanent magnet. The resulting force causes the diaphragm to move, thus reproducing the original sound.

5. The ring potential (90 V_{rms} at 20 Hz) is superimposed on the existing on-hook dc potential (−48 V dc) because the telephone circuit, at this time, is an open circuit for dc. The −48 V dc on-hook potential will remain present until the telephone is answered or goes off-hook.

7. The *ring signal*, 90 V_{rms}, causes the telephone set to ring, whereas the *ring-back* signal is an acknowledgment tone to the party placing the call that the telephone at the destination end is ringing. The two signals occur simultaneously but not necessarily in sync with each other.

9. **(a)** D1 acts as the dial switch.
   (b) D2 shunts the receiver off when dialing. This prevents clicking pulses from being heard at the receiver.
   (c) Switches S1 and S2 are open when the telephone is in the on-hook position.
   (d) Varistors are used to regulate voice amplitude.

11. A dial pulse has a period of 100 ms, a repetition rate of 10 Hz, and a pulse width of 40 ms (60 ms on-hook and 40 ms off-hook).

13. The *interdigit time* is the time interval between the digits dialed.

15. **(a)** 1.6 to 1.8 MHz or the 46-MHz band for more recent cordless phones.
   (b) 49.8 to 49.9 MHz or the 49-MHz band.

17. There are 10 pairs of carrier frequencies used between the base unit and portable handset of a cordless telephone.

19. -48 V dc

21. The *congestion tone* sounds like the busy tone but is twice as fast. It is on for 0.2 sec and off for 0.3 sec. Its frequencies are the same as the busy tone: 480 and 620 Hz.

23. 3400 Hz

25. (a) 0.42 dB/mi
    (b) 430 Ω
    (c) 3440 Hz

27. (a) $I_{sc} = 48$ V/800 $\Omega = 60$ mA
    (b) $V_{off\text{-}hook} = 60$ mA $\times$ 142 $\Omega = 8.52$ V dc
    (c) distance $= [(800\ \Omega - 142\ \Omega)\ (1000\ \text{ft}/8.33\ \Omega)] \times (1\ \text{mi}/5280\ \text{ft}) \times \frac{1}{2} = 7.48$ miles

29. (a) 1500 μs
    (b) -2 to $+3$ dB

31. Cellular technology permits many mobile telephone users to operate in heavily populated areas that are divided up into cells. Since transmitters are of low power, frequency channels can be reused in other cells. This is a major advantage over conventional mobile telephone.

33. A *handoff* occurs in a cellular telephone when a mobile unit's signal strength diminishes as it exits a cell. The mobile unit is handed a new frequency channel to operate with the new cell site, thus improving signal strength.

CHAPTER 8

1. (a) Interoffice trunk
   (b) Tandem trunk
   (c) Toll trunk
   (d) Intertoll trunk

3. Consistent line characteristics; less prone to noise; less attenuation and delay distortion; higher data transfer rates; available 24 hours a day.

5. For a signal to propagate back and forth between two telephone sets separated by a distance of 2900 miles, it would take 31.2 ms. An echo suppressor or canceller would not be necessary.

7. Signal-to-noise ratio (SNR).

9. When the central office is temporarily overburdened with calls and cannot provide further switching, a call can be *blocked*. A fast busy tone is sent to the caller, indicating that the call must be placed again at a later time when traffic subsides.

11. *Space-division multiplexing (SDM):* the combining of physically separate signals into a bundled cable. Space is shared between the signals. *Frequency-division multiplexing (FDM):* the simultaneous transmission of signals within a designated band of frequencies. Bandwidth is shared between the signals. *Time-division multiplexing*

(TDM): the interleaving of two or more signals in the time domain. Time is shared between the signals.

13. 6800 Hz

15. **(a)** 648 ns

 (b) 8 kHz

 (c) 1.5 ms

17. An alternating pattern of 1's and 0's is used for the framing bits in odd-numbered frames. For even-numbered frames, the pattern 001110 is used.

19.

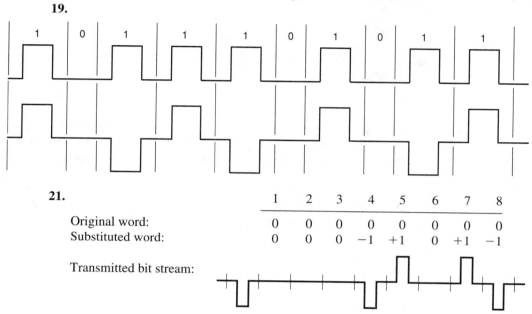

21.

	1	2	3	4	5	6	7	8
Original word:	0	0	0	0	0	0	0	0
Substituted word:	0	0	0	−1	+1	0	+1	−1

Transmitted bit stream:

CHAPTER 9

1. Modulator–demodulator.

3. The acoustically coupled modem suffers a major drawback in terms of transmission speed. The rubber fittings may not always fit the telephone's handpiece snugly. Even if they do, ambient noise levels may still leak through, thus limiting speed and performance.

5. A *dibit* is the encoding of two bits into a single phase or signal change.

7. 2225 Hz

9. *Simplex:* transmission, in one direction only. *Half-duplex:* two-way transmission, one station at a time. *Full-duplex:* simultaneous two-way transmission.

11. 1600

13. 4

15. An *analog loopback test* (ALB) is a test feature in modems that loops back the modem's transmitter output to its receiver input. The modem is therefore tested without the need for a remote station to echo back the character. A *digital loopback test* (DLB) is a test feature in modems that permits data from a remote terminal to be looped back by a local modem through each modem's interface. Both modems and their interface are tested.

17. The original Xmodem protocol uses a 128-byte fixed-length data field. In contrast, 1K-Xmodem/G uses a 1-kbyte data field. The "/G" signifies the use of *Microcom Networking Protocol* (MNP) for error control.

19. Microcom Networking Protocol (MNP).

21. 9146 bps

23. A 2:1 compression ratio can be achieved with a Huffman-based data compression technique.

CHAPTER 10

1. A *protocol* is a set of rules and procedures that governs communications.

3. *Applications layer (7):* specifies the application program for the end user. *Presentation layer (6):* addresses any code or syntax conversions necessary to present data to the network in a common format. *Session layer (5):* defines the management of a session. *Transport layer (4):* ensures the reliable and efficient transportation of data on a network. *Network layer (3):* defines how messages are broken down into data packets and sent on a network. *Data link layer (2):* defines the successful transmission of data between devices. *Physical layer (1):* defines the electrical and mechanical specifications of the network.

5. Physical layer.

7. **(a)** DLE STX
   (b) DLE ETX or DLE ETB
   (c) DLE DLE

9. When a time-out error occurs, an ENQ is issued by the transmitting station to the receiving station. An ACK 0, ACK 1, or NAK in response will allow the transmitting station to determine whether or not the last message was received.

11. **(a)** ENQ
    (b) ACK 0 and ACK 1
    (c) NAK

13. In BISYNC, a *tributary* is a terminal or station, other than the host computer, that is polled or selected by the host to transmit or receive data, respectively.

15. Circuit switching, message switching, and packet switching.

17. A circuit-switched connection between two stations permits exclusive use of the channel between the two stations until it is relinquished by one of the stations.

19. The *Telex* network is an example of message switching.

21. In packet switching, a PAD assembles a terminal's packet and sends it to the network nodes. It also disassembles received packets from the network nodes for the terminal.

23. HDLC is a bit-oriented protocol.

25. For a two-byte address field in HDLC, the LSB of the first address byte must be a logic 0. This informs the receiver that the address field is more than one byte. The LSB of the second byte must be a logic 1, indicating the final byte of the address field.

27. Zero-bit insertion or "bit stuffing."

29. In HDLC, if a receiving device receives five consecutive 1-bits, after detecting the opening flag, the sixth bit is tested for a 1 or 0. If it is 1, it is assumed to be the closing flag pattern (or an abort pattern). If the sixth bit is a 0, it automatically deletes it and continues to receive the remainder of the message block. The deleted 0 was inserted at the transmitter as part of zero-bit insertion.

31. In HDLC, only the Information Transfer Frame and Unnumbered Frame types can have an information field.

33.

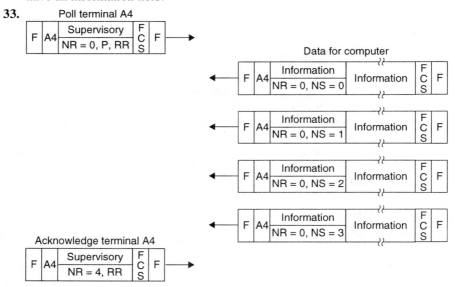

35. In ATM, VP stands for *virtual path* and VC stands for *virtual channel*. A VC is a connection between communicating entities. A VP is a group of VCs carried between two points.

37. *Header field:* includes address and control information. *Information field:* contains digitized intelligence (e.g., voice, data, image, and video).

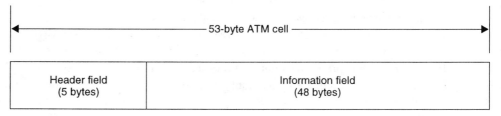

39. *Generic flow control (GFC):* The first four bits of the ATM header control the flow of traffic across the user–network interface (UNI) and into the network. *Virtual path identifier (VPI)/ Virtual channel identifier (VCI):* The next 24 bits after the GFC make up the VPI and VCI. These bits make up the ATM address field. *Payload type (PT):* This three-bit field specifies one of eight types of message payloads. *Cell loss priority (CLP):* This single-bit field specifies the eligibility of the cell for discard by the network under congested conditions. *Header error control (HEC):* The fifth and final byte of the header field is designed for error control of the header field only.

41. ISDN services provided to residential customers has been slow, due primarily to the existing voice-grade lines. These lines have been designed for voice and must be modified for digital signals.

43. An ISDN B (Bearer)-channel is a 64-kbps channel that carries end-user voice, audio, video, or data. An ISDN D (Delta)-channel carries packet-switched user data and call control messaging information at 16 or 64 kbps.

45. R, S, T, and U

47. 51.84 Mbps

49. 12

51. 100 Mbps

53. The *Fibre Channel Systems Initiative* is a joint effort between Hewlett-Packard and Sun Microsystems to advance the Fibre Channel as an affordable high-speed inter-connection standard for workstations and peripherals.

CHAPTER 11

1. LAN stands for *local area network.*

3. LAN *topology* refers to the geometric pattern or configuration of intelligent devices and how they are linked together for communications.

5. (a) The critical resource for a bus LAN is the bus cable.
   (b) The distinguishing feature of the bus LAN is that control of the bus is distributed among all the nodes on the bus.
   (c) A major disadvantage with the bus LAN is that each node must contend for use of the channel. In heavy volumes of traffic on the bus, collisions occur, thus hindering communications.

7. A *collision* is said to occur on a bus LAN when two or more stations assert their packets onto the channel at the same time.

9. CSMA/CD stands for *carrier sense multiple access with collision detection.*

11. The round-trip propagation delay time between the farthest two nodes on a bus LAN is referred to as the *slot time.*

13. In slotted ring, *time slots* are circulated around the ring.

15. Radio-frequency (RF) and infrared (IR) technology.

17. CATV stands for *community antenna television,* or simply, *cable TV.*

19. (a) Bus (branching nonrouted tree)
   (b) 10 Mbps
   (c) 1024 nodes
   (d) 500 m
   (e) 51.2 μs
   (f) 50 m
   (g) Repeaters are used to join segments together to extend the area of the network. The repeater is simply tapped into the Ethernet cable in the same manner as a node. A transceiver cable is used.

21.

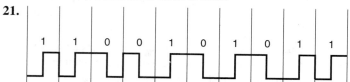

23. Cheapernet uses BNC-type connectors and RG-58 U coaxial cable as a transmission medium.

25. The major advantage of 10BaseT over 10Base5 is that it uses inexpensive UTP and modular wall jacks.

CHAPTER 12

1.

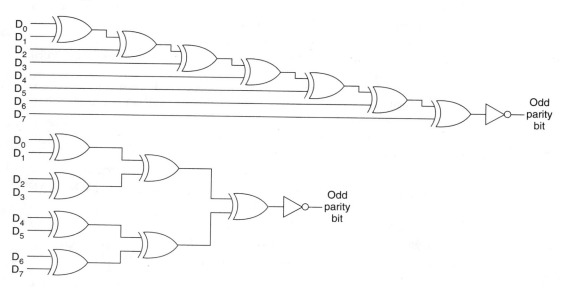

3. A simple parity check, even or odd, will not detect if an even number of bit errors has occurred. For example, if the seven-bit data word below with an odd parity bit were transmitted and bits 3 and 4 were corrupted (changed from 1 to 0), the final parity bit would still agree with the data word received (10001010). The same would be true if four or six bits (i.e. an even number) were inverted from the original. The odd parity bit would still indicate that the total number of 1-bits received, including the parity bit, is an odd quantity.

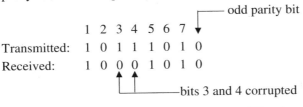

```
                              ┌── odd parity bit
              1 2 3 4 5 6 7 ↓
Transmitted:  1 0 1 1 1 0 1 0
Received:     1 0 0 0 1 0 1 0
                  ↑ ↑
                  └─┴──────bits 3 and 4 corrupted
```

5.

	J	u	s	t	(SP)	d	o	(SP)	i	t	!	BCC (LRC)	
LSB (D0)-	0	1	1	0	0	0	1	0	1	0	1	1	
	1	0	1	0	0	0	1	0	0	0	0	1	
(D2)-	0	1	0	1	0	1	⊗	0	0	1	0	1 ← row error	
	1	0	0	0	0	0	1	0	1	0	0	1	
	0	1	1	1	0	0	0	0	0	1	0	0	
	0	1	1	1	1	1	1	1	1	1	1	0	
MSB (D7)-	1	1	1	1	0	1	1	0	1	1	0	0	
VRC-	1	1	1	0	1	1	0	1	0	0	0	0	

└─column error

7. **(a)** 16
   (b) 11010001011010010000000000000000
   (c) 1101110010111001
   (d) 1101101110011001
   (e) 11010001011010011101101110011001

9. Single-precision; double-precision; Honeywell; Residue.

11. 07H

13. 0197H

15. B1D0

17. 83

19. ARQ (automatic repeat request) is an error-correction technique used in full- and half-duplex communication systems. When a block is received in error, a request for retransmission is made by the receiver to the transmitter. In FEC, this is not possible since communications is simplex. Instead, special error-correcting codes are placed into the message stream so that the receiver can extract the code and predict and correct the error.

21. **(a)** $C_0 = 0$
$C_1 = 1$
$C_2 = 0$
$C_3 = 0$
(b) 10111001010110
(c) 0100
(d) 1101

CHAPTER 13

1. $\text{SNR} = 10 \log (3.58 \text{ mW}/45 \text{ } \mu\text{W}) = 19 \text{ dB}$
3. **(a)** 20.97 dB
(b) 19.5 dB
(c) 1.4
(d) 1.46 dB
5. 12.5×10^{-6}
7. 30 kbps
9. 15.6 μV
11. 69.3 nA
13. 12 bits

CHAPTER 14

1. Bandwidth; less weight and volume; less loss; security; flexibility; safety.
3. The two classic theories of light are the *particle theory* and the *wave theory*.
5. Infrared.
7. **(a)** 187.5 m
(b) 3.41 m
(c) 7.5 cm
(d) 5.0 cm
(e) 1546 nm
9. 51.2×10^{-6} in.
11. $n = 1.5$
13. $\phi = 40.6°$
15. $\theta = 70.1°$
17. In transparent mediums, the *critical angle* is the angle above which a ray of light will experience total internal reflection.
19. 81.9°
21. 572 modes

23. *Graded-index fiber* is characterized by having a core whose index of refraction is graded from the center of the core out to its cladding interface.

25. (a) 175.9 μW
    (b) -0.395 db/km

27. 0.227

29. The two commonly used methods for splicing fiber ends are the *mechanical* splice and the *fusion* splice.

31. Gallium-aluminum-arsenide is used for LEDs that emit light at 820 nm.

33. *Laser* stands for *light amplification by stimulated emissions of radiation.*

35. The sensitivity of a photodiode is typically specified by its *responsivity* rating.

References

AT&T TECHNICAL REFERENCE, Pub. *54075, 56 Kbps Subrate Data Multiplexing,* AT&T Communications, Florham Park, NJ, 1985.

BELL COMMUNICATIONS RESEARCH INC., *The Extended Superframe Format Interface Specification,* Morristown, NJ, 1985.

BELL LABORATORIES, *A History of Engineering and Science in the Bell-System–Transmission Technology,* AT&T Bell Laboratories, Indianapolis, Ind., 1985.

BELL LABORATORIES, *A History of Engineering and Science in the Bell System–Communications Science,* AT&T Bell Laboratories, Indianapolis, Ind., 1984.

BELL LABORATORIES, *A History of Engineering and Science in the Bell System–Switching Technology,* AT&T Bell Laboratories, Indianapolis, Ind., 1982.

BELL SYSTEM TECHNICAL REFERENCE Pub. 62411, *High Capacity Digital Service Channel Interface Specification,* Basking Ridge, NJ, 1983.

CARR AND WEIDNER, *Physics From the Ground Up,* McGraw-Hill, Inc., New York, NY 1971.

CCITT RECOMMENDATION V.22bis, *2400 bps Duplex Modem Using the Frequency Division Technique Standardized for Use on the General Switched Telephone Network.*

CCITT RECOMMENDATION V.32, *A Family of 2-Wire Duplex Modems Operating at Data Signaling Rates of Up to 9600 bps for Use on the General Switched Telephone Network and on Leased Telephone-Type Circuits,* 1984.

CCITT RECOMMENDATION V.42, *Error Correcting Procedures for DCEs Using Asynchronous-to-Synchronous Conversion.*

CONCORD DATA SYSTEM'S *All About Adaptive Data Compression for Asynchronous Applications,* Concord Data Systems, Inc., Marlborough, MA.

CONCORD DATA SYSTEM'S *All About Error Protection Protocols for Asynchronous Data Transmission,* Concord Data Systems, Inc., Marlborough, MA.

COUCH, L. W. II, *Digital and Analog Communications Systems,* 2nd ed., Macmillan Publishing Co., New York, NY, 1983.

DIGITAL, *Introduction to Local Area Networks,* Digital Equipment Corp., Santa Clara, CA, 1982.

EIA RS-232-C, *Interface Between DTE and DCE Employing Serial Binary Data Interchange,* Washington, D.C., 1969.

EIA RS-422-A, *Electrical Characteristics of Balanced Voltage Digital Interface Circuits,* Washington, D.C., 1978.

EIA RS-423-A, *Electrical Characteristics of Unbalanced Voltage Digital Interface Circuits,* Washington, D.C., 1978.

EIA RS-449, *General Purpose 37-Position and 9-Position Interface for DTE and DCE Employing Serial Binary Data Interchange,* Washington, D.C., 1977.

FIKE, J. L. AND FRIEND, G. E., *Understanding Telephone Electronics,* Texas Instruments, Dallas, Texas.

FINK, D. G. AND CARROL, J. M., *Standard Handbook for Electrical Engineers,* McGraw-Hill, Inc., New York, NY, 1969.

HAMMING, R. W., *Coding and Information Theory,* 2nd ed., Prentice-Hall, Inc., Englewood Cliffs, NJ, 1986.

HEWLETT-PACKARD, *Fiber Optics Handbook,* Hewlett-Packard GmbH, Boeblingen Instruments Division, Federal Republic of Germany, 1988.

HEWLETT-PACKARD, *HP Smart Wand Bar Code Reader User's Manual,* Hewlett-Packard, 1993.

HEWLETT-PACKARD, *Introduction to SONET Networks and Tests,* MCG/Queensferry Telecommunications Division, 1992.

HEWLETT-PACKARD APPLICATION STAFF, *Optical Communication Application Seminar,* Hewlett-Packard Optical Communication Division, San Jose, CA, 1988.

JOHNSON, B., *Analysis of Fault Tolerant Systems,* Addison-Wesley, 1989.

KENNEDY, G., *Electronic Communications Systems,* 3rd ed., McGraw-Hill, Inc., New York, NY, 1985.

LANE, J., *Asynchronous Transfer Mode: Bandwidth for the Future,* Telco Systems, Inc., Norwood, MA, 1992.

MCNAMARA, J. E., *Local Area Networks, An Introduction to the Technology,* Digital Press, Burlington, MA, 1985.

MCNAMARA, J. E., *Technical Aspects of Data Communications,* 2nd ed., Digital Equipment Corporation, Bedford, MA, 1982.

MARTIN, J., *Telecommunications and the Computer,* 3rd ed., Prentice-Hall, Inc., Englewood Cliffs, NJ, 1989.

MOODY, D., *ISDN,* Northern Telecom, Issue 2, 1991.

NORTHERN TELECOM, *SONET 101: An Introduction to SONET,* Fiber World.

PEARSON, G., *MNP Error-Correcting Modems,* Microcom, Norwood, MA.

PECAR, J., O'CONNER, R., GARBIN, *Telecommunications Fact Book,* McGraw-Hill, 1993.

PRENTISS, S., *Introducing Cellular Communications,* Tab Books, Inc., Blue Ridge Summit, PA, 1984.

RACAL-VADIC, *Data Communications, A User's Handbook,* Racal-Vadic, Sunnyvale, CA.

ROWE, R. I., *Data Compression Algorithm for V.42 Modems,* Adaptive Computer Technologies, Inc., Santa Clara, CA. 1988.

SPRINT CENTRAL TELEPHONE, *Nevada First Source Phone Book,* Las Vegas Telephone Directory, Reuben H, Donnelley and Centel Directory Company, July 1993.

STALLINGS, W., *Data and Computer Communications,* 3rd ed., Macmillan, New York, NY, 1988.

STALLINGS, W., *Data and Computer Communications,* 4th ed., Macmillan, New York, NY, 1994.

STEIN, D. H., *Introduction to Digital Data Communications,* Delmar Publishers Inc. Albany, NY, 1985.

SINNEMA, W. AND MCGOVERN, T., *Digital, Analog, and Data Communications,* 2nd ed., Prentice-Hall, Inc., Englewood Cliffs, NJ, 1986.

Index

Bar codes (*cont.*)
common industrial bar codes, 27
continuous code, 35
definition of, 25
intercharacter space (CS), 28, 31
left and right hand characters,
32–34
start/stop/center guard patterns,
32–33
structure of, 26–27
UPC, 32–35
Baseband LANs, 351–53
Base unit, cordless telephone,
158–61
Basic Rate Interface (BRI),
317–19
Baud, 16
Baudot Code, 8, 16–17
deficiencies of, 17
Baudot, Emile, 16
Baud rate, 47–48, 108
ASCII terminals, 47–48
BCC, *See* Block check character
BCD, *See* Binary coded decimal
BEL control character, 24, 52
Bell, Alexander Graham, 2, 145,
412
Bell 500 telephone, 150–53
Bell System:
divestiture, 183
FDM groups, hierarchy of, 218–
21
L5 coaxial carrier, 197
T1 carrier, 232–39
Bell System modems, 249–56
103/113 modem, 249–50
201 B/C modem, 254–255
202 modem, 250–51
disabling the echo suppressors,
251–52
pseudo full-duplex operation,
251
208A/208B modems, 255
209A modem, 255–56
212A modem, baud rate versus
bit rate, 254
212A modem, 252–54
Binary coded decimal (BCD),
8–10
BCD addition, 9–10
excess-3 code, 10–11
Binary phase shift keying
(BSPK), 247
Binary to Gray conversion, 12–13
Bipolar pulse train, 237–39

Birth defects, VDTs and, 41
BISDN, *See* Broadband Integrated
Services Digital Network
BISYNC, 284–94
BCC errors, 289
data link control characters,
285–87
device polling, 290–91
device selecting, 290–91
handshaking, 288–89
message block format, 285–86
multipoint line control, 290–94
point-to-point line control,
288–90
time-out errors, 289–90
transparent text mode, 286–87
Bit error rate (BER), 403–04
Bit-oriented protocols, 283–84
Bit time, 47–48, 73, 107–08
Block-by-block synchronization,
63
Block check character (BCC),
286, 379, 381
Block mode, ASCII terminals, 38,
49, 51–52
BPSK, *See* Binary phase shift
keying
BRI, *See* Basic Rate Interface
Bridge, 373–74
Broadband Integrated Services
Digital Network (BISDN),
311, 320
Broadband local area networks,
351–53
Buffer jackets, 433
Bus topology, LAN, 334–45
Busy bit, 345
Busy tone, 167
Byte-oriented protocols, 283–84

C

Cable construction, optical fibers,
432–34
Cable television, 352
Caller ID, 216
Caller ID block, 216
Call forwarding, 215
Call progress tones, telephone set,
166–68
Call trace, 216
Call waiting, 216
CAN control character, 24

Carrier Detect (CD), 80–82,
85–88
Carrier sense multiple access with
collision detection
(CSMA/CD), 340–42
CATV, *See* Community antenna
television
CCITT modems, 256–63
V.22bis specification, 256–58,
266
V.29 specification, 258
V.32 specification, 259–60
V.32bis/turbo, 260–61
V.33, 261–62
V.42, 261
V.42bis, 261–62
V.34 (V. Fast), 262–63
C-type/D-type line conditioning,
174–75
Cells, 177–79
Cellular technology, 177–81
Central office, 146, 162, 163, 165,
166, 185
switching systems, 204–16
crossbar switches, 206–10
step-by-step switching (SXS),
206–07
Channel access:
contention:
Aloha, 338–39
carrier sense multiple access
with collision detection
(CSMA/CD), 340–42
slotted Aloha, 339–40
LANs, 336–45
polling, 337
token ring passing, 343–45
slotted ring, 344–45
token on a bus, 345
Channel capacity, 404
Channel groups, 219–20
Character mode:
ASCII terminals, 49–50
half-duplex (HDX) versus
full-duplex (FDX), 50
Cheapernet, 365
Checkbits, 392–95
Checkerboard character, 49
Checksums, 387–90
double precision, 388–89
Honeywell, 389–90
residue, 389–90
single-precision, 388
Christensen, Ward, 267
Chromatic dispersion, 432

Type field, 363